Geographic Information Systems

Geographic Information Systems

Geographic Information Systems
An Introduction

Third Edition

Tor Bernhardsen

Asplan Viak
Arendal, Norway

JOHN WILEY & SONS, INC.

This book is printed on acid-free paper. ∞

Copyright © 2002 by John Wiley & Sons, New York. All rights reserved.

Published simultaneously in Canada.

This publication is designed to provide accurate and authoritative information in regard to the subject matter covered. It is sold with the understanding that the publisher is not engaged in rendering professional services. If professional advice or other expert assistance is required, the services of a competent professional person should be sought.

Wiley also publishes its books in a variety of electronic formats. Some content that appears in print may not be available in electronic books. For more information about Wiley products, visit our web site at *www.wiley.com*.

Library of Congress Cataloging-in-Publication Data:

Bernhardsen, Tor.
   Geographic information systems : an introduction / Tor Bernhardsen.—3rd ed.
     p. cm.
   Includes bibliographical references (p.).
   ISBN 0-471-41968-0 (cloth : alk. paper)
    1. Geographic information systems. I. Title.
  G70.212.B473 2001
  910'.285—dc21   2001046661

Printed in the United States of America.

10 9 8 7 6 5 4 3 2 1

CONTENTS

Foreword   x

Preface   xiii

1   Geographical Information Systems and
    Graphical Information   1
    1.1   Basic Concepts   1
    1.2   Socioeconomic Challenges   12
    1.3   Benefits of Computerizing Information   21
    1.4   Users of GIS   25

2   Historical Development: Geographical Data and GIS   27
    2.1   Early Developments   27
    2.2   First Automatic Processing of
          Geographical Information   31
    2.3   The Microprocessor   32
    2.4   Recent Developments   34

3   From the Real World to GIS   36
    3.1   The Real World   36
    3.2   Real-World Model   36
    3.3   Data Model   37
    3.4   Levels of Measurement   45
    3.5   From Database to GIS to Map   46
    3.6   Extension of the Traditional GIS Data Model   46
    3.7   Conceptual Generalization   50
    3.8   Role of Maps in Data Modeling   51
    3.9   Extension of the Reality Concept   52

4   Basic Data Models   54
    4.1   Introduction   54
    4.2   Vector Data Model   55
    4.3   Raster Data Models   73
    4.4   Automatic Conversion Between Vector and
          Raster Models   84
    4.5   Vector Versus Raster Models   86
    4.6   Attribute Data and Computer Registers   87
    4.7   Linking Digital Map and Attribute Information   91

                                                                 v

5 Advanced Data Models **94**
5.1 Terrain Surface Representation 94
5.2 Three-dimensional Objects 102
5.3 Representation of Time 105
5.4 Models for Movable Objects 108
5.5 Combination of Models 111

6 Georeferencing Systems **112**
6.1 Datum 113
6.2 Coordinate Systems 115
6.3 Map Projection 118
6.4 UTM 119
6.5 Coordinate Conversion and Transformation 121
6.6 Elevation Referencing 123
6.7 Relative Georeferencing 125
6.8 Discrete Georeferencing Systems 127

7 Hardware and Communication Technology for
GIS Applications **130**
7.1 Computers 130
7.2 Networks 133
7.3 Displays 137
7.4 Quantizers 139
7.5 Plotters and Other Output Devices 142

8 Basic Software and Databases for GIS **146**
8.1 The Foundation Stones of GIS Software 146
8.2 Operating Systems 148
8.3 Communications Between Users and Computers 148
8.4 Database Management Systems 152
8.5 Computer-Aided Design 164
8.6 Multimedia 164
8.7 World Wide Web 166
8.8 User Requirements 168
8.9 Working Environment 169

9 Data Collection I **170**
9.1 Introduction 170
9.2 Digitizing Maps 171
9.3 Scanning 176
9.4 Manual Digitizing or Scanning 184
9.5 Aerial Photographs and Photo Interpretation 185
9.6 Remote Sensing 187

10 Data Collection II **199**
10.1 Surveying 199
10.2 Satellite Positioning System 201

10.3	Photogrammetric Mapping	207
10.4	Collection of Attribute Data	210
10.5	Text Data	214

11 Data Quality **215**

11.1	Selection Criteria	215
11.2	Measuring Accuracy and Precision	218
11.3	Resolution and Sampling Rate	222
11.4	Data Storage Precision	223
11.5	Positional Accuracy	223
11.6	Attribute Data Accuracy	225
11.7	Temporal Accuracy	227
11.8	Logical Consistency	228
11.9	Completeness	230
11.10	Data Quality Overview Elements	231
11.11	Accessibility	232
11.12	Probable Sources of Error	232
11.13	Quality Control	234

12 Database Implementation and Spatial Indexing **235**

12.1	Database	235
12.2	Distributed Databases	239
12.3	Databases for Map Data and Indexing	240
12.4	Partitioning and Indexing	241
12.5	Database Design	244

13 Housekeeping Tools **251**

13.1	Introduction	251
13.2	Data-Entry Functions	253
13.3	Importing Existing Digital Data	253
13.4	Organization of Data Storage Operations	254
13.5	Functions for Correcting and Adapting Geometric Data for Further Use	256
13.6	Editing and Correcting Attribute Data	263

14 Basic Spatial Analysis **265**

14.1	Analysis of Spatial Information	265
14.2	Logic Operations	266
14.3	General Arithmetic Operations	267
14.4	General Statistical Operations	268
14.5	Geometric Operations	268
14.6	Search and Report Generation from Attribute Data	273
14.7	Geometric Data Search and Retrieval	274
14.8	Complex Operations of Attribute Data	275
14.9	Classification and Reclassification	277

14.10	Integrated Processing of Geometry and Attributes	278
14.11	Overlay	279
14.12	Buffer Zones	283
14.13	Raster Data Overlay	284
14.14	Procedures in Integrated Data Analyses	285

15 Advanced Analysis **288**
15.1	Network and Raster Connectivity Operations	288
15.2	Spatial Interpolation and Proximity Operations	292
15.3	Fuzzy Analysis	297
15.4	GIS Analytic Models	297
15.5	Practical Application of GIS Analytical Functions	300
15.6	Digital Terrain Models	304
15.7	Hydrologic Modeling	309
15.8	Functions for Engineering GIS	311

16 Visualization **313**
16.1	Theoretical Foundation	314
16.2	Graphic Generalization	322
16.3	Selecting Map Symbols	323
16.4	Limitations and Potentials of GIS in Cartographic Communications	324
16.5	Final Comments	327

17 Choosing a GIS: Organizational Issues **329**
17.1	Technology and Organization	329
17.2	Phases in Organizational Issues	333
17.3	Development of a Business Concept and the Identification of Goals	334
17.4	Appraisal of Current Tasks, Users, Data, and Data Flow	335
17.5	Review of Others' Experience with GIS	336
17.6	Identification of User Requirements	337
17.7	Financial Evaluations—Cost-Benefit Analyses	339
17.8	Developing a Strategic Plan	345
17.9	Developing a Logical Data Model	357
17.10	Creating National Geographic Databases and Developing New Business Sectors	358

18 Choosing a GIS: Technical Issues **361**
18.1	Pilot Project	361
18.2	Choosing Hardware and Software for GIS	362
18.3	Contracts	371
18.4	Technical Database Design	372
18.5	Creating a Database	373
18.6	System Operation and Maintenance	374

18.7 Safekeeping and Security Routines   374
18.8 Evaluating New Applications   376

19 Standards and Geospatial Infrastructure   377
19.1 Introduction to Standards   377
19.2 Elements for Standardization   380
19.3 Standard Transfer Formats   382
19.4 Special Standardization Elements   387
19.5 Standardization on the Application Level   389
19.6 Metadata   392
19.7 Infrastructure for Georeferenced Data   397
19.8 Data Access and Digital Libraries   401

20 Formal Problems in Establishing GIS   403
20.1 Ownership and Copyright   403
20.2 Cost Recovery and Pricing   404
20.3 Public or Private Organization of Geodata   406
20.4 Example of Strategy for a National Map Service   407
20.5 Data Security   408

21 A Vision for the Future   409

References   414

Index   417

FOREWORD

In approaching global problems, the people of the world are benefiting from more than 50 years of major developments in the science and technology used for studying the Earth, its human population, and the natural and cultural resources that make up the global environment. Especially important among these have been the rapid developments in remote sensing of the Earth environment, in computer science and technology, in global electronic communications, and in the gathering of spatial information about the Earth and its inhabitants into vast new electronic storehouses. In sum, as we enter the twenty-first century, we are in the fortunate position of having both the technical and the political means of bringing the people of the world together to deal with the world's problems; we must seize this opportunity.

One of these newly developed technologies, and one of the most important components of any approach to global problem solving, is geographic information systems (GIS) technology. Developed in just the last 30 years or so, GIS technology already represents a billion dollar industry worldwide, growing at perhaps 25% per year and serving about one million persons on a daily basis in more than 100 countries.

GIS has been used on problems at widely varying scales and for geographic areas ranging from a few hectares or a single city block up to those encompassed by the global databases just now beginning to become available for extensive use. GIS has been applied by a number of disciplines to a correspondingly wide range of problems. Governments, nongovernmental organizations, businesses, and educational institutions all now use GIS technology. Members of the general public will continue to become GIS users in the years just ahead. The fact that GIS technology has proven its value to its users is indicated by the rapid growth in its use, the rapid growth in expenditures for GIS technology, and the very large amounts of resources now being devoted throughout the world to creating digital data for use in GISs of various kinds.

Another indication of the value of GIS technology is that the number of GIS educational programs is growing faster than the use of the technology. For this important educational enterprise to succeed, sound textbooks in GIS technology and science are needed. This volume is just such a textbook. Until just a few years ago there were few introductory textbooks dealing with GIS technology. Those who learned the technology learned chiefly or exclusively through experience. Fortunately, that is now changing. I think this book is well designed, solidly constructed, and finely crafted; those who depend on it as they set out to explore our spatial world will be well served.

In my own work in the GIS field over the last 30 years or so, I have become convinced of the importance of providing GIS technology that works for its users. Many persons, especially those new to the field, can be overwhelmed by the enormous amount of information about GIS available. Students and others new to the GIS field need the guidance of textbook authors with extensive experience in the field, discernment, and careful judgment—authors who can separate the sound and essential from the ephemeral, authors who can indicate what actually works. That experience is reflected in these pages.

It is also important to realize that a GIS is considerably more than just technology; most of the problems with GIS now have more to do with the people and the procedures they use (system design, applications programming, effective use of GIS in solving real problems, funding new systems, GIS planning and implementation, and so on) than with computer hardware and software. The reality is that although one can now buy a reliable GIS hardware/software system and can often obtain a great deal of GIS data through purchase or conversion, there is no similarly reliable way to create the organizational structures and experienced staff necessary to support an effective GIS.

Some years ago, in thinking about how GIS use could best be developed throughout the world, I made a list of the traits that I thought GIS technical advisers would require if they were to be successful in spreading GIS use into new disciplines, new organizations, and new areas in the world. The list of intellectual and personal traits was a formidable one, chiefly because I believe that GIS technology can be applied to so many kinds of problems and in so many situations. The list, in turn, led me to reflect on how one could possibly find, recruit, and train and educate such people. How does one go about developing people with boundless enthusiasm and extraordinary patience, coupled with the appropriate technical, scientific, managerial, and political skills; those who are technicians as well as humanitarians; persons sympathetic to the ills of humankind and not afraid to wade in and try to improve things even when, often, they will not be successful?

In the end, I have concluded that such persons will, naturally and inevitably, be drawn to GIS technology, bringing with them the many skills and abilities they have already acquired elsewhere. When they understand just what GIS technology is capable of doing for many kinds of problem solving, they will embrace it and take it with them wherever they go. Books like this one will foster that process and so contribute to the improvement of the world in which we live.

I hope that the readers of this book will profit from what it has to teach them. If we are to solve many of the problems facing us—in the cities, in the wild areas of the Earth, in the atmosphere and the oceans—problems of the Earth as a whole, we shall need the help of skilled users of GIS technology. If readers can master what is in this volume, they will be well started on this enterprise.

Jack Dangermond

PREFACE

It is no more than three years since the previous edition of this book was published; however, technological developments in the field of GIS occur rapidly. In addition, there is always room for improvement. In this third edition I have attempted to maintain the original philosophy that the book should be at the introductory level while at the same time covering all relevant aspects linked to GIS. No major changes have been made in the chapter structure or in the choice of material. However, technological references have been updated and slightly extended. I have attempted to simplify the more difficult and obscure material, while going into greater depth with some of the material, such as georeferencing systems and geometric operations on geodata. The standardization work being carried out by such organizations as the International Standardization Organization (ISO), Open GIS Consortium (OGC), Federal Geographic Data Committee (FGDC), and others has lately become a central theme in the GIS world, and I have attempted to incorporate this material more actively in the text. Moreover, as in earlier editions I have supplemented the text with simple, but I hope instructive, illustrations.

It is not possible to write this kind of book without assistance from others. In this connection I would particularly like to thank the following from the National Mapping Authority, Norway: Bjoern Geirr Harsson for his contribution to Chapter 6, "Georeferencing Systems," Morten Borrebæk for his contribution to Chapter 3, "From the Real World to GIS," and Chapter 19, "Standards and Geospatial Infrastructure," as well as Hans Roenning for his contribution to Chapter 9, "Remote Sensing."

I would like to thank Mahala Mathiassen for her excellent translation to English of all the new material in this edition. I would also like to thank James Harper of my publisher John Wiley & Sons for his pleasant and professional cooperation. Finally, I would like to thank my wife and family for support and great patience while I spent much of my time working on the book.

Tor Bernhardsen
Arendal, Norway

Geographic Information Systems

Geographic Information Systems

CHAPTER 1

Geographical Information Systems and Geographical Information

Society is now so dependent on computers and computerized information that we scarcely notice when an action or activity makes use of them. Over the past few decades we have developed extremely complex systems for handling and processing data represented in the only form acceptable to computers: strings of zeros and ones, or bits (binary digits). Yet is has proved possible to represent not only numbers and letters, but sound, images, and even the contents of maps in this simple, universal form. Indeed, it may be impossible to tell whether the bits passing at high speed down a phone line or stored in minute detail on a CD-ROM (compact disk—read-only memory) represent a concerto by Mozart or the latest share prices. Unlike most of its predecessors, computer technology for processing information succeeds in part because of its ability to store, transmit, and process an extremely wide range of information types in a generalized way.

Computerization has opened a vast new potential in the way we communicate, analyze our surroundings, and make decisions. Data representing the real world can be stored and processed so that they can be presented later in simplified forms to suit specific needs. Many of our decisions depend on the details of our immediate surroundings and require information about specific places on the Earth's surface. Such information is called *geographical* because it helps us to distinguish one place from another and to make decisions for one place that are appropriate for that location. Geographical information allows us to apply general principles to the specific conditions of each location, allows us to track what is happening at any place, and helps us to understand how one place differs from another (Figure 1.1). Geographical information, then, is essential for effective planning and decision making in the modern society.

We are used to thinking about geographical information in the form of maps, photos taken from aircraft, and images collected from

1

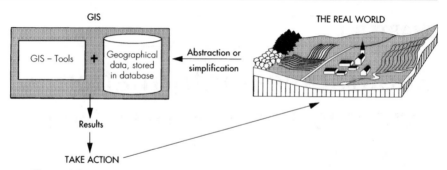

Figure 1.1
GIS is intended to be a means of improving everyday life. It is therefore
important that the information that results from data processing be applied to
guide the real world in the right direction.

satellites, so it may be difficult at first to understand how such information can be represented in digital form as strings of zeros and ones.
That problem is one of the central issues of this book, and the fact that
many alternatives exist is one of the reasons why that book is as long
as it is. If we can express the contents of a map or image in digital
form, the power of the computer opens an enormous range of possi-
bilities for communication, analysis, modeling, and accurate decision
making (Figure 1.2). At the same time, we must constantly be aware
of the fact that the digital representation of geography is not equal to
the geography itself—any digital representation involves some degree
of approximation.

Since the mid-1970s, specialized computer systems have been developed to process geographical information in various ways, which
include the following:

- Techniques to input geographical information, converting the information to digital form
- Techniques for storing such information in compact format on
  computer disks, compact disks (CDs), and other digital storage
  media
- Methods for automated analysis of geographical data, to search
  for patterns, combine different kinds of data, make measurements,
  find optimum sites or routes, and a host of other tasks
- Methods to predict the outcome of various scenarios, such as the
  effects of climate change on vegetation
- Techniques for display of data in the form of maps, images, and
  other formats
- Capabilities for output of results in the form of numbers and
  tables

The collective name for such systems is *geographical information systems* (GISs). The acronym GIS has come to signify much more than a
software system that processes, stores, and analyzes geographical data.

Figure 1.2
By use of geographical information systems, a simplified world can be brought into the computer. Different techniques can be applied to analyze and simplify the data, and the foundation is laid for the decision-making pyramid. Today, geographical information systems are in the process of filling the upper half of the pyramid. (Figure freely adapted from Grossman 1983.)

GIS is a "hot" application area for digital technology. Its software industry has been growing at more than 20% a year for many years, and recent figures for total annual sales of GIS software exceed $800 million. The term *GIS* has come to be associated with any activity involving digital geographical data; we now talk about GIS data, GIS decisions, and even GIS systems.

Although it is very easy to purchase the constituent parts of a GIS (the computer hardware and basic software), the system functions only when the requisite expertise is available, the data are compiled, the necessary routines are organized, and the programs are modified to suit the application. A computer system can function at what may appear to be lightning speed, yet the entire time span of a GIS project can stretch to months and even years. These facets of an overall GIS are interlinked (Figure 1.3). In general, procurement of the computer hardware and software is vital but simple. The expertise required is often underestimated, the compilation of data is expensive and

GIS CHAIN

Figure 1.3
A GIS system cannot be bought off the shelf. The system has to be built up within an organization. When planning to introduce GIS, it is important that equal attention be given to all four links in the GIS chain.

time-consuming, and the organizational problems can be most vexing. These facets of an overall GIS are discussed in detail later.

Traditionally, geographical data are presented on maps with the use of symbols, lines, and colors. Most maps have a legend in which these elements are listed and explained—a thick black line for main roads, a thin black line for other roads, and so on. Dissimilar data can be superimposed on a common coordinate system. Consequently, a map is both an effective medium for presentation and a bank for storing geographical data. But herein lies a limitation. The stored information is processed and presented in a particular way, usually for a particular purpose. Altering the presentation is seldom easy. A map provides a static picture of geography that is almost always a compromise between many differing user needs. Nevertheless, maps are a substantial public asset. Surveys conducted in Norway indicate that the benefit accrued from the use of maps is three times the total cost of their production.

As compared with maps, GIS has the inherent advantage that data storage and data presentation are separate. As a result, data may be presented and viewed in various ways. Once they are stored in a computer, we can zoom into or out of a map, display selected areas, make calculations of the distance between places, present tables giving details of features shown on the map, superimpose the map on other information, even search for the best locations for retail stores! In effect, we can produce many useful products from a single data source (Figure 1.4).

GIS defined

The term *geographical information system* (GIS) is now used generically for any computer-based capability for the manipulation of geographical data. A GIS includes not only hardware and software, but also the special devices used to input maps and to create map products, together with the communication systems needed to link various elements. The hardware and software functions of a GIS are as follows:

- Acquisition and verification
- Compilation

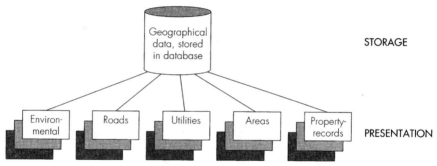

Figure 1.4
A map can be both a presentation medium and a storage medium, with
resulting limitations. With GIS, storage and presentation are separated, thereby
enabling a wide variety of products to be created from the same basic data.

- Storage
- Updating and changing
- Management and exchange
- Manipulation
- Retrieval and presentation
- Analysis and combination

These actions and operations are applied by a GIS to the geographical
data that form its database.

All of the data in a GIS are georeferenced, that is, linked to a specific location on the surface of the Earth through a system of coordinates. One of the commonest coordinate systems is that of latitude and
longitude; in this system location is specified relative to the equator
and the line of zero longitude through Greenwich, England. But there
are many other systems as well, and any GIS must be capable of transforming its georeferences from one system to another.

Geographical information attaches a variety of qualities and characteristics to geographical locations (Figure 1.5). These qualities may be
physical parameters such as ground elevation, soil moisture level, or
atmospheric temperature, as well as classifications according to the
type of vegetation, ownership of land, zoning, and so on. Such occur-
rences as accidents, floods, or landslides may also be included. We use
the general term *attributes* to refer to the qualities or characteristics of
places and think of them as one of the two basic elements of geographical information, along with locations.

In some cases, qualities are attached to points, but in other cases
they refer to more complex features, either lines or areas, located on
the Earth's surface; in such cases the GIS must store the entire mapped
shape of the feature rather than a simple coordinate location. Examples of commonly mapped features are lakes, cities, counties, rivers,
and streets, each with its set of useful attributes. When a feature is
used as a reporting zone for statistical purposes, a vast amount of

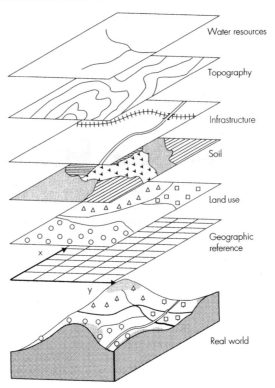

Figure 1.5
One can visualize the data stored as theme layers in the computer, with each layer linked to a common georeferencing system.

information may be available to be used as attributes for the zone in a GIS. In market research, for example, it is common for postal codes to be used as the basis for reports on demographics, purchasing habits, and housing markets.

The relationships between geographical features often provide vital information. For example, the connections of a water supply pipe network may be critical for firefighters, who need to know which valves to close in order to increase water pressure while extinguishing a fire. The details of properties bordering a road are necessary if all property owners affected by roadwork are to be properly notified. Connections between streets are important in using a GIS to assist drivers in navigating around an unfamiliar city. The ability of a GIS to store relationships between features, in addition to feature locations and attributes, is one of the most important sources of the power and flexibility of this technology. Some GISs can even store flows and other measures of interaction between features to support applications in transportation, demography, communication, and hydrology, among other areas.

Stored data may be processed in a GIS for presentation in the form of maps, tables, or special formats. A major GIS strength is that geographical location can be used to link information from widely scattered sources (Figure 1.6). Because the geographical location of every item of information in a GIS database is known, GIS technology makes it possible to relate the quality of groundwater at a site to the health of its inhabitants, to predict how the vegetation in an area will change as the climate warms, or to compare development proposals with restrictions on land use. This ability to overlay gives the GIS unique power in helping us to make decisions about places and to predict the outcomes of those decisions. The only requirement is that the geographical information from each source be expressed in compatible georeferencing systems.

A GIS can process georeferenced data and provide answers to questions involving, say, the particulars of a given location, the distribution of selected phenomena, the changes that have occurred since a previous analysis, the impact of a specific event, or the relationships and systematic patterns of a region. It can perform analyses of georeferenced data to determine the quickest driving route between two points and to help resolve conflicts in planning by calculating the suitability of land for particular uses.

Many recent GISs can process data from a wide range of sources, including data obtained from maps, images of the Earth obtained from space satellites, video film of the Earth taken from low-flying aircraft, statistical data from published tables, photographs, data from computer-assisted design (CAD) systems, and data obtained from archives by electronic transmission over the Internet and other networks. Data integration is one of the most valuable functions of a GIS, and the data that are integrated are increasingly likely to be obtained from several distinct media—multimedia is an active area for research and development in GISs (Figure 1.7).

Technically, a GIS organizes and exploits digital geographical data stored in databases. As mentioned earlier, the data include information on locations, attributes, and relationships between features. But a

DIGITAL MAP DATABASE   COMPUTERIZED TABULAR DATA

Building ID	Property	Owner	Year	Type
589	44/110			
610	44/50			
955	44/99			

Property ID	Owner	AREA	ADDRESS
44/99			
44/50			
44/110	Nils Nilsen	6,51	9999 Toppen

Figure 1.6
How GIS functions, based on the interaction between a digital map database and computerized tabular data.

Figure 1.7
GIS is in the process of becoming a typical data integration machine. Some modern systems can receive, process, and transmit data of widely varying origin. (Figure freely adapted from ESRI.)

database can only approximate the real world, inasmuch as the storage capacity of a database is minuscule in comparison with the complexity of the real world, and the cost of building a database is directly related to its complexity. The contents of a book of 100,000 words can be stored in digital form in roughly 1 million bytes (the common unit of computer storage is a byte, defined as 8 bits; 1 megabyte is slightly more than 1 million bytes). The information on a topographic map is comparatively dense, and it commonly takes 100 megabytes to capture it in digital form. A single scene from an Earth-observing satellite may contain 300 megabytes, the information content of 300 books. Thus, even crude approximations to the complexity of real-world geography can rapidly overtake the capacity of our digital storage devices.

Although we often think of the contents of a GIS database as equivalent to a map, there are important differences. On a map, a geographical feature such as a road or a power line is shown as a symbol, with the use of a graphic that will readily be understood by the map reader (Figure 1.8). In a geographical database, a road or power line will be represented by a single sequence of points connected by straight lines, and its symbolization will be reattached when it is displayed. A windmill will be represented by a single point, with the attribute "windmill," and will be replaced by a symbol when displayed. This approach is economical, in that the geometric form of the windmill symbol will be stored only once rather than repeated at each windmill location, and it also allows analysis to be more effective.

MAP SYMBOL

TABLE
INFORMATION

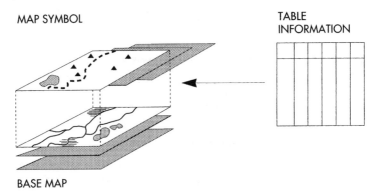

BASE MAP

Figure 1.8
For presentation purposes, the table information is translated into map symbols, which are then superimposed on existing map data.

Databases are vital in all geographical information systems, because they allow us to store geographical data in a structured manner that can serve many purposes. Many GISs impose further structure by using a database management system (DBMS) to store and manage part or all of the data in a largely independent subsystem within the GIS itself (Figure 1.9). A DBMS is a general-purpose software product, and GISs that use this approach are often able to function in conjunction with a wide range of DBMS products.

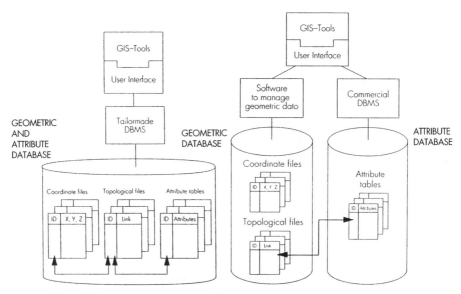

Figure 1.9
Alternative DBMS solution for GIS.

9

The database underlying a GIS achieves many objectives. It ensures that data are:

- Stored and maintained in one place
- Stored in a uniform, structured, and controlled manner that can be documented
- Accessible to many users at once, each of whom has the same understanding of the database's contents
- Easily updated with new data

This system contrasts with the traditional way of organizing and storing data on paper in filing cabinets, in which data are often:

- Stored in a way that is understandable to one person only
- Easily corrupted by use, or edited in a way that is meaningful only to the editor
- Inaccessible to anyone other than the creator of the system
- Stored in formats and at scales that are so diverse that they cannot be compared or collated
- Difficult to update

The constituent parts and modes of operation of GISs are discussed in detail in Chapter 3.

GIS diversity

Although the general definition of GIS given here is quite valid, in practice the diversity of GIS has spawned various definitions. First, users have contrived working definitions suited to their own specific uses. Thus, definitions may vary according to whether operators are planners, water supply and sewage engineers, support service personnel, or perhaps professional and public administrators or Earth scientists. Second, those with a more theoretical approach, such as researchers, software developers, or sales and training staff, may use definitions that are different from those used in practical applications. Systems can be tailor-made by assembling them from available software tool kits of semi-independent modules, assorted computer hardware components, and other interoperable devices. Many applications can be addressed by acquiring a single generic GIS product and a standard configuration of hardware. GISs can be viewed in a number of ways, including:

- A data processing system designed for map production or visualization
- A data analysis system for examining conflicts over plans or optimizing the design of transport systems
- An information system to be used in responding to queries about land ownership or soil type
- A management system to support the operations of a utility company, helping it to maintain its distribution network of pipes or cables

- A planning system to aid the design of road systems, excavations, or forest harvest operations
- An electronic navigation system for use in land or sea transport

GISs are often designated according to application. When used to manage land records, they are sometimes called land information systems (LISs); in municipal and natural resource applications, they are important components of urban information systems (UISs) and natural resource information systems (NRISs), respectively. The terms *spatial* and *geospatial* are often used almost interchangeably with *geographical*, although *spatial* is also used to refer more generally to any two- or three-dimensional data whether or not they relate directly to the surface of the Earth. The term *automatic mapping/facility management* (AM/FM) is frequently used by utility companies, transportation agencies, and local governments for systems dedicated to the operation and maintenance of networks. Nonetheless, *GIS* is now accepted internationally as an umbrella term for all digital systems designed to process geographical data, and is used in that sense in this book.

As befits this breadth of applications, the field of GIS involves many disciplines, applications, types of data, and end users, examples of which are:

- *Disciplines:* computer sciences, cartography, photogrammetry, surveying, remote sensing, geography, hydrography, statistics, information sciences, planning
- *Applications:* operation and maintenance of networks and other facilities, management of natural resources, real estate management, road planning, map production
- *Data:* digital maps, digital imaging of scanned maps and photos, satellite data, ground truth data, video images, tabular data, text data
- *Users:* water supply and sewage engineers, planners, biologists and cartographers, surveyors

The software capabilities required for a GIS often overlap those needed by other computer applications, particularly image processing and computer-assisted design (CAD). Image processing systems are designed to perform a wide range of operations on the images captured by video cameras, still cameras, and remote-sensing satellites. Today, the distinction between image processing and GIS is becoming increasingly blurred as images become ever more important sources of GIS data. Broadly, though, it is convenient to think about image processing systems as concerned primarily with the extraction of information from images, and GIS as concerned with the analysis of that information.

CAD systems have been developed to support design applications in engineering, architecture, and related fields. In general, CAD systems emphasize design over analysis and often lack the capabilities needed to process the complex attributes and information of georeferenced data or to integrate georeferenced data from many sources. Nevertheless, the distinction between CAD and GIS has increasingly diminished in recent years; by adding appropriate features, many former vendors of CAD systems are now able to compete effectively in the GIS market.

1.2   Socioeconomic Challenges

Various surveys have been conducted to determine the importance of geographical data by assessing the proportion of all data that are geographical, as defined earlier. In local government bodies and in utility companies the proportion is often over 80%; in mapping agencies it may come very close to 100%. These surveys have also shown that major users of geographical data—the construction sector, public administration, agriculture, forestry and other resource management, telecommunications, electricity supply, transportation—spend 1.5% to 2% of their annual budgets in acquiring geographical data. In relation to gross national product (GNP), annual expenditures average 0.5% in industrialized countries and 0.1% in developing countries. In the United States, a 1993 survey by the Office of Management and Budget (OMB) found annual expenditures on geographical information in federal agencies totaling more than $4 billion, suggesting that the total in all areas of U.S. society is well over $10 billion. The major challenges to system developers and users today are very different from those at the time of the OMB survey, and are related to the comparative ease of use of the technology, the problems of finding and accessing suitable data, and the lack of trained personnel able to exploit the technology's potential to its full.

Our complex society

Modern societies are now so complex, and their activities so interwoven, that no societal problem can be considered in isolation or without regard for the full range of its interconnections. For example, a new housing development will affect the local school system. Altered age distribution in a village will affect expenditures for health and social services. The volume of city traffic will put constraints on the maintenance of buried pipe networks, thus affecting citizens' health. Street excavations may drastically reduce the turnover of local retail shops. The noise of traffic on a new road or motorway may well drive people from their homes. The actions needed to solve such problems are best taken on the basis of standardized information that can be combined in many ways to serve many users (Figure 1.10). GISs have this capability.

Populations are now extremely mobile; changing jobs and moving

Forestry or holiday cabins

Water source
or swimming area

Agriculture
or industry

Pollution
Emission
permit

Cultural monuments

Road safety
Road access permission

Figure 1.10
Today's society has a growing need for information. Conflicts of interest have to
be brought into the open, and information has to be made available to the
parties concerned. It has been shown that at least 50% of the public
administration sector needs geographical data in one form or other. (Courtesy of
the Norwegian Cartographers Union et al.)

to another location have become commonplace. When key personnel
leave a company, they take their expertise with them; if that expertise
involves specific knowledge of, say, the water supply and sewage network of a community, the loss can be serious if the information is otherwise inadequately documented. Here, too, GIS has an advantage in
that it can act as an effective filing system for dissimilar sectors of a
complex society.

Operation and maintenance
Another aspect of developed industrial societies is that emphasis has
shifted from the planning and development of infrastructure to its operation and maintenance, particularly for such facilities as transportation systems and buried pipes and cables. At the beginning of the
twenty-first century the infrastructure of many industrial societies is
deteriorating and the costs of maintenance are increasing accordingly.
It is estimated that the average age of the water supply pipes of many
cities of eastern North America is now more than 100 years, and some
20% to 30% of the water supplied to these systems is now lost through
leakage. The costs of replacing this aging infrastructure over the next
decades will be truly astronomical. In Sweden the costs of operation

and maintenance of the water supply system in the 1990s were double those of the 1980s (Figure 1.11).

In Norway in 1990, municipal budgets for operation and maintenance were nine times those allocated to building new facilities. The large sums involved indicate that even minor gains in efficiency in this area can result in considerable savings. Yet more efficient operation and maintenance are possible only when full information is available on a facility's location, condition, age, construction material(s), and service record.

As many as 35 different types of piping and cable may be buried beneath the surface of a major city street. These include telecommunications and optical cables, high-voltage, local power, signal, heating, and TV cables, as well as piping for water supplies, sewerage, gas, and remote heating (Figure 1.12). Indeed, such is the confusion of pipes and cables that it is impossible to include all the relevant information on an ordinary map. Since the mid-1970s, therefore, piping and cable data have increasingly been computerized.

Computerization is the only realistic way of systematizing and standardizing the enormous amounts of data involved. For instance, even in a small city (population 500,000) there may be several million meters of buried pipes and cables. Even though the cities of industrialized countries have long had such problems, the overall picture seems equally serious in the cities of developing countries. In some cases, construction workers have been electrocuted when high-voltage cables, whose locations were not pinpointed, have been cut accidentally. The situation with water supply and sewage networks is equally adverse. Far too many human tragedies may be ascribed to an inade-

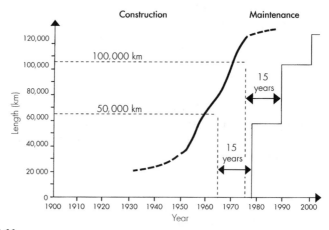

Figure 1.11

Total length of the water and sewerage system in Sweden from 1935 to date. Maintenance is normally carried out about 15 years after the pipes have been laid. This means that maintenance costs in the 1990s are double those of the 1980s.

Figure 1.12
Complexity of the systems under a city street. There can be up to 35 different utility systems under a single street (standard specifications for piping cartography).

quate, leaky water supply, deplorable sewerage, lack of maintenance, and lack of knowledge of the networks involved. The prime requirement for the information involved is, of course, that it must be manageable. This means that now, and in the future, it must be computer based.

Environmental and resource management
Decision making is becoming increasingly complex as dwindling natural resources and more demanding economic priorities diminish the chances of today's decision being right tomorrow. Furthermore, environmental awareness is constantly increasing among the general public, particularly among the younger generation. The pressing global challenges today include the following:

- Uncontrolled desertification
- Erosion, particularly of productive agricultural soils
- Monoculture, particularly in areas devoted to cereal production such as the American Midwest, and its associated pollution and soil deterioration
- Loss of endangered animal and plant species
- Acid rain and associated deterioration of forest and aquatic environments
- Pollution of rivers, lakes, and oceans
- Environment-related illnesses
- Contaminated groundwater and other environment-related stresses

15

- The greenhouse effect and resultant climatic change
- Reduction in the ozone layer
- Repeated environmental catastrophes, such as oil spills, poison leaks, and radioactive releases

Despite many comprehensive studies, the global environment is still not well understood, because nature is complex and most effects are interrelated. For example, a small decrease in the atmospheric ozone layer permits more ultraviolet radiation to reach the surface of the Earth; this kills the marine algae on which fish feed. Fish populations dwindle, which reduces catches and threatens the economy of fishing villages in major coastal areas. To help us map and monitor such changes, and plan appropriate responses that take account of the complex interactions of the Earth's systems, many countries now have comprehensive programs to capture and archive information on existing natural resources and known sources of pollution, using technologies such as satellite remote sensing and GIS (Figure 1.13). Worldwide databases have been established by agencies such as the United Nations Environmental Program, whose Global Resource Information Database (GRID) program makes such data available to a wide range of users in a GIS format.

Environmental data may be used both to expose conflicts and to examine environmental impacts. Impact analyses and simulated alternatives will probably become increasingly important. GIS is playing a key role in:

- Monitoring and documenting natural conditions and detecting change

Figure 1.13
Today the world's natural resources are under great pressure, and it is important to ensure sustainable development built on a strong foundation. GIS can be instrumental in initiating such a process.

- Documenting the suitability of resources and land areas for various uses
- Exposing conflicts and conflicting interests
- Revealing cause–effect relationships
- Modeling the interaction of various components of the Earth's environment to predict the effects of changes—for example, the effects of continued burning of fossil fuels on atmospheric carbon dioxide or the effects of global climate change on forests and agriculture

The ability to address environmental problems is complicated by a sparsity of information. Even the data that have been compiled are of limited use in decision making because of nonuniform storage and filing, lack of verification, lack of a systematic approach to data acquisition, and even obsolescence. GIS techniques and databases are ideally suited for the optimal manipulation of such environmental data.

Planning and development

As discussed earlier, the planning and development of new housing, roads, and industrial facilities requires data on the terrain and other geographical information. Development often involves building on marginal terrain, increasing the density of building in areas already built up, or both. Yet the new structures must fit within the existing technical infrastructure; here computerization is a great aid. For example, when Stockholm's Central Railway Station was expanded, all technical infrastructure facilities above and below ground were registered, and the entire project managed with the use of computers and GIS. In the United States, GIS has been used to plan and manage the massive Inner Relief Project in Boston, Massachusetts. One of the benefits GIS holds for such projects is a minimization of disruption to the existing infrastructure (Figure 1.14).

Escalating construction costs have made the optimizing of building and road location extremely important. Minimizing blasting and earth-moving is a significant aspect of minimizing costs. Flexibility is vital: Plans should be amenable to rapid change as decisions are made. The influence of special-interest groups and of individual citizens requires that initial plans be presented efficiently and in a manner that is easily understood. Simplified, visualized plans are instrumental in conveying both the content of a scheme and the nature of any likely impact on those concerned.

Management and public services

In modern societies, decisions should be made quickly, using reliable data, even though there may be many differing viewpoints to consider and a large amount of information to process. Today, the impact of development decisions is ever greater, involving conflicts between society and individuals, or between development and preservation.

Figure 1.14
Today's requirements for physical planning are stringent. Projects that involve encroachments on nature have to be carried out in the most environmentally friendly, yet economical way. (Courtesy of Asplan Viak.)

Information must therefore be readily available to decision makers; most such information is likely to be geographical in nature and best handled with the use of GIS.

Overviews of administrative units and properties are crucial in the development of both virgin terrain and built-up areas, in both developing and developed nations. In many countries, property registration is extensive; even in smaller states, 2 or 3 million properties may be involved. Moreover, property is an economic factor in taxation and as security for loans, so comprehensive overviews are essential to a well-ordered society. Computerized registers based on GIS technology are now well established in many countries. They are playing a vital role in restructuring the system of landownership of former communist countries and in modernizing land records in developing countries.

Safety at sea

In many waters, the volume of shipping has increased considerably. Offshore oil rigs and other fixed facilities have been erected in customary shipping channels. As alterations along channels often affect navigation, there is an urgent need in shipping circles for up-to-date information. Today, a ship sailing in international waters may carry as many as 2000 navigational charts, each of which may need updating as often as weekly (Figure 1.15).

Figure 1.15
A prerequisite for safe navigation at sea is rapid access to relevant navigational data. The risk of collision at sea is heightened by the increasing speed of vessels and complexity of shipping lanes. Electronic sea chart systems function in the same way as GIS at sea and improve safety.

Technology now permits the sending of information to mariners via public telecommunications networks, and electronic charts may be combined with positioning and radar displays on a ship's bridge. This means that ships can sail more safely and at greater speeds in hazardous waters. Computers are widely used for the production of traditional paper charts as well as for the production of updates.

Land transportation
In many countries, the greater part of transportation has shifted from rail and water to roads; at the same time, the use of private cars has greatly increased. These developments have created traffic problems, which cause loss of time and money. Large (and sometimes hazardous) goods are now transported by road. In most countries the annual costs of traffic accidents have become extremely high.

The transportation sector has always been a major consumer of maps and geographical data, so new technologies may enable considerable savings. The automobile industry is now investing heavily in the development of in-vehicle navigation systems (Figure 1.16), and several systems are now on the market. In principle, all of them

19

Figure 1.16
The transport sector has always made extensive use of geographical data. The introduction of new technology has opened a new range of applications in the fields of transport planning and traffic information.

involve simple GIS functions with digital maps and supplementary information. In many urban areas, where traffic is complex, increased driver information can ease congestion and improve safety. Such innovative programs to address problems of traffic congestion, known collectively as *intelligent transportation systems* (ITSs), are the subject of much research, particularly in the United States and Europe. They rely heavily on database representations of road networks and related information of importance to drivers and travelers, and on GIS.

Military uses
Modern military planning, training, and warfare are extensively computerized. In many ways, military computer systems include GIS functions. Digital data representing the form of the land surface are used in flight simulators and as the basis of automatic navigation and targeting in many weapons systems, such as the Cruise missile. Digital information on shorelines, water depth, land cover, and roads, now widely used by the military, is processed with the use of military forms of GIS technology.

The *global positioning system* (GPS), another increasingly useful form of geographical information technology, was initially developed by the U.S. military but is now widely used for both military and civilian applications around the world, often in conjunction with GIS. It consists of a constellation of 24 satellites whose signals can be analyzed by a very simple hand-held device to determine position on the Earth's surface. A single receiver, available at a retail price of less than $300, can provide positioning in civilian applications to about 20 m. By using a combination of a roving receiver and a base station, it is possible to achieve much higher accuracy, down to the centimeter level. The importance of geographical data and information technologies in the Gulf War was such that it was described on more than one occasion as the "first GIS war." The implications for the GIS community of such statements, and of the role of GIS in warfare in general, have been the subject of comment. As a rule, military operations are relatively independent of civilian activities. However, the increasingly demanding military specifications pertaining to equipment and data access will undoubtedly benefit civilian GIS in due course, and trends in modern warfare indicate that military GIS activities can be expected to increase.

Conclusions

This is the age of information, and information technologies such as GIS and GPS will have increasing impact on our lives. GIS technology can be used to build detailed information databases on individual citizens, and thus threaten individual privacy, and it can be used to improve our monitoring of the global environment and the decisions we make about the future of the planet. It can be used to make either reliable decisions based on good data, or unreliable decisions based on bad data or inappropriate methods of analysis. However, the computer itself is no more than a mechanical processor. The responsibility for wise use lies entirely with the user and with those who provide its data, develop its software, and build its hardware. The technology has enormous potential, both for use and for abuse, both conscious and unconscious. There is sometimes a tendency to give computer results more credibility than they deserve, and the wise use of a powerful technology like GIS often requires greater responsibility on the part of its users than did the earlier techniques of map use and analysis.

1.3   Benefits of Computerizing Information

Almost all aspects of modern society use digital information, and the total amount that flows through our communication networks daily is truly staggering. GIS offers its users the ability to process quantities of data far beyond the capacities of manual systems (Figure 1.17). Data in GIS are stored in a uniform, structured manner, as opposed to manual systems in which data are stored in archives and files, within agencies, on file cards, on maps, or in long reports. Data may be retrieved from GIS databases and manipulated far more rapidly and reliably than data in manual systems. In addition, data are quickly compiled into documents using techniques that include automatic mapmaking and direct report printouts. The potential gains to be derived from switching from manually prepared maps and ordinary files to computerized GIS are considerable, in both the public and private sectors.

Figure 1.17
There are considerable gains to be made by converting to computerized
information. However, unlike the results obtained with slot machines, the
winnings are not based on chance.

A joint Nordic research project (Nordic KVANTIF) has evaluated the
gains realized by implementing GIS through studies of its use in 50 to
60 organizations in the United States, Canada, Italy, and the five Nordic
countries—Norway, Sweden, Denmark, Finland, and Iceland. The
study showed that considerable benefits may be achieved, provided
that the strategy used to implement GIS is suitably chosen. The study
also showed that benefits are often related to objectives and that the
following benefit/cost ratios may be attained by introducing GIS (Figure 1.18):

1. If computerized GIS is used for automated production and maintenance of maps, the benefit/cost ratio is 1:1.

Objective level	Map production	Map production and internal use of data	Map production, internal use of data and shared use of data
Tasks	• storage • manipulation • maintenance • presentation	• map production • planning • facility maintenance • project management	• map production • project • planning • facility maintenance • coordination • general service • facility management • economic planning • service and information
Benefit/cost ratio:	1:1	2:1	4:1

Figure 1.18
The benefit/cost ratio of transferring to GIS depends on how the system and
information are applied. The ratio will normally be 1:1 for pure map
production; 2:1 for other applications such as planning, and as much as
4:1 for a multiple-user information system.

2. If the system is also used for other internal tasks such as work manipulation and planning, the benefit/cost ratio may be 2:1.
3. The full benefit of the system is first realized when information is shared among various users. The benefit/cost ratio may then be 4:1.

The benefit/cost ratios quoted in the preceding list refer to municipal services. Studies have shown that corresponding ratios for nationwide uses are somewhat lower, ranging from 2:1 to 3:1. Similar research in Australia and New Zealand (Price Waterhouse, 1995) shows a benefit/cost ratio of 4:1 in making geographical information available through a national infrastructure. Nonetheless, it is obvious that investment in GIS is at least as productive as investment in other sectors, such as road construction, kindergartens, etc. These benefits are not automatic. They depend largely on proper choice of an acquisition and implementation strategy, following careful study of the objectives and requirements of GIS investment, and careful selection of the appropriate system. Without these safeguards, many GIS projects eventually fail to deliver the promised benefits and may eventually fail entirely, at considerable cost to the institution. Even with a carefully selected strategy, it is difficult to estimate benefits precisely. The ratios discussed here are averages over many projects varying widely in scale and scope. Some figures, however, are impressive, with benefit/cost ratios of up to 8 to 10:1 or more. The following are benefit/cost ratios for various sectors and activities (Tveitdal 1987).

National data

Forestry	2.0:1
Transportation	4.0:1
Environmental data	1.8:1
Documentation	2.0:1

Municipal data

Map production	1:1
Agency uses	2:1
Joint uses	4:1

A number of GIS projects have been terminated because benefits did not meet expectations; in other cases, a project might not have been terminated had benefits been evaluated in a rigorous, objective manner. The Nordic KVANTIF study showed that GIS projects almost always involve high financial risk but that successful projects often result in benefits greater than anticipated.

Benefits are a function of many factors, including the goals and objectives of the project, the strategy adopted in its implementation, and the structure of the system built to serve the objectives (Figure 1.19). Systematic planning and implementation often set profitable GIS

Figure 1.19
High benefit/cost ratio figures are not easy to achieve, and the level of difficulty of the project tends to cause the ratio to increase proportionately.

projects apart from those that are unprofitable. Projects based on carefully estimated cost and benefit calculations are often more profitable than projects driven by pure technology. Profitable projects are user oriented rather than production oriented. Profitable projects start by being defined so clearly and convincingly that they are funded outside the ordinary operating budget.

The measurable benefits of GIS are usually expressed as gains in efficiency in terms of time saved, but there are also many cases of direct increases in income and reductions in costs. Measurable benefits may include:

- Improved efficiency due to more work being performed by the same staff, or the same work performed by a smaller staff
- Reduction in direct operating costs through better bases for financial management, less costly maintenance of facilities, and joint uses of available data
- Increases in income due to increased sales or sales of new products and services

Experience indicates that when GIS makes some traditional jobs superfluous, staff are not made redundant but are instead put to tasks in the GIS environment that create more value.

Intangible benefits may also accrue. They cannot be expressed directly in monetary terms, but attempts should always be made to include them when benefits are evaluated. Intangible benefits may include:

- Improved public and private decision making in administration, planning, and operations
- Improved information and service to the public

- Increased safety and reduction in the impact of disasters through better-planned evacuation and more efficient management of emergency services
- An improved environment for future generations
- Better presentation of plans and their associated effects
- Improved decisions regarding new development and better analysis of market and site conditions

The greatest long-range global benefits of GISs are probably in the sectors where decisions have an environmental impact. The environment and the natural relationships within it are complex and not yet fully understood. It is widely known, however, that environmental degradation is implicated in the causes of many modern illnesses, such as asthma and cancer, the annual costs of which are enormous.

Increases in safety at sea are associated with the introduction of electronic navigation systems, such as the *electronic chart display and information system* (ECDIS). The benefits realized can best be appreciated by considering what the ECDIS may help to prevent. Among the more serious environmental disasters are oil spills from tankers that run aground. The costs of subsequent cleanups may have a cost of $40 per kilogram of oil, summing up to totals of $2 billion for one accident. If improved navigation using ECDIS can prevent one such disaster per year, the benefits are considerable, albeit difficult to assess directly.

1.4   Users of GIS

This chapter has reviewed some of the basic concepts of GIS and a number of the issues that arise from its use. Today, the widespread acquisition of digital computers by businesses, schools, and households has allowed technologies such as GIS to penetrate many aspects of our lives. Nevertheless, computer processing of geographical data remains problematic, and GISs are widely regarded as difficult to learn about and to use. We hope that subsequent chapters of this book will provide a conceptual and technical understanding of GIS that will allow readers to make effective use of its capabilities in one or more of the many areas of its application.

Users of GIS naturally fall into two groups. Some are professional operators of GIS, who spend much of their lives working with the technology in their jobs. They are well trained in the particular software they use and are well aware of its capabilities. In many cases they do not use the results of their work themselves, but pass them to end users. The results may be maps, designed and produced by the GIS operator, results of analysis to be used in planning harvesting of trees, or work orders for maintenance staff in a major utility company.

The second group of users spend a relatively small proportion of their lives using GIS. They may maintain a GIS capability on their personal workstation in order to produce an occasional map, to find a restaurant in an unfamiliar city, to plan a driving route for a vacation, or to carry out analysis of map data in connection with a research project. In these cases the opportunities for lengthy training are much less, so the GIS must be simple and easy to use.

This second group also comprises end users and primary users who make professional decisions based on GIS products, such as the following:

- Operation and maintenance engineers; a typical decision may be whether to replace or repair a damaged water main.
- Regional planners; characteristic tasks involve presentations of plans to municipal authorities in a realistic, varied, visual manner.
- Building authority functionaries; representative jobs include processing building permit applications involving access roads, water supply, or sewage.
- Revenue officials, typically dealing with tax assessment and taxpayer addresses.
- Road engineers, whose responsibilities include locating new roads to minimize cut-and-fill operations.
- Information officers; information produced may include complete packages for newly established firms, with details on industrial areas, schools, and transportation.
- Local officials, who may require updated overviews on the effects of effluents on water quality at public beaches or the effects of zoning on school capacities.
- Fire brigades, for whom rapid, reliable information on the locations of fires and the presence of hazards such as explosives would be invaluable.
- Forest managers planning harvest operations, computing volumes of annual growths, estimating road costs, and identifying sensitive wildlife areas.
- Bank officials, perhaps wishing to verify ownership of properties offered as collateral.
- Oil tanker captains, in maneuvering a ship in hazardous waters.
- Truck drivers seeking to minimize the problems of transporting an extrawide load between two points.

Such a list could be endless; only imagination sets a limit. Indeed, GIS may spawn information technologies applicable to completely new sectors, such as optimum warehousing, or even brain mapping.

CHAPTER 2

Historical Development: Geographical Data and GIS

2.1 Early Developments

Geographical information systems evolved from centuries of map-making and the compilation of registers. The earliest known maps were drawn on parchment to show the gold mines at Coptes during the reign (1292–1225 B.C.) of Rameses II of Egypt. Perhaps earlier still are Babylonian cuneiform tablets that describe the world as it was then known (Figure 2.1). At a later date, the Greeks acquired cartographic skills and compiled the first realistic maps. They began using a rectangular coordinate system for making maps around 300 B.C. About 100 years later, the Greek mathematician, astronomer, and geographer Eratosthenes (ca. 276–194 B.C.) laid the foundations of scientific cartography. One of the earliest known maps of the world was constructed by Claudius Ptolemaeus of Alexandria (ca. A.D. 90–168).

The Romans were more concerned with tabulations and registers. The terms *cadastre* (an official property register) and *cadastral* (of a map or survey that shows property or other boundaries) originate from the late Greek *katà stíkon*, which means "by line." But it was the Romans who first employed the concept of recording properties, in the *capitum registra*, literally, "land register." In many countries, the term *cadastre* designates map and property registers. In France it also applies to the administrative agency in charge of such documents.

Throughout history, as societies organized, it became necessary to establish a mean of meeting the expenses entailed. Some of the better known earlier examples include taxation levied by emperors and kings to meet military expenses. These direct levies are the foundations of today's complicated revenue systems involving the taxation of income, property, and goods. Because both the ancient Egyptians and the Romans taxed property, property registration was early systematized to ensure tax revenues.

The earliest maps were drawn almost exclusively to facilitate commercial sea voyages (Figure 2.2). On these maps, coasts were meticulously detailed and harbors were plumbed, whereas interiors remained uncharted, apart from details of important trade and caravan routes.

Figure 2.1
One of the oldest known maps in the world, a Babylonian cuneiform tablet dating from about 1400 B.C.

The Arabs were the leading cartographers of the Middle Ages. European cartography degenerated as the Roman Empire fell. But in the fifteenth century, old skills were revived and Claudius Ptolemaeus's *Geographia* was translated into Latin to become the then-existent view of the world. Although cartography was neglected, in many countries property registry thrived. The best known example is the Domesday Book, the record of the lands of England compiled in 1086 for the first Norman king, William the Conqueror (1027–1087). The data included specifications of properties and their value and a count of inhabitants and livestock, as well as incomes earned and taxes paid.

The travels and explorations of Marco Polo, Christopher Columbus, Vasco da Gama, and others resulted in increased trade. In turn, maps of previously unmapped seas and coasts were needed. As the European countries and the newly discovered regions evolved to more organized societies, the need for geographical information increased. Ordnance developments, such as the introduction of artillery, made maps important in military operations, and military agencies became

Figure 2.2
One of the oldest known maps of the northern European seas, dating from 1570.
The cartographer was obviously more familiar with the southern than with the
northern tracts.

the leading mapmakers. In many countries, the military mapmakers
became responsible for both topographic land maps and navigational
charts (Figure 2.3). Vestiges of this trend remain; map agencies, particularly nautical chart agencies, seem characteristically military. For example, the official mapmaking agency of Great Britain is the Ordnance
Survey. The introduction of mass printing techniques enabled maps to

Figure 2.3
In the eighteenth century, vast land areas were surveyed using relatively exact
methods. This task was often carried out by military mapmakers.

be produced as consumer articles rather than as works of art, as was often the case earlier when maps were drawn by hand.

Until the nineteenth century, geographical information was used mostly for trade and exploration by land and sea and for tax collection and military operations. New needs arose in step with evolving infrastructures, such as roads, railways, telegraph and telephone lines, and gas and water supplies. Planning these facilities required information about the terrain beyond that commonly available. The accurate location of towns and cities, lakes and rivers, mountains and valleys became increasingly important. Detailed topographic information was needed to lay out railway and road gradients and curve radii. Then, as now, foundations were a major challenge, so maps showing the type of soil and the quality, location, and properties of bedrock were required. As planning advanced, specialized maps became more common. The first geological map of Paris was produced in 1811. In 1838, the Irish government compiled a series of maps for the use of railway engineers, which may be regarded as the first manual geographical information system.

Development became increasingly dependent on socioeconomic factors. The rights of property owners entered the picture, because the construction of roads and railways often necessitated the expropriation of private lands. New applications arose for property registers and maps as builders needed to compile overviews of affected properties in order that their owners might be justly compensated.

As cities grew larger and more complex, accurate urban planning became a necessity. Many countries began compiling statistical information relating to urban planning in the early nineteenth century. By 1837, the British Registrar General's Office had amassed extensive population statistics.

On the Continent, traditional village property ownership became a hindrance to effective farming. Many properties had become fragmented over the years, owing to inheritance settlements. In some cases a single property might comprise several hundred dispersed parcels of land. Sometimes the ownership of, or rights to, a parcel were divided: One owner could have timber rights, another grazing rights, a third hunting rights, and so on. Therefore, property mapping in the late nineteenth century aimed to wrest order from chaos. With reference to available land registers, the various parcels were assembled into properties that were easier to work. Borders were consolidated, clarifying ownership and facilitating the taxation of property.

Aerial photography accelerated the progress of mapmaking. The first aerial photograph was used for mapmaking, and the first mapmaking instrument devised, in 1909. Photogrammetry, the technique of making measurements from photographs, developed rapidly in the 1920s and 1930s, and the two world wars also hastened developments. After World War II, photogrammetry became widely used in mapmaking, mostly for maps in scales from 1:500 to 1:50,000. Aerial photographs themselves became important sources of quantitative information in evaluating such features as vegetation and geological formation.

2.2   First Automatic Processing of Geographical Information

Although Blaise Pascal is credited with devising the first true calculating machine in 1647, large amounts of data were first processed automatically in 1890, when a tabulating device conceived by Hermann Hollerith was used in compiling the U.S. census. In Hollerith's first apparatus, census data were punched on cards, which were then read electromechanically to compile data in separate registers. In the first half of the twentieth century, Hollerith's various mechanisms were developed further. Data processing using punch cards became an industry.

During World War II, data processing again advanced, primarily to meet the military need for predicting ballistic trajectories. One of the most famous computers developed for that purpose was ENIAC, an acronym for *electronic numerical integrator and calculator*. After the war, computer development continued. In 1953, IBM launched the model 650, which became the "Model T of the Computer Age" by virtue of being the first electronic computer not to be handmade. More than 100 were produced—in those days an amazing quantity. In today's computer terms, ENIAC, Whirlwind, the IBM 650, and other early electronic computers are referred to as first generation. All first-generation computers suffered from a common drawback: They used vacuum tubes, which, like light bulbs, gave off heat and had limited lifetimes. That alone restricted their application. One 25,000-tube computer of the period was continuously manned by a staff of ten, two of which were technicians assigned to continuous replacement of burned-out tubes. Nonetheless, computerization was the established technology for processing large amounts of data. By 1952, all U.S. governmental statistical data were processed by electronic computers.

By the late 1950s and early 1960s, second-generation computers using transistors became available, outperforming their vacuum-tube predecessors. Suddenly, computers became affordable in disciplines other than those of major governmental agencies. Meteorologists, geologists, and other geophysicists began using electronic mapmaking devices. Initially, the quality was poor, not least because automatic drawing machines had yet to be developed.

As the uses of second-generation computers spread, theoretical models were evolved to use statistical data. Then, as now, public and private decision making was often based on analyses of various classes of geographical data. These included demographic trends,

31

cost-of-living variations, the distribution of natural resources, wealth, and social benefits, and the demography of employment. The first geographical information system was constructed by the government of Canada in the late 1960s, which by modern standards was both unbelievably crude and expensive. It required a large mainframe computer, and its output was entirely in the form of tables. It was thus limited, in part, because no computer-controlled devices were available at that time to draw maps and, in part, because of the system's emphasis on analysis. Later, in the United States, a similar system, MIDAS, began processing data on natural resources.

The need for reliable geographical data multiplies with the expansion of road, rail, telecommunications, and sewage networks, airports, electricity and water supplies, and other essential services vital to the infrastructure of urban areas. Terrain information on maps is now a vital planning tool, from the first conceptual stage to the final, legally binding plan. Burgeoning road networks have mandated extensive analyses of transport patterns. Indeed, since the mid-1950s computers have been used in the United States to simulate traffic flows in relation to population distribution.

Often, conflicts of interest arise between developers and conservationists, or between municipal and regional planners and individuals. Most countries now have laws intended to resolve such conflicts—planning and building statutes, environmental and historical preservation laws, and the like. Enforcement of these laws requires an improved overall view of property ownership. Many countries began computerizing property registers in the mid-1970s; such systems may be viewed as the initial forms of land information systems (LISs).

2.3 The Microprocessor

In the 1960s and early 1970s, integrated circuits were developed and computer programs refined. The result: third-generation computers, which brought computerization to virtually all professional disciplines, especially those processing large amounts of data.

The next major breakthrough came in 1971–1972 with the development of the microprocessor. In 1974, a microprocessor was used to build the first fourth-generation desktop computer. Seven years later, the first microprocessor-based desktop computer was launched as a personal computer (PC). By the mid-1980s, the computer field was divided into three categories according to the size of the computer: mainframes, the descendants of the original large computers, intended for major data processing and computational tasks; PCs, the increasingly ubiquitous desktop computers; and minicomputers/workstations (Figure 2.4), which were smaller than mainframes but larger than desktop PCs.

By the early 1990s, mainframes had become physically smaller and

Figure 2.4
Early workstation, with graphics display, digitizer, secondary storage, and data processor. (Courtesy of SysScan.)

computationally more capable. That trend was reflected strongly in PCs, which by 1990 were outperforming minicomputers built only a few years earlier. This development signaled the demise of the mini-computer, an event that was further hastened by the introduction of PC networks, in which processing and storage capacity may be shared and distributed. The development of powerful workstations in the mid-1980s, however, led to an acceleration in the use of GIS. The over-all trend is best illustrated in terms of the costs of computing: A computer's processing and storage capabilities which, in 1960, cost $100,000, could be purchased for $10 in 1984 and $0.0005 in 2001. In other words, cost efficiency increased by a factor of 10 every two to three years.

In the 1970s and 1980s, various systems were evolved to replace manual cartographic computations. Workable production systems became available in the late 1970s, and system development continued through the 1980s. Nonetheless, by the beginning of the twenty-first century, elegant approaches to some cartographic tasks had yet to be found, and computerized cartographic research and development remains a continual challenge. The spread of PCs spurred user-friendly operations and programs capable of processing in ways previously not possible, for example, by considering the logical connections in geographical data.

Increases in microprocessor computing capacity also made the processing of digital and satellite images and other types of raster images commercially available in the mid-1980s. Software systems have developed apace. Relational database systems, such as dBase and Oracle, which first appeared in the late 1980s, are particularly useful in processing geographical data. Commercially available relational databases are now used routinely in GIS systems.

In the late 1980s, computing capability became widely accessible, and microprocessors were used for a multitude of devices, from household appliances and automobiles to an extensive range of specialized instruments, including those used in GIS. For GIS users, microprocessors have improved such devices as:

- Surveying instruments
- GPS (global positioning system)
- Digitizing table
- Scanners
- Environmental monitoring satellites and data presentation systems, including graphic displays, electrostatic plotters, and laser printers

2.4 Recent Developments

The 1990s produced even faster and more powerful computer equipment and peripherals. However, new developments in the field of data networks and communications are of equal importance—specifically, local area network (LAN), wide area network (WAN), and, of course, Internet and World Wide Web (WWW). The development of the Internet was initiated by the U.S. Department of Defense as long ago as the late 1960s. The World Wide Web was developed at CERN (Conseil Européen pour la Recherche Nucléaire) in Switzerland in 1990.

Data networks have opened up a whole new range of opportunities for geographical data search and distribution, thereby considerably increasing the value of GIS, particularly since common data have become

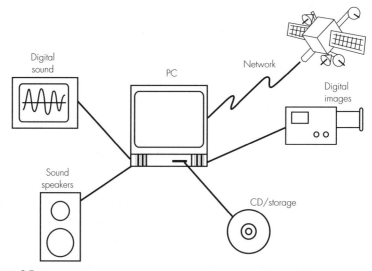

Figure 2.5
PC set up for multimedia.

more easily accessible. The most spectacular development in the GIS arena has occurred in the field known as multimedia. Multimedia techniques are based on the combination of elements such as figures, text, graphics, pictures, animation, sound, and video (Figure 2.5). Multimedia brings geospatial information into living maps and makes complex information understandable to those who are not technically sophisticated. Multimedia technology is available on the Internet and has proved to be eminently suitable as an information tool in planning (city, roads), tourism, and the distribution of environmental information.

Flight simulators are perhaps the best-known application of data technology to create near real-life situations, thus making them ideal for use in training. The concept behind flight simulators has now been adopted for other activities and is known as virtual reality. *Virtual reality* is a term used in multimedia; it was marketed as fiction in 1984 and became commercial technology in 1992. The most extreme virtual reality experience is attained by "dressing up" in a computer and moving into a world where almost all external impulses are artificial. Virtual reality and GIS have many features in common and are becoming increasingly integrated.

CHAPTER 3

From the Real World to GIS

In many ways GIS presents a simplified view of the real world. Since the processes involved are seldom straightforward because realities are irregular and constantly changing, perception of the real world depends on the observer. For example, a surveyor might see a road as two edges to be surveyed, the roadwork authority might regard it as an asphalt surface to be maintained, and the driver will see it as a highway. Moreover, the real world may be described in terms of countless phenomena, from basic subatomic particles up to the dimensions of oceans and continents. The complexity of the real world, as well as the broad spectrum of its interpretations, suggests that GIS system designs will vary according to the capabilities and preferences of their creators. This human factor can introduce an element of constraint, as data compiled for a particular application may be less useful elsewhere.

The systematic structuring of the data determines its ultimate utility and, consequently, the success of the relevant GIS application. This aspect is also characteristic of the data available in traditional maps and registers. The real world can be described only in terms of models that delineate the concepts and procedures needed to translate real-world observations into data that are meaningful in GIS. The process of interpreting reality by using both a real-world and a data model is called *data modeling*. As illustrated in Figures 3.1 and 3.2, the elements involved are:

- The real world
- Real world model
- Data model
- Database
- Maps and reports

3.2   Real-World Model

Each person creates his or her own subjective model of the real world. The point with GIS is to establish a common model of the real world. The arrangement of the common real-world model determines which

Figure 3.1
To bring the real world into GIS, one has to make use of simplified models of the real world. Uniform phenomena can be classified and described in the real-world model, which is converted into a data model by applying elements of geometry and quality. The data model is transferred to a database that can handle digital data, from which the data can be presented.

data need to be acquired. The basic carrier of information is the *entity*, which is defined as a real-world phenomenon that is not divisible into phenomena of the same kind. An entity consists of:

- Type classification
- Attributes
- Relationships

Although in theory the modeling process can be divided into two phases: (1) establishment of a common real-world model and (2) establishment of a common data model, in practice they will be carried out in one and the same process. Therefore, the emphasis will be on data modeling in the further treatment of this subject.

3.3 Data Model

A real-world model facilitates the study of a selected area of application by reducing the number of complexities considered. Those outside the selected area are considered insignificant. However, if a real-world model is to be of any use, it must be realized in a database. A data model makes that possible.

Unlike humans, computers cannot "learn" the essentials of manholes, property lines, lakes, or other types of objects. What they can do is to manipulate geometric objects such as points, lines, and areas, which are used in data models. The carriers of information in data models are known as *objects* (Figure 3.3) or *features*.* These correspond

---

* The terms *entity, feature,* and *object* are often interchangeable. In this book *entity* has been chosen as a term for a real-world thing or phenomenon. The term *feature* is used very little in the book but is synonymous with the term *object type* and is defined as a group of spatial elements that together represent a real-world entity. In international standardization work (ISO, etc.) the term *feature* is used instead of *object type*. *Object* is defined in object modeling as a digital representation of a discrete spatial entity and may describe the spatial aspect of *feature*. In this book *object* normally covers both *feature* and *object*.

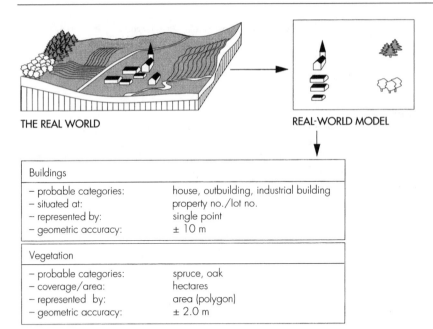

THE REAL WORLD   REAL-WORLD MODEL

Buildings	
– probable categories:	house, outbuilding, industrial building
– situated at:	property no./lot no.
– represented by:	single point
– geometric accuracy:	± 10 m

Vegetation	
– probable categories:	spruce, oak
– coverage/area:	hectares
– represented by:	area (polygon)
– geometric accuracy:	± 2.0 m

DATA MODEL

ID	Type	Property No.	X	Y	Accuracy
1	House	44 113	350	575	± 10.0
2	Outbuilding	45 6	375	600	± 10.0
3	Industrial	45 11	345	630	± 10.0

ID	Type	Area	Coordinates	Accuracy
10	Spruce	100	250,420 250,455 370,475 360,420 250,420	± 2.0
20	Oak	50	360,420 370,475 425,395 425,420 360,420	± 2.0

DATA BASE

MAP WITH SYMBOLS

Figure 3.2
Modeling process. The transformation of the real world into GIS products is achieved by means of simplification and models in the form of maps and reports.

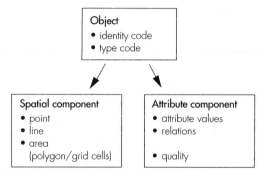

Figure 3.3
In a data model, objects are categorized as object classifications, geometric elements (point, line, area), attributes, relations between the entities, and quality definitions of these descriptive elements.

to entities in real-world models and are therefore regarded as abstractions of real-world phenomena.

Objects are characterized by:

- Type
- Attributes
- Geometry
- Relations
- Quality

Real-world models and entities cannot be realized directly in databases, partly because a single entity may comprise several objects. For instance, the entity "Church Road" may be represented as a compilation of all the roadway sections between intersections, with each of the sections carrying object information. Multiple representations produced by such divisions may promote the efficient use of GIS data (Figure 3.4). This means that information-carrying units and their magnitudes must be selected before the information is entered in a database. For example, the criteria for dividing a roadway in sections must be selected before the roadway can be described.

Each single object is identified with an *identity (ID) code,* which may be designated by numbers and is unique: No two objects have the same identity.

Figure 3.4
A single entity can be described by several objects (i.e., there are many relationships between entities).

3.3.1 Object type (feature type)

An object may belong to an object class. A class contains a set of objects with a specific theme or type. Each object is identified with a type code or *theme code,* which is based on object classification. An object may be classified under one type code only. *Object type* corresponds to the term *feature.* In this book, as previously mentioned, the term *object (type)* is used instead of *feature.*

During the classification process, each object type must be uniquely defined to preclude ambiguity. For example, "house" must be defined in such a way that "detached house at No. 10 Church Road" is classified under "house" and not under "industrial building."

Some user organizations may need to classify object types into categories as well as according to type. For example, national highways, country roads, urban roads, and private roads may come under the "roadways" category; alternatively, all objects within a specific geographical area may belong to a unique category of that area. In geographical data an object type is also known as the nominal scale or qualitative data (Figure 3.5).

Data models may be designed to include the following main categories of object types:

- Physical discrete objects, such as roads, water mains, and properties
- Continuously changing objects, such as elevation contours and temperature limits
- Classified objects, such as types of vegetation, climatic zones, or age groups
- Events, such as accidents or water leaks
- Artificial objects for a selected representation and database (raster)

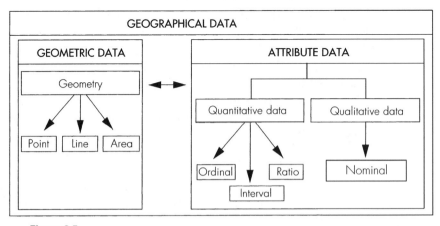

Figure 3.5
Geographical data can be divided into geometric data and attribute data. Attribute data can in turn be subdivided into qualitative data and quantitative data.

3.3.2 Object attributes

Object type code and object ID code are really general *attributes*, but each object may incorporate one or more attributes that describe the fundamental characteristics of the phenomena involved. For example, objects classified as "buildings" may have a "material" attribute, with legitimate entries *"wooden"* and *"brickwork"* and a "number of stories" attribute with legitimate values of 1 to 10, and so on.

In principle an object may have any number of attributes. For example, a lake may be described in terms of its name, depth, water quality, or fish population, as well as its chemical composition, biological activity, water color, algae density, potability, or ownership.

Attributes describe an object's features and may thus be regarded as a computer's "knowledge" of the object. In practice, object attributes are stored in tables (Figure 1.6), with objects on lines and attributes in columns. Theoretically, attribute values connected to grid data can be presented in the same way. Each grid cell corresponds to an object (Figure 3.6).

Although identity code and object type code (theme code) in principle are attributes, they are often considered to be more closely linked to geometry than other types of attributes.

3.3.3 Geometric representation of objects

Geometric information about objects may be entered in terms of (Figures 3.3 and 3.5):

- Points (no dimensions)
- Lines (one dimension)
- Areas (two dimensions)

Points

A *point* is the simplest geometric representation of an object (Figure 3.7). Points have no dimensions but may be indicated on maps or

Cell	Attributes				
no.	A	B	C		X
01	///	///			
02	///	///	///		///
03	///				
11					
12					
–					
nn	///		///		

Figure 3.6
In principle, the difference between vector data and raster data is not great. Raster data could well be arranged in tabular form with each cell number representing a line and each attribute (layer or raster values) a column.

Point: A zero-dimensional object that specifies geometric location specified throught a set of coordinates.

Node: A zero-dimentional object that is a topological junction and may specify geometric location

Line segment (vector): A one-dimensional object that is a direct line between two endpoints.

String: A sequence of line segments.

Area/polygon: A two-dimensional object bounded by at least three one-dimensional line segments.

Raster cell/pixel: A two-dimentional object (not bounded area) that represents an element of a regular tesselation of a surface.

Figure 3.7
Point, line, area. (Adapted from NCDS.)

displayed on screens by using symbols. The corner of a property boundary is a typical point, as is the representative coordinate of a building. It is, of course, the scale of viewing that determines whether an object is defined as a point or an area. In a large-scale representation a building may be shown as an area, whereas it may be only a point (symbol) if the scale is reduced.

Lines

Lines connect at least two points and are used to represent objects that may be defined in one dimension (Figure 3.7). Property boundaries are typical lines, as are electric power lines and telecommunications cables. Roads and rivers, on the other hand, may be either lines or areas, depending on the scale.

Areas

Areas are used to represent objects defined in two dimensions (Figure 3.7). A lake, a portion of woodland, or a township may typically be represented by an area. Again, physical size in relation to the scale determines whether an object is represented by an area or by a point. An area is delineated by at least three connecting lines, each of which comprises points. In databases, areas are represented by polygons (i.e., plane figures enclosed by at least three straight lines intersecting at a like number of points). Therefore, the term *polygon* is often used instead of *area*.

Physical reality is often described by dividing an area into regular squares or rectangles so that all objects are described in terms of areas (Figure 3.8). This entire data structure is called a *grid*. Population density is well suited to grid representation; each square or rectangle is

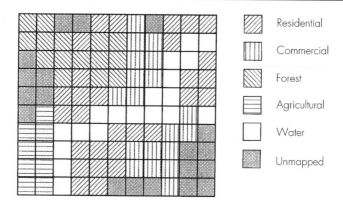

Residential

Commercial

Forest

Agricultural

Water

Unmapped

Figure 3.8
Land use in a town, in the form of a raster map. The land use is registered in a raster system with cells of varying size. Each category is given its own symbol on the map.

known as a *cell* and represents a uniform density or value. The result is a generalization of physical reality. All cells of a grid in a data model or a database are of uniform size and shape but have no physical limits in the form of geometric lines.

Point, lines, and areas are often defined as the simple geometric primitives for objects.

3.3.4 Object relations

Relations between objects are often encountered. Typically, these include (Figure 3.9):

Relation	Typical example(s)
Pertains/belongs	A depth figure *pertains* to a specific shoal, or a pipe *belongs* to a larger network of contiguous pipes.
Comprises	A country or a state *comprises* counties, which in turn *comprise* townships.

Figure 3.9
The computer cannot see the real world, so it is necessary to specify the various relations between entities, such as *belong to, comprise, are located in/on,* and *border on.*

43

Located in/on	A particular building is *located on* a specific property.
Borders on	Property A *borders on* property B.

Some relations may be unwieldy. One cause may be complexity, as in networks where the states of switches or contacts, open or closed, determine which parts of the network may be viewed as constituting a logical entity. In such situations, one may differentiate between actual and potential relations. Differentiation is made between:

1. Relations that may be calculated from:
   a. The coordinates of an object: for example, which lines intersect or which areas overlap
   b. Object structure (relation), such as the beginning and end points of a line, the lines that form a polygon, or the locations of polygons on either side of a line
2. Relations that must be entered as attributes, such as the division of a county into townships or the levels of crossing roads that do not intersect

Although the relations are intuitively obvious on ordinary maps, computers have no intuition. The computer processing of relations therefore requires either further descriptive information or instructions on how the data may be compiled. This aspect of map reading—the human ability to see what a computer needs to be "told" to "see"—highlights pivotal differences between human and computer processing. As a sign in a certain international computer research laboratory states, Computers: Fast, Accurate, Stupid. Humans: Slow, Slovenly, Smart.

3.3.5 Object quality

The true value of any description of reality depends on the quality of all the data it contains: geometry, attributes, and relations. Geometric data accurate to ± 0.1m obviously describe reality more faithfully than data accurate to ± 100m. Similarly, recently updated data are preferable to five-year-old data (which bring in temporal factors).

In the initial data modeling stage, the assessment of the data quality should include:

- Positional accuracy
- Attribute accuracy
- Temporal accuracy
- Logical consistency
- Data completeness

In addition, there are overview elements, such as:

- Purpose
- Usage
- Lineage

Information on the quality of data is important to users of the database. Requirements for data quality are discussed in greater detail in Chapter 11.

3.4 Levels of Measurement

Attributes may be grouped in different ways. A common method of division, as previously mentioned, is to differentiate between qualitative and quantitative attributes. *Qualitative* attributes represent the type of data, such as type of vegetation or type of language. We can therefore say that qualitative data is the name of the attribute. On the other hand, *quantitative* attributes are related to quantities and are normally represented by figures, such as number of inhabitants in a city and the acreage of a property. In addition, data that can be ranked in a series, such a good, better, best, can be placed in a separate group known as *ordinal* data. In cartography is it usual to divide data into qualitative, quantitative, and ordinal.

Attributes may also be ranked in four levels of accuracy: ratio, interval, ordinal, and nominal (Figure 3.10). The most accurate are *ratio* or *proportional* attributes, such as length and area, which are measured in respect to a fixed origin or starting point and on a continuous scale at

Figure 3.10
Attribute data consist of qualitative or quantitative data. Qualitative data specify the type of object, whereas quantitative data can be categorized into ratio data, data measured in relation to a zero starting point; interval data, data arranged into classes; and ordinal data, which specify quality by the use of text. Interval data are illustrated by day of birth of four persons on a segment of a number line, but we do not know whether they are recorded in Gregorian calendar or in Islamic calendar.

fixed intervals. *Interval* data comprise numerical data that are less accurate. An *interval scale* places the data on a number scale and, in principle, with a randomly selected starting point and arbitrary interval. Temperature, for example, can be measured in Celsius and Fahrenheit and years according to different calendars. A segment of these number lines without the origin and without definition of the unit of measure gives only the intervals between the observations and thus yields data of limited accuracy. *Ordinal data,* even less accurate than interval data, are characterized by rank such as "good," "better," and "best." Normally, one would know only the order and not the numerical difference between observations. Ratio, interval, and ordinal attributes represent quantitative data. *Nominal* is the least accurate level of measurement, and attribute data are here placed in groups only according to type and given a "name," such as type of vegetation, type of language. This means in practice that nominal data are also qualitative data.

3.5 From Database to GIS to Map

Once a data model is specified, the task of realizing it in a computer is technical and the task of entering data is simple and straightforward, albeit time-consuming. A database need seldom be made to suit a data model, as many databases compatible with GIS applications are now on the market. The problem at hand is more one of selecting a suitable database with regard to:

- Acquisition and control
- Structure
- Storage
- Updating and changing
- Managing and exporting/importing
- Processing
- Retrieval and presentation
- Analyses and combinations

Needless to say, a well-prepared data model is vital in determining the ultimate success of the GIS application involved. But the data in the database will not be meaningful before it is possible for the users to physically see them. Users view reality using GIS products in the form of maps with symbols, tables, and text reports. The dissemination of information via maps is discussed further in Chapter 16.

3.6 Extension of the Traditional GIS Data Model

3.6.1 Overlapping polygons

In the real world, one specific area or field may have many different characteristics; one area may in reality represent a number of entities or object types, such as coniferous forest, protected area, Property No. 44/133, and so on. We experience, on a daily basis, that it is the area as an entity that carries the information. However, in our real-world model we split phenomena into entities (*entity:* a real-world phenomenon that is not divisible into phenomena of the same kind) and allow the entities to be bearers of information. This model will allow an entity to represent only one phenomenon (e.g., only coniferous forest or only protected area). To adapt the model to reality, overlapping phenomena (entities/objects) are separated into different layers (Figure 3.11).

Reality is thus adapted to fit into a layer system, which is also traditionally used in map presentation. Because this type of model does not have overlapping polygons, each polygon in the different layers may be represented with one single set of coordinates, such as the center of gravity of the polygon. The polygons' attributes can thus be connected to this point. This is a very practical solution in regard to data management; thus many GISs make use of it. In the real world, areas are not divided into any form of horizontal two-dimensional physical layers. However, modern GISs have developed data structures that can handle overlapping polygons, so that the problem of sorting data in layers with nonoverlapping polygons will gradually be eliminated.

We can say that geometry—in which coordinates define points, lines, and areas—is in many ways an artificial concept and an unnatural way to describe reality. Coordinates are not tangible and are never used in our everyday description of reality. Instead, we define a phenomenon's location in relation to other phenomena with which the recipient of the information is familiar. We can therefore establish that our model of reality is not perfect.

During the last decade, new models were developed, known as *object-oriented models,* which to a certain extent can allow for the fact that the entity bearing the information can represent many phenomena. Object-oriented database systems are currently used in a few commercial GISs (Laser-scan, Smallworld) and appear to have many advantages over traditional database systems.

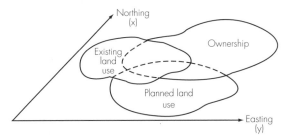

Figure 3.11
In the traditional discrete GIS-model, the bearers of information often consist of areas that overlap. There are systems available today that can handle overlapping areas in the same layer.

47

3.6.2   Continuous phenomena—field

To regard the real world as consisting of geometric constructs (points, lines, areas) means viewing objects as discrete data model representations—that is, to perceive that all objects have clearly defined physical limits. These limits are most obvious on maps, where lines imply sharp demarcations and no smooth, continuous transitions. Lines limit objects, not continuous transitions. For example, heights cannot be seen on the ground and changes in contours have, in addition, a typical continuous function. On a map, on the other hand, they are represented with discrete lines, and height differences between the contours will, in principle, remain unknown.

As previously mentioned, geographical phenomena can be divided into two main groups (Figure 3.12):

- Continuously varying phenomena, where the geography can be observed as a set of spatial continuous phenomena, represented by a function that gives unique attribute values in space. These continuous phenomena are also described as a *geographic field*.
- Clearly delineated phenomena, where the real world is covered by limited (discrete) objects in an otherwise empty world. A point can lie in one or more overlapping discrete units/surfaces. Each object is characterized by the attributes that are linked to it.

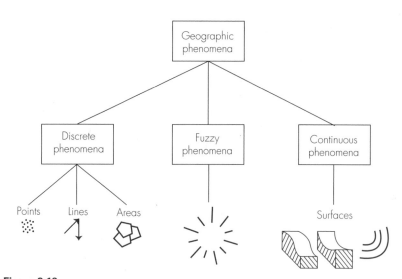

Figure 3.12
Points, lines, and areas represent discrete phenomena. Terrain surface and other mathematical surfaces represent continuous phenomena—known as geographical fields—where attribute values can be calculated on the basis of mathematical functions. The difference between discrete and continuous phenomena can be diffuse. The fuzzy theory can, to a certain extent, be used to represent diffuse phenomena.

The discrete data model is the model used so far to describe the real world, but it does not always suit reality. Difficulties arise in depicting phenomena that lack clear physical demarcation, such as soil types, population densities, or prevailing temperatures. Continuous phenomena can be represented by mathematical surfaces in the form of digital terrain models, isolines, and similar structures. The principle for terrain models can be used to describe the terrain surface and other continuous phenomena, where, for example, population density values or temperature values can lie on the z axis for different observation points in the area. However, such surface models are only an approximate description and there will always be some uncertainty linked to the interpolation of new values. As we will se later, raster data are more suitable to represent continuous phenomena than vector data. By making the cell size small enough, the raster model will be able to capture the greater part of the continuous variations, and it is relatively easy to interpolate new attribute values for random points inside the raster.

In practice, many geographical phenomena can be represented by both mathematical surfaces and discrete objects, depending on what is most appropriate in each case.

3.6.3 Uncertainty

In the discrete model, entities or objects are defined as being either within specific classes or outside them, and thus operate only within areas that are homogeneous in respect to limitation and classification. In reality, phenomena often vary even within small, limited areas. For example, coniferous forest often contains deciduous trees and population density is variable. There can also be uncertainty in other attribute values such as measurement values, time range, and the like. It can be difficult, even with discrete objects, to create data models that have a uniform and clear definition of the objects' classes. For example, does a pedestrian area that is accessible to emergency vehicles classify as a road? In practice, a large degree of uncertainty can be linked to the geographical borders between objects—borders that are usually based on individual evaluation. We can determine that uncertainty can be linked to:

1. Assignment of attribute values
2. Geographical limits

Once again, we have established that our traditional real-world model is not perfect and that it is (too) closely linked to traditional mapping concepts.

The "fuzzy" theory is an attempt to solve the problems of representing uncertainty in attribute values and in limitation of geographical units (Figure 3.12). In a discrete model objects are stated to be outside or inside defined classes. This is not suitable for phenomena of a continuous nature. For example, if areas with a population density of 1,000 persons per km^2 are defined as high density, then areas with 999 persons per km^2 fall into the category of medium density, even though the real difference is very small. Problems of this type can be partly solved by using the *fuzzy set theory*, which allows an object to belong only partially to a class. It is possible to give each object one or more additional values ranging from 0 (nonmember) to 1 (full member). A value of 0.8 will thus specify how a certain object is placed in a class. Where there are several attributes within an area, more additional values will be used. To date, fuzzy theory has had little use in commercial GIS software.

A great deal of further research is needed to determine an appropriate way to handle the uncertainty linked to geographical data. The significance of this type of deficiency in the data model is left to the person interpreting the results (maps and reports) of the GIS process. This means, in practice, that users will often have to make their own evaluations of uncertainty.

3.7 Conceptual Generalization

Normally, the aim of information management is to ensure that the recipient picks up the message quickly and with as little uncertainty as possible. This objective is mainly achieved in GIS by simplifying the real world through a systematic selection of information to be entered into the system. This is known as *conceptual generalization* or *data structuring*. When presenting the information, an exaggeration technique (broad lines, large symbols, etc.) is used in the map image. This technique, known as *cartographic generalization,* is discussed in Chapter 16 (Figure 3.13).

As mentioned earlier, the real world is interpreted differently because of subjective differences between observers, but the real world itself is objective. In conceptual generalization it is important to differentiate between the objective reality and the significance that reality has for the task to be performed. The only information to be registered is that which is relevant to carrying out the task. When establishing the data model, iformation is divided into two main groups: interesting and uninteresting information/object types. Only interesting object types will be put into the model; all other object types found in the real world are uninteresting and are ignored. This may sound simple, but it is often quite difficult to decide how interesting information has to be for it to be included in the model. Certain techniques have been developed that can give support to such processes; these are discussed in greater detail in Chapter 12, Section 12.5.1, "Conceptual database design."

The need to divide objects into classes also results in generalizations. For example, an area of forest that is mainly coniferous, but with some

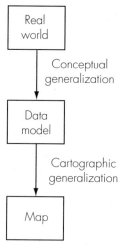

Figure 3.13
Several levels of generalization have to be carried out in order to represent the
real world on a map.

deciduous trees, will often be generalized and classified as coniferous,
not as a combination. Thus, conceptual generalization is also a method
for handling uncertain elements, whereby one can ignore some uncertainty or variation.

When points, lines, and polygons are selected as the geometric representation of objects, this very often results in a *generalization* of the
real world; a town can be represented by a point rather than a polygon, and a road is frequently represented by a center line rather than
two curbs. This kind of generalization lies on the borderline between
conceptual generalization and cartographic generalization, and the
map scale that the data show will, to a large extent, determine what
one chooses.

It will always be necessary to make choices about such generalizations in relation to the real world when making data models. This may
be seen as a problem, but generalization is also a technique that makes
it possible to obtain an overview of our complex reality.

3.8   Role of Maps in Data Modeling

Maps are, in general, good sources for describing objects and their attributes. However, maps always represent particular models of the real
world, and GIS should represent the real world, not the maps that depict it. For instance, ferry routes are often shown by dotted lines on
maps, whereas in transport planning, data models should form integral parts of a contiguous road network. As a rule of thumb, therefore,
always look at a map as a data source, not as a data model for GIS.

3.9 Extension of the Reality Concept

The traditional model for transformation from the real world to GIS, as described earlier, has its obvious faults. In addition, it describes only flat and unchanging reality. Models for describing objects in three-dimensional space and terrain have not yet been discussed. In the traditional layer-based data model heights are treated as attributes to the objects, not as a part of the geometry. But the real world is three-dimensional (Figure 3.14). Three-dimensional data models are discussed in Chapter 5. The fourth dimension—time—and its inroads into a geographical data model must also be considered. The same applies to models for dealing with objects (traffic) moving along defined networks.

Here it is also most practical to use the same basic concept: a geometry consisting of points, lines, and polygons, and attributes that describe the objects or phenomena. Elevation values can be linked to points, lines, and polygons and thereby give the objects a position in space. The surface of the terrain can be described with the help of sloping areas or with the help of horizontal surfaces with an elevation value linked as an attribute. Elevation values can also be linked to objects such as towers, wells, and buildings as attributes. The time factor can be accommodated by storing *all* historical data for the objects, such as changes in the geometry or attribute values. The movement of objects (traffic) along a road network can be simulated by assigning attribute values to elements in the network. These should be values that are relevant to the transfer speeds, and the sum of attribute values for different routes will be the measurement of passage in time or dis-

Figure 3.14
The world is three-dimensional, with phenomena having a location and surface area in both elevation and ground plane.

tance. A technical description of how this can be realized in GIS is given in Chapter 5, and the basic data models are described in Chapter 4.

Undoubtedly, the traditional data model concept has definite drawbacks in describing these new real-world elements. We must accept that the real world is too complex to be described in full at present, although researchers are continuously engaged in developing improved models.

CHAPTER 4

Basic Data Models

As discussed in Chapter 3, GIS depicts the real world through models involving geometry, attributes, relations, and data quality. In this chapter, the realization of models is described, with the emphasis on geometric spatial information, attributes, and relations. Data quality is discussed further in Chapter 11.

A prerequisite for describing the real world by use of GIS is that the different types of geographical information can be stored in a computer. All the operations in a computer are based on the storage and handling of numbers. This is why the data stored in a computer are known as digital data. In GIS there is a need to store graphical figures, images, numerical values, and plain text. All these forms of data must thus be able to be converted to digital representation.

Digital representation of data in a computer

In principle, only two different numerical symbols or signals can be stored in a computer—0 and 1. The numerical system based on 0 and 1 is known as the binary system. In order to separate different numbers form each other, the stream of 0's and 1's is divided into groups of 8 bits. Each group is known as a byte. The highest number in a binary system with bytes of 8 bits is 255. Numbers stored in this way are known as integers. In practice, a computer can today work with larger units than 8 bits, such as 16, 32, and 64 bits. Using 32 bits (double integer), one can at the most represent a whole number of 4 294 967 295. In order to give negative numerical values, the minus sign will need a place, and the maximal values will be reduced to +2 147 488 647 and −2 147 483 647.

Decimal figures are stored with the help of 4 bytes (32 bits), based on logarithmic notation, by separate storage of the number's mantissa and exponent. Of the 32 bits, 1 is used for storing a minus sign, 7 for storing the exponent of the number in the decimal system, and 24 for storing the mantissa. Numbers stored in this way are known as floating points. Very large decimal figures can, in principle, be stored by the floating point technique. The largest decimal number that can be stored with 32 bits in $3.4*10^{38}$. However, number precision is limited by the number of significant figures, in this case 7. Noticeable rounding-off errors can occur when large decimal numbers are multiplied over and over again. By using 64 bits (double precision), it is possible to store numbers with 15 significant figures, but double precision at the same time increases the calculation time and the storage volume. Coordinate values are often high decimal numbers and are therefore usually stored with floating point double precision.

Text is stored digitally with the help of a code system called ASCII (American Standard Code for Information Interchange). Each number between 0 and 127 corresponds to one sign on the computer's keyboard. For example, uppercase letters A through Z are represented by numbers between 97 and 90, and lowercase a through z by numbers between 97 and 122. To handle special national signs, a variation of the system has been developed to accommodate up to 255 different signs.

Geometric presentations are commonly called *digital maps*. Strictly speaking, a digital map would be peculiar because it would comprise only numbers (digits). By their very nature, maps are analog, whether they are drawn by hand or machine, or whether they appear on paper or are displayed on a screen. Technically, GIS does not produce digital maps—it produces analog maps from digital map data. Nonetheless, the term *digital map* is now so widely used that the distinction is well understood. Spatial information is presented geometrically in two ways: as vector data in the form of points, lines, and areas (polygons), or as raster data in the form of uniform, systematically organized cells. The vector model and raster model are discussed in the following sections.

4.2 Vector Data Model

The basis of the vector model is the assumption that the real world can be divided into clearly defined elements, each of which consists of an identifiable object with its own geometry of points, lines, or areas (Figure 4.1). In principle, every point on a map and every point in the terrain it represents is uniquely located using two or three numbers in a coordinate system, such as in the northing, easting, and elevation Cartesian coordinate system. On maps, coordinate systems are commonly displayed in grids with location numbers along the map edges (Figure 4.2). On the ground, coordinate systems are imaginary, yet marked out by survey control stations. Data usually may be transformed from one coordinate system to another. (See Chapter 6 for further details on coordinate systems.) Points, lines, and areas are mathematical figures that are used to describe the positions and extensions of geographical objects.

Points

Points are the fundamental and simplest form of geographical objects and are zero-dimensional because they have no extension. Each point

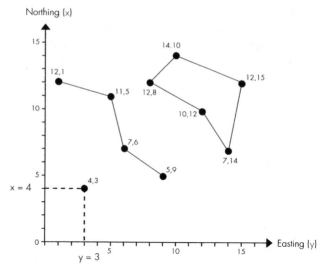

Figure 4.1
In the vector data model, the objects are described as points, lines, or areas (polygons). These three geometric phenomena may be described individually by a single point in a coordinate system and with connected lines (for lines and areas). Therefore, each object comprises one or more sets of coordinates.

Figure 4.2
On maps, coordinate systems in an x- and y-direction format are generally displayed in grids, with location numbers along the map edges.

is represented by a coordinate pair. As discussed later, we also operate in a GIS with a special type of point called a node (Figure 4.3). A point has no boundary.

Lines

Lines consist of points linked together with line segments. The points, represent start, break, or end points of the line. A line has two points as a boundary; a start point and an end point. Lines are one-dimensional, as they stretch in only one direction. Two is the lowest number of coordinate pairs representing a straight line between two points. Mathematically, a vector is a straight line having both magnitude and direction. Therefore, a straight line between two coordinate points on a digital map is a vector—hence the concept of vector data used in GIS and the designation of vector-based systems. There are a number of different line types, such as string, link, chain, arc, and so forth; see Figure 4.3. *Link, chain,* and *arc* are terms often used with the same meaning, that is for lines linked together in a network structure.

Lines can also include certain lines or parts of lines (e.g., those representing road curves or property boundary curves) that may be described by mathematical functions, such as those for circles, parabolas, or a polynomials. In these cases, GIS data include equation parameters: for example, the radii of the circles used to describe parts of lines, or third-degree polynom that describe a cubic curve and are expressed with the formula:

$$f(x) = a_0 + a_1 x + a_2 x^2$$

Such smooth curves fitted mathematically to a sequence of points are known as splines and are used, among other things, for interpolation and making gentle curves in drawings.

Together with the coordinate data, instructions are entered as to which points are connected to lines. These instructions can subsequently be used to create lines and polygons and to trigger "pen up" and "pen down" functions in drawing.

Area

An area is represented by a single line that encloses a space, thus forming a closed polygon. The surrounding line, called a *ring*, has to start and end at the same point in order for the area to be closed and defined. Areas are two-dimensional because they stretch in two dimensions. The term *polygon* is often used to designate an area. Area is the most complicated of the geometric primitives. It is possible to have complex areas with different variants of inner rings defining holes, and holes in holes (Figure 4.3). Thus, an area has a closed line in the outer boundary, but can also have several inner closed boundaries. We shall see later that there are different methods of storing polygons. A

57

ZERO-DIMENSIONAL OBJECTS:

Point: A zero-dimensional object that specifies geometric location specified throught a set of coordinates.

Node:.A zero-dimensional object that is a topological junction and may specify geometric location.

ONE-DIMENSIONAL OBJECTS

Line segment (vector): A one-dimensional object that is a direct line between two end points.

Link: A one-dimensional object that is a direct connection between two nodes.

Directed link: A link between two nodes with one direction specified.

String: A sequence of line segments.

Chain: A directed sequence of nonintersecting line segments with nodes at each end.

Arc: A locus of points that forms a curve that is defined by mathematical function. Also defined as a string or chain.

Ring: A sequence of any line segments with closure.

TWO-DIMENSIONAL OBJECTS

Simple area/polygon: An area defined by an outer ring that may not have inner rings (holes).

Complex area/polygon: An area defined by an outer ring with optional inner rings defining holes.

Raster cell/pixel: A two-dimentional object (not bounded area) that represents an element of a regular tesselation of a surface.

Figure 4.3
Geometric objects. (Adapted from NCDS.)

frequent question is whether the border between two polygons should be stored only once by decomposing the geometry, or whether the geometry of each polygon should be stored separately, resulting in double storage of the boundary.

In a vector model, points, lines, and areas (polygons)—often called the geometric primitives—are the homogeneous and discrete units that carry information. As discussed earlier, these three types of objects may be represented graphically with the use of coordinate data. However, as we shall see, the objects may also carry attributes that can be digitized, and all digital information can be stored.

4.2.1 Coding digital data for map production

The question now is how coordinates representing points, lines, and areas can be converted to maps on a computer monitor or plotter. Anyone familiar with maps knows that map data are traditionally coded. Roads, contour lines, property boundaries, and other data indicated by lines are usually shown in lines of various widths and colors. Symbols designate the locations of churches, airports, and other buildings and facilities. In other words, coordinates and coding information identify all objects shown on a map.

Not surprisingly, then, the digital data used to produce maps are also coded, usually by the assignment of numerical codes used throughout the production process—from the initial data-to-computer manipulation and on to the drawing of the final map. Each numerical code series contains specific codes assigned to the objects in the group. This code, often called a *theme code*, is normally represented by numbers in a certain system For example, the codes for boundaries may be as illustrated in Figures 4.4 and 4.5.

Digital data for map production comprise sequences of integers, such as

$-5314401112345678912340678 0 - 5314401112333678812330670 0$

Use of the format permits the numerical sequence to be divided into groups and read

$-5/314/4011/12345/6789/12340/6780$

$-5/315/4011/12333/6788/12330/6700$

The figure designations are as follows:

Figure   **Designates**

−5   Start of a continuous sequence of data (i.e., if there are several coordinates, they are to be connected in a line: *pen down*)

Numerical code series	Object group
1 000	Survey control stations
2 000	Terrain formation
3 000	Hydrography
4 000	Boundaries
5 000	Built-up areas
6 000	Buildings and facilities
7 000	Communications
8 000	Technical facilities

Figure 4.4
Digital map data often use numerical coding, in the form of different numerical series, to identify object groups.

Numerical code	Object type
4 001	National border
4 002	County boundary
4 003	Township boundary
4 011	Property boundary
4 022	National park border
	etc.

Figure 4.5
By using a numerical coding system, codes can be assigned to all levels of detailed information on the objects.

314	Serial number of data sequence (such as of a unique line)
4011	Property boundary (such as might produce a final line width of 0.3mm)
12345	First easting coordinate
6789	First northing coordinate (pen moves to next coordinate set)
12340	Last easting coordinate
6780	Last northing coordinate
−5	End of data sequence, start of next sequence (*pen up*—moved and set down for a sequence of new coordinates, etc.)

In thematic coding, which may be compared to the overlay separation of conventional map production (Figure 4.6), data are divided into single-topic groups, such as all property boundaries. Information on symbol types, line widths, colors, and so on, may be appended to each thematic code, and various combinations of themes may be drawn. Data may be presented jointly in this way only if all objects are registered, using a common coordinate system.

Coding digital data for GIS

More coding is required to develop digital map data into part of an information system. Point objects may easily be realized in a database because a given number of attributes and coordinates is associated with each point. Line and polygon objects are more difficult to realize in a database because of the variation in the number of points composing them. A line or a polygon may comprise two points or 2000 or more points, depending on the extent of the line and the complexity of the area, which is delineated by a boundary line that begins and ends at the same point (Figure 4.7).

Object spatial information and object attributes are often stored in different databases to ease the manipulation of lines and areas, but in some systems they are stored together. The separate storage of at-

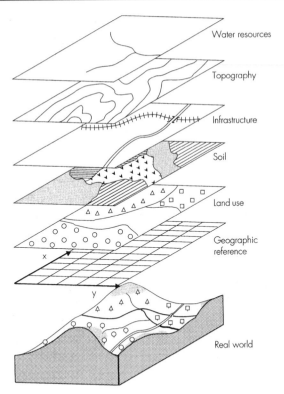

Water resources

Topography

Infrastructure

Soil

Land use

Geographic reference

Real world

Figure 4.6
Theme codes in the digital map data can be used to separate data into thematic layers.

tribute and spatial information data requires that all objects in the attribute tables be associated with the corresponding spatial information. This association is achieved by inserting spatially stable and relevant attribute data or codes from the attribute table into the spatial information, or vice versa. In other words, identical objects have the same identities in both databases.

The identity (ID) codes used to label and connect spatial information and attribute table data are most often numerical, but may be alphanumerical. Typical identity codes include building numbers, property numbers, and street numbers and addresses. If the data are ordered in a manuscript map, each object may be assigned a serial number used in both the spatial information and the attribute databases. Polygons for vegetation mapping can, for example, be numbered from 1 onward, but pipes, manholes, and so on, are usually numbered according to an administrative system. ID codes allow differentiation between objects, whereas theme codes allow for differentiation between different groups of objects. Identity codes and thematic codes are, in principle, both attribute data. However, they are very closely tied to geometry and are therefore often treated as such, as described earlier.

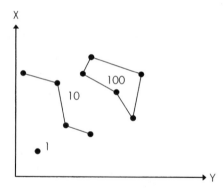

Geometry	ID	Thematic codes	Attributes		Coordinates
Point	1	1000	A	B	4,3 (single point)
Line	10	7000	C	D	2,1 11,5 · · 5,9 (string)
Area	100	5000	C	E	14,10 12,14 · · · · 14,10 (closed polygon)

Figure 4.7
Each object is assigned attributes and coordinates. However, the number of coordinates for lines and polygons will vary considerably, depending on the length of the line and circumference of the polygon. This may make it inappropriate to store attributes and geometry together and is one of the reasons that many systems store them separately.

Identifiers are normally entered together with the relevant data, but they may also be entered later, with the use of an interactive human–machine process such as keying in identifiers for objects pointed out on the screen.

Some systems tie a polygon's ID code to a characteristic point in the polygon, known as the *representative point* or *label point*. Representative points may be computed or identified interactively on-screen, and codes may be entered manually for the relevant polygons. The attribute values of the polygon are then linked to this label point. Today, systems are available that treat polygons as independent objects without label points. The representative points can also be used for placement of labels (attribute values, text, etc.) in systems based on graphic presentation.

Typical digital geometric data for GIS are illustrated in Figure 4.8.

Plotting may be controlled by appending drawing instructions to the thematic code, to the individual identifiers, or to other object attribute values. In a finished map, tabular data appear on a foreground map against the background of a base map derived from the remaining map data. Look-up tables are usually used to translate tabular data to map symbols (Figure 4.9). Map presentations are discussed further in Chapter 16.

I.D.	Thematic code	X-coordinates	Y-coordinates
11	30	23 999.80	10 008.55
	30	23 990.50	10 015.10
34	40	24 876.30	11 122.86
	40	24 890.10	11 150.30
–	–	–	–
	–	–	–
122	20	24 870.25	11 130.23

Code list	
20	Topography
30	Infrastructure
40	Soil

Figure 4.8
Typical section of digital map data with relevant code list.

4.2.2 Spaghetti model

As illustrated in Section 4.2.1, digital map data comprise lines of contiguous numerals pertaining to spatially referenced points. *Spaghetti
data* are a collection of points and lines with no real connection (Figure 4.10). What appears as a long, continuous line on the map or in the
terrain may consist of several line segments that are to be found in odd
places in the data file. There are no specific points that designate where
lines might cross, nor are there any details of logical relationships
between objects. Polygons are represented by their circumscribing

Figure 4.9
Drawing instructions are often designated in look-up tables. Thematic code
values or attribute values are often input values in the tables, whereas output
values can be symbol types, colors, line thickness, and so on.

63

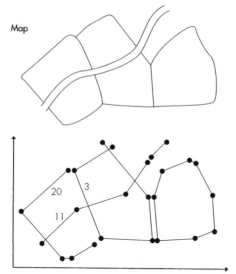

Figure 4.10
Spaghetti data is a term often used to describe digital map data with crossing lines, loose ends, double digitalization of common boundaries between adjacent polygons, and so on. These data lie in a pile, just like spaghetti. Several line segments are to be found in odd places in the data file.

boundaries, so that common boundaries between adjacent polygons are registered twice (often with slightly differing coordinates, which can lead to overlaps and gaps between the polygons). The lines of data are unlinked and together are a confusion of crossings (Figure 4.10).

Unlinked (spaghetti) data usually include data derived either from the manual digitizing of maps or from digital photogrammetric registration. Consequently, spaghetti data are often viewed as raw digital data. These data are amenable to graphic presentation—the delineation of borders, for example—even though they may not form completely closed polygons. Otherwise, their usefulness in GIS applications is severely limited.

One drawback is that both data storage and data searches are sequential. Hence, search times are often unduly long for such routine operations as finding commonality between two polygons, determining line intersection points, or identifying points within a given geographical area. Other operations vital in GIS, such as overlaying and network analysis, are intractable. Furthermore, unlinked data require an inordinate amount of storage memory because all polygons are stored as independent coordinate sequences, which means that all lines common to two neighboring polygons are stored twice. The typical memory required for unlinked data is illustrated in Figure 4.11.

Complex polygons with holes cannot be handled in this simple polygon structure. However, it is possible to extend the simple polygon

Map scale	Map sheet	MBytes
1 : 250,000	50 x 60 cm	25 – 50
1 : 50,000	50 x 60 cm	15 – 25
1 : 5000	48 x 64 cm	2,5 – 10
1 : 1000	60 x 80 cm	1 – 3

Figure 4.11
Digital map data require an inordinate amount of storage memory. Normally, elevation data comprise 25 to 90% of the total data volume.

structure by making the joint border between polygons identical and storing it only once. Storing a joint border can be achieved by adding to the coordinate list a separate table with an overview of which polygons belong to the different points. It is still not possible to handle complex polygons with holes.

4.2.3 Topology model

Topology is the branch of mathematics that deals with geometric properties that remain invariable under certain transformations, such as stretching or bending. The *topology model* is one in which the connections and relationships between objects are described independent of their coordinates; their topology remains fixed as geometry is stretched and bent. Hence, the topology model overcomes the major weakness of the spaghetti model, which lacks the relationships requisite to many GIS manipulations and presentations.

The topology model is based on mathematical graph theory and employs nodes and links. A *node* can be a point where two lines intersect, an endpoint on a line, or a given point on a line. For example, in a road network the intersection of two roads, the end of a cul-de-sac, or a tunnel adit may generate a node. A *link* is a segment of a line between two nodes. A link is also designated in the literature as an arc or chain. The term *edge* is often used in data modeling to mean *link*. Links have a start node and an end node and therefore have a direction in a topology model. Several links can share a node, and a collection of such links and nodes is known as a *network*. A closed polygon consisting of alternating nodes and links forms an area (Figure 4.12). A closed polygon is also known as *ring* or *boundary*. The expression *face* is often used in data modeling to mean the area formed by a closed polygon. A single point can be regarded as a degenerate node and as a link with zero length (Laurini and Thompson 1992). Theme codes should be taken into consideration in creating nodes to ensure that they are created only between relevant themes (e.g., at the junction between a national highway and a county road, not between roads and property boundaries).

Unique identities are automatically assigned to all links, nodes, and polygons, and attribute data describing connections are associated

65

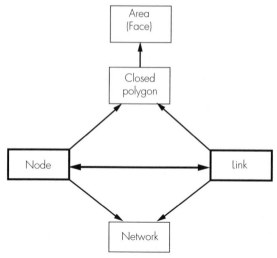

Figure 4.12
Elements that build up topological structures: node, link (arc, chain, edge),
network, closed polygon (ring, boundary), and area (face, surface, polygon).

with all identities. Topology can therefore be described in three tables
(Figure 4.13):

1. The polygon topology table lists the links comprising all polygons, each of which is identified by a number.
2. The node topology table lists the links that meet at each node.
3. The link topology table lists the nodes on which each link terminates and the polygons on the right and left of each link, with
   right and left defined in the direction from a designated start
   node to a finish node.

The system creates these tables automatically.

A table with coordinates to the links ties these features to the real
world and permits computations of distances, areas, intersections, and
other numerical parameters. The geometry of the links is stored in its
own subordinate table (see Figure 4.13). In general, one can say that
vector data consists of geometric primitives (point, lines, and areas)
and topological primitives (links and nodes). These primitives are
used, separately or in combination, to construct objects that express
the spatial characteristics of geographic objects.

In the topology model, border lines that separate adjoining polygons are stored only once. Complex polygons with holes are not handled in any special way, but the holes can be identified by the fact that
they are surrounded by the same adjoining polygon. Because every
polygon is individually identified, it is also possible to link attributes
directly to each by means of its ID.

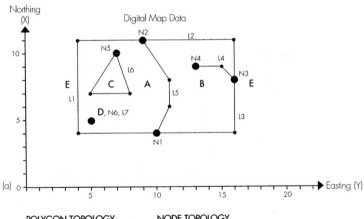

(a)

POLYGON TOPOLOGY

Polygon	Links
A	L1, L5
B	L2, L3, L5
C	L6
D	L7
E	L1, L2, L3

(b)

NODE TOPOLOGY

Node	Links
N1	L1, L3, L5
N2	L1, L2, L5
N3	L2, L3, L4
N4	L4
N5	L6
N6	L7

(c)

LINK TOPOLOGY

Links	Start node	End node	Left polygon	Right polygon
L1	N1	N2	E	A
L2	N2	N3	E	B
L3	N3	N1	E	B
L4	N3	N4	B	B
L5	N2	N1	B	A
L6	N5	N5	A	C
L7	N6	N6	A	A

(d)

LINK COORDINATES

Link	Coordinates			
L1	4,10	4,4	11,4	11,9
L2	11,9	11,16	8,16	
L3	8,16	4,16	4,10	
L4	8,16	9,15	9,13	
L5	11,9	8,11	6,11	4,10
L6	10,7	7,8	7,5	10,7
L7	5,5			

(e)

Figure 4.13
The idea behind the topology model is that all geometric objects *(a)* digital map data can be represented by nodes and links. The objects' attributes and relationships can be described by storing nodes and links in three tables: *(b)* a polygon table, *(c)* a node topology table, and *(d)* a link topology table. *(e)* An additional table gives the objects' geographical coordinates.

Numerous spatial analyses may then be performed, including:

- Overlaying
- Network analyses
- Contiguity analyses
- Connectivity analyses

Topological data may be used directly in contiguity analyses and other manipulations with no intervening, time-consuming geometric operations.

Once the topology has been created, a map can be plotted with solid colors. This is not always possible with spaghetti data. Thematic layers of topological data can also be used to steer the plotting sequence. The sequence influences what becomes visible on the map. For example, a green area superimposed on a white house will render the house invisible on the map unless the house creates a complex polygon with a *hole* in it.

Topology requires that all lines should be connected, all polygons closed, and all loose ends removed. Even gaps as small as 0.001mm may be excessive, so errors should be removed either prior to or during the compilation of topological tables. Coordinates are stored in GIS as *floating points*. Therefore, it is difficult to get two chance points to lie exactly one on top of the other. A function known as *snap* or *fuzzy tolerance* can be used to define two points as identical if they are close enough to each other. Using the snap function with a defined tolerance of, say, 0.01mm, a search can be made around the end of a line or around an existing point that is assumed to have the same coordinates as the last point registered. When this point is found, the two points will be snapped together to form a common node, thereby closing the polygon. The same procedure can be carried out automatically on existing data. A node can also be created in existing data by calculating the point of intersection between lines. Meaningless loose ends can be removed by testing with a given minimum length.

The automatic building of topology includes the following:

- Links are established.
- Isolated points are converted to nodes.
- Nodes are established at all ends of links.
- Nodes are established in all links that cross each other.
- Nodes that are very close to each other are merged.
- Very short links are removed.

Topological information permits automatic verification of data consistency to detect such errors as the incomplete closing of polygons during the encoding process. The graph theory contains formulas for the calculation of such data errors. There has to be a fixed relationship between the number of nodes, lines, and polygons in one data set. A run-through of the data in positive and negative directions will produce the same result.

The topological model has a few drawbacks. The computational time required to identify all nodes may be relatively long. Uncertainties and errors may easily arise in connection with the closing of polygons and formation of nodes in complex networks (such as in road interchanges). Operators must solve such problems. When new data are entered and existing data updated, new nodes must be computed and the topology tables brought up to date.

Topological data may require a longer plotting time than spaghetti data because of the separation of lines into nodes and links. However, the overall advantages of the topology model over the spaghetti model make it the prime choice in most GISs. Today, efficient software and faster computers enable topology to be established on-the-fly; thus the disadvantages of topological data as compared with spaghetti data have become less important.

4.2.4 Data compression

The amount of memory needed can be reduced by using data compression techniques. Most of these automatic techniques are based on removing points from continuous lines (contour lines, etc.). Good data compression techniques, therefore, are those that preserve the highest possible degree of geometric accuracy. The most basic technique involves the elimination of repetitive characters: for example, the first character of all coordinates along a particular axis. The repetitive character needs to be entered only once; subsequently, it may be added to each set of coordinates. This particular technique has no effect on the geometry. Savings in characters stored are illustrated in Figure 4.14.

There are other automatic methods of removing points. One simple means is to keep only every nth point on a line. The lower the value of n, the greater the number of points that will be removed. This method does not take into account geometric accuracy; however, this can be compensated for by testing the curvature of the line. The simplest and most popular of the more advanced methods is the *Douglas–Peucker algorithm* (Figure 4.15). The routines used in this method are as follows:

Original data		Compacted data	
Northing	Easting	10,000	70,000
10,234	70,565	234	565
10,245	70,599	245	599
10,167	70,423	167	423
–	–	–	–
–	–	–	–
etc.	etc.	etc.	etc.

Figure 4.14
Simple data compression. The volume of data to be stored can be reduced to a single entry, giving the value common to all coordinate values.

69

Figure 4.15
Douglas–Peucker method, to save storage space, t = tolerance. See the text.

1. A minimum distance, tolerance *(t)*, is defined.
2. A trend line is drawn between the start and end points on the original line.
3. The angular distance between the trend line and all break points on the original line is calculated.
4. Should no break points lie farther from the trend line than the tolerance *(t)*, the original line is replaced by the trend line.
5. Where some points are at a greater distance than the tolerance value, the point that lies farthest from the trend line is identified.
6. Lines are drawn from each of the trend line's end points to this maximum point.
7. Steps 3 to 6 are repeated until all line segments have been handled.

Points of little or no value in describing a line may also be eliminated by moving a corridor step by step along a line and deleting points that are closer to the neighboring point than a given value or where the vectors create an angle that is smaller than the given value (Figure 4.16). Contours and other lines can also be replaced with mathematical functions, such as straight lines, parabolas, and polynomials. A spline function comprises segments of polynomials joined smoothly at a finite number of points so as to approximate a line. As shown in Figure 4.17, a spline function can involve several polynomials to build a complex shape. It has been reported that a spline function representing nautical chart data has reduced data volume by 95%.

The amount of memory required to store a given amount of data often depends on the format in which data are entered. Some formats contain more administrative routines than others, some have vacant space. Thus, the gross volumes stored are frequently related to format.

4.2.5 Storing vector data

The manner in which digital map data are stored in a record is determined by a format, a set of instructions specifying how data are arranged in fields. The latter are groups of characters or words, which, in turn, are treated as units of data. The format stipulates how the GIS will read data into the fields: total number of fields specified, number

Figure 4.16
Reducing lines by use of a corridor. The number of points needed to describe a line may be reduced by moving forward a corridor of a given width until it touches the digitized line. All the points on the line in the corridor, apart from the first and last, are thereby deleted. This process is repeated until the entire line has been trimmed.

of characters permissible in each field, number of spaces between fields, which fields are numeric and which are text, and so on. An example is shown in Figure 4.18. The way files are stored on the hard disk is dependent upon the operating system.

The information content of the data is designated not in the format but ancillary to it—for example, in a heading. Typical specifications for information content may include field assignments, such as the point number in the first field, the thematic code in the second, easting in the third, northing in the fourth, and elevation in the fifth. The meanings of the numeric codes used, the georeferencing system, and so

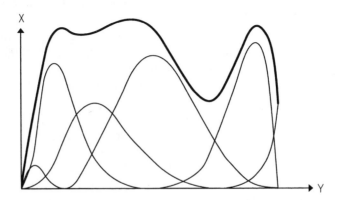

Figure 4.17
A complicated line course can be replaced by a few simple polynomials (spline). This reduces considerably the volume of data emanating from line courses.

	Field				
Record	1	2	3	4	5
1	10	4003	1556.3	2303.1	131.0
2	11	4003	1590.2	2370.2	135.5
3	12	4003	1610.6	2415.4	139.7
4	13	4011	1605.3	2541.6	133.2
5	14	4011	1540.1	2623.3	139.2
6	15	4011	1510.5	2759.9	145.6
¦					
¦					
¦					
¦					

Figure 4.18
In a single data file, data are stored in records and each record is divided into fields where the values are stored. The format determines how the data are arranged in fields; according to the number of characters, whether they are text or numeric, and so on.

forth, must also be given. This kind of information is usually placed in the header of the file. Spaghetti data are stored in a simple file structure and in the order in which the data have been registered. Topological data are stored in different tables/files.

Users of conventional maps know the frustrations of extracting information from maps produced by various agencies using differing map sheet series, varying scales and coordinate systems, and, frequently, different symbols for the same themes. Moreover, the cartographic version of Murphy's law dictates that the necessary information is all too often located in the corners where four adjoining map sheets meet.

Database storage of cartographic data can overcome these problems because it involves standardization of data through common reference systems and uniform formats. Cartographic data from various sources can, with few limitations, be combined. The results are then independent of map sheet series and scales.

Standardized storage makes the presentation of data compiled from dissimilar sources much easier. For example, uniform storage formats permit the combination of telecommunications administration network data with property survey data, or of geological information from 1:50,000 scale maps with vegetation data from 1:20,000 scale maps.

Digital map data are stored in databases, the computerized equivalent of conventional file drawers and cabinets. Data entries in a database can be updated far more rapidly than data printed on map sheets on file. "Smart" programs, known as database management systems (DBMSs), have been compiled to maintain, access, and manipulate databases. The various DBMSs differ primarily in the ways in which the data are organized. Their selection and use are vital in GIS applications because they determine the speed and flexibility with which data may be accessed.

It is usual to split topological data into different thematic layers to simplify storage and to improve access to data. This division is done so that no overlap occurs between polygons within each thematic layer. For example, property boundaries are stored in one layer while other data overlapping the property, such as roads, buildings, and vegetation boundaries, are stored in another. The disadvantage of this system is that common lines between objects (e.g., roads and properties) that are stored in different layers have to be removed several times. This problem can be avoided by using object-based storage.

4.2.6 Comments on spaghetti and topology models

When digitizing lines such as those on land-use maps, the borders of surfaces are digitized both as spaghetti data and as separate objects. When creating topology, this model is converted to a layer model. The discussion of spaghetti and topology is very much based on the assumption that a class of area entities is always a tiling of the plane in which every point lies in exactly one polygon. (See the discussion on entities and fields in Section 3.6.2.) However, the problems related to spaghetti and topology have changed somewhat during recent years with the advent of new GIS software that treats polygons as independent objects that may overlap and need not fill the plane, and with systems permitting shapes. Many of the traditional arguments for area coverage/layer model and use of topology are based on the assumption of needing to avoid computation. New and more powerful computers eliminate the need for reduction in calculation time. Today, topology can easily be built on-the-fly.

4.3 Raster Data Models

Raster data are applied in at least four ways:

1. Models describing the real world
2. Digital image scans of existing maps
3. Compiling digital satellite and image data
4. Automatic drawing driven by raster output units

In the first method, raster data are associated with selected data models of the real world; in the second and third, with compilation methods; and in the fourth, with presentation methods. The respective computer manipulations may have much in common; as discussed later, satellite data may be entered in a raster model. For ease of explanation, digital raster maps are treated in Section 9.3, satellite data are discussed in Section 9.6, and output units generally are the subject of Chapter 7.

Raster model
The raster model represents reality through selected surfaces arranged in a regular pattern. Reality is thus generalized in terms of uniform,

regular cells, which are usually rectangular or square but may be triangular or hexagonal. The raster model is in many ways a mathematical model, as represented by the regular cell pattern (Figure 4.19). Because squares or rectangles are often used and a pictorial view of them resembles a classic grid of squares, it is sometimes called the *grid model*. Geometric *resolution* of the model depends on the size of the cells. Common sizes are 10×10m, 100×100m, 1×1km, and 10×10km. Many countries have set up national digital elevation models based on 100×100 m cells. Within each cell, the terrain is generalized to be a flat surface of constant elevation. Each cell represents an area, although the area has no direct digital limitation. The limitation is given only indirectly through cell size, in contrast to the area in vector data, which is limited by a sequence of line segments that form a closed polygon.

The quadric raster cells, of uniform size throughout a model, affect final drawings in two ways. First, lines that are continuous and smooth in a vector model will become jagged, with the jag size corresponding to cell size. Second, resolution is constant: Regions with few variations are as detailed as those with major variations, and vice versa.

The cells of a model are given in a sequence determined by a hierarchy of rows and columns in a matrix, with numbering usually starting from the upper left corner (Figure 4.20). The geometric location of a cell, and hence of the object it represents, is stated in terms of its row and column numbers. This identification corresponds to the directional coordinates of the vector model. The cells are often called *pixels* (picture elements), a term borrowed from the video screen technology used in television and computer displays. A pixel is the smallest ele-

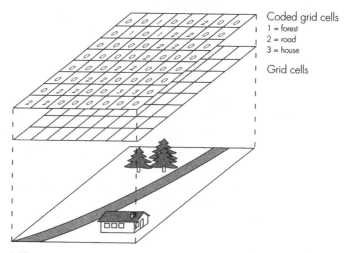

Figure 4.19
Raster data can be visualized as a grid lying over the terrain. Each grid cell has a code stored in the database describing the terrain within that particular cell.

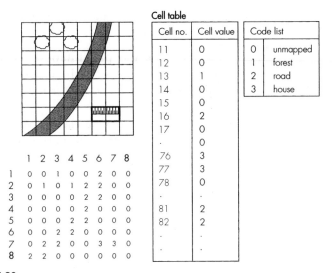

Cell table

Cell no.	Cell value
11	0
12	0
13	1
14	0
15	0
16	2
17	0
.	0
76	3
77	3
78	0
.	.
81	2
82	2
.	.
.	.

Code list

0	unmapped
1	forest
2	road
3	house

	1	2	3	4	5	6	7	8
1	0	0	1	0	0	2	0	0
2	0	1	0	1	2	2	0	0
3	0	0	0	0	2	2	0	0
4	0	0	0	0	2	0	0	0
5	0	0	0	2	2	0	0	0
6	0	0	2	2	0	0	0	0
7	0	2	2	0	0	3	3	0
8	2	2	0	0	0	0	0	0

Figure 4.20
A line number and column number define the cell's position in the raster data.
The data are then stored in a table giving the number and attribute value of
each cell.

ment of an image that can be processed and displayed individually.
The raster techniques used in GIS are siblings of the rasters long used
to facilitate the manipulation and display of information and, consequently, are suited to computerized techniques.

Because cells form a regular pattern, relatively simple rules can be
made to calculate new attribute values for random points in the grid.
In addition, the size of cells can initially be selected so that variations
in the terrain can be captured satisfactorily. This means that the raster
model is relatively well suited to represent phenomena with continuous variation—known as a geographic field.

4.3.1 Realizing the raster model

Raster models are created by assigning real-world values to pixels
(Figure 4.19). The assigned values comprise the attributes of the objects that the cells represent—and because the cells themselves are in
a raster, only the assigned values are stored. Values, usually alphanumeric, should be assigned to all the pixels in a raster. Otherwise, there
is little purpose in drawing empty rows and columns in a raster.

Consider a grid of cells superimposed on the ground or on a map.
Assigning the values/codes of the underlying objects/features to the
cells creates the model. The approach is comprehensive because everything covered by the raster is included in the model. *Draping* a ground
surface in this way regards the ground or map as a plane surface.

Some GISs can manipulate both numerical values and text values (such as types of vegetation). Hence, cell values may represent numerous phenomena, including:

- Physical variables, such as precipitation and topography, with amounts and elevations, respectively, assigned to the cells
- Administrative regions, with codes for urban districts, statistical units, and so on
- Land use, with cell values from a classification system
- References to tables of information pertaining to the area(s) the cells cover, such as references to attribute tables
- Distances from a given object
- Emitted and/or reflected energy as a function of wavelength—satellite data

A single cell may be assigned only one value, so dissimilar objects and their values must be assigned to different raster layers, each of which deals with one thematic topic (Figure 4.21). Hence, in raster models as in vector models, there are thematic layers for topography, water supply systems, land use, and soil type. However, because of the differences in the way attribute information is manipulated, raster models usually have more layers than those in vector models. In a vector model, attributes are assigned directly to objects. For instance, a pH value might be assigned directly to the object "lake." In a raster model, the equivalent assignment requires one thematic layer for the lake, in which cells are assigned to the lake in question, and a second thematic layer for the cells carrying the pH values. Raster databases may therefore contain hundreds of thematic layers (Figure 4.21)

In practice, a single cell may cover parts of two or more objects or values. Normally, the value assigned is that of the object taking up the

Figure 4.21
Only one attribute value may be assigned to each cell. Objects that have several attributes are therefore represented with a number of raster layers, one for each attribute.

greater part of the cell's area, or of the object at the middle of the cell, or that of an average computed for the whole of the cells.

Cell locations, defined in terms of rows and columns, may be transformed to rectangular ground coordinates—for example, by assigning ground coordinates to the center of the upper left cell of a raster (cell no. 1,1). If the raster is to be oriented north–south, the columns are aligned along the northing axis and the rows along the easting axis. The coordinates of all cell corners and centers can then be computed using the known cell shapes and sizes.

Object relations, which in the vector model are described by topology, are only partly inherent in the raster structure. When the row and column numbers of a cell are known, the locations of neighboring cells can easily be calculated. In the same way, cells contained in a given polygon may be located simply by searching with a stipulated value. It is much more difficult, however, to identify all the cells located on the border between two polygons. Polygon areas are determined merely by adding up constituent cells. Some operations, though, are more cumbersome. An example of this is computation of a polygon's perimeter length, which requires a search for, and identification of, all the cells along the polygon's border.

Overviews of phenomena in a given area are obtained from a raster model quickly and easily by searching all the thematic layers for cells with the same row and column numbers. The relevant overlay analysis is described in Section 14.11.

4.3.2   Coding raster data

A raster has no meaning until the cells are filled with attribute values. Normally, cell values are represented with numerical codes, either as whole numbers (integer) or decimal numbers (real).

Cell values are entered from word processing files, databases, or other sources in the same sequence in which they are registered (Figure 4.24). When coding cells, it will be the case that a whole cell will not be covered by one attribute. Thus, as previously mentioned, some doubt could arise as to which value the cell should represent (see Figure 4.22).

Raster data may be available from a variety of dissimilar sources, ranging from satellite data and data entered manually to digital elevation data. Their collocation requires that cells from differing sources and thematic layers correspond with each other. In other words, cells having the same row and column numbers must refer to the same ground area in order to allow operations to be performed on several layers (Figure 4.23). Various computations such as transformation and resampling may be necessary to accommodate any differences in cell shape and size.

Cells may also contain values referenced to attribute tables. The cells of a thematic layer may be coded so that their values correspond to

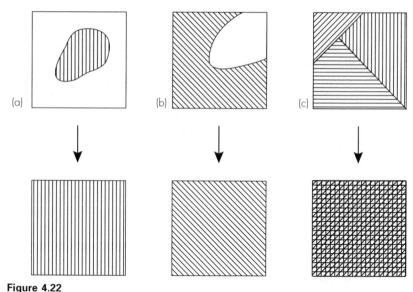

Figure 4.22
When an attribute does not cover a whole cell, different principles can be selected to give a cell value: *(a)* the cell is given the value that lies in the center of the cell, *(b)* the cell is assigned the attribute that covers the greater parts of the cell, *(c)* the cell gets its value from the average (or another interpolation method) of its attributes.

identities in a given attribute table. Attribute data or tabular data may be coded independently, irrespective of whether the geometry is represented with the use of vector data or raster data. Generally, raster data are based on the division of the area covered into small, uniform units according to a tessellation of the space, and the assignment to each unit of an attribute value.

4.3.3 Storage of raster data

If the cell values of a raster model are entered in fixed matrices with rows and columns identical to those of the registered data, only the

Figure 4.23
It is possible to make direct calculations on the raster layers.

cell values need to be stored; row and column numbers need not. Even when only cell values are stored, the volumes of data can easily become unwieldy. Typical operations may involve 200 thematic layers, each containing 5000 cells.

The total number of cell values stored is thus 200 × 5000 = 1 million. A Landsat satellite raster image contains about 7 million pixels, a LANDSAT TM image about 35 million pixels. In addition, data volume is dependent on how many bytes are used to store a single cell value. This again is dependent on whether the numerical value is expressed as an integer, real, single, or double precision.

Different GISs organize raster data in different ways (format). The most visual method is to store cell values continuously in the file—that is, to start with the first cell and the first row in the matrix and fill up the file, column by column and row by row, without marking the division between columns and rows (Figure 4.24). Thus, information on the number of columns and rows has to be entered separately at the beginning of the file. An less used method is to enter values that mark the end of each row.

When storing raster data, some general information on the data set has to be stored, such as the following:

1. Numbers of rows and columns
2. Cell size
3. Georeferencing—for example, specification of coordinates of the upper left corner of the upper left cell
4. An explanation of what the different cell values/codes represent

These meta-data are stored as a header in the raster file or in a separate file.

4.3.4 Compression of raster data

Because raster data often represent very large volumes of data, methods of compressing data have been developed. Various devices may be

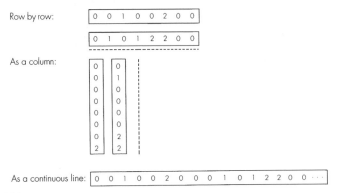

Figure 4.24
Typical cell input.

employed to reduce data volume and, consequently, storage memory requirements. The most common devices are:

- Run-length encoding
- Chain codes
- Block codes
- Quad-tree model

Run-length encoding

Cells of the same value are often neighbors because they pertain to the same soil type, the same population density of an area, or other similar parameters. Thus, cells of the same value in a row may be compacted by stating the value and their total. This type of compacting, called *run-length encoding,* is illustrated in Figure 4.25.

However, data volume is doubled in a case where a single cell has a different value from the neighboring cells in the row. On the other hand, where three or more identical cells follow each other, a reduction will occur. An even greater reduction can be achieved for binary (black and white) raster data by storing only the cells that represent a change in cell values. Further compacting may be achieved by applying the same process recursively to subsequent lines.

Chain codes

Raster data can also be compressed by storing only the cells that lie on the outer border of homogenous areas/polygons. These are called *chain codes.* The cells are linked together by specifying a start cell, the sequence/number of similar cells in the same direction, and a direction code (North, South, East, and West). Thus, the entire polygon bor-

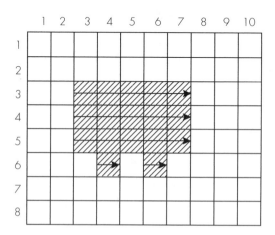

Figure 4.25
Run-length encoding is the method used to save data storage space, by reducing a row of cells with the same value to a single unit having a specific value and quantity. Result of the run-length encoding: Row 3: 3,8, Row 4: 3,8, Row 5: 3,8, Row 6: 4,4 6,6.

Figure 4.26
Chain coding. Staring in the upper left corner and choosing the codes East = 0, South = 1, West = 2, and North = 3, the following chain coding will result: 1^6, 2^1, 3^2, 2^2, 0^1, 3^2, 2^1, 0^1, 3^1, 0^1. The exponent represents the length of the segment and the mantissa represents the direction code.

der is covered (see Figure 4.26). For large areas with a simple border it is possible to achieve considerable compression, but for small and complex areas double storage can easily occur because the same cells are described twice (both East/West or both North/South) when the chain goes in and out of narrow passages and corners.

Block codes
The basis of the *block code* method is that homogenous quadratic areas are stored as one unit (block). Homogenous quadratic blocks must first be identified. Each block is allocated three values linked to position and size. The position of the block's center or bottom left corner is given, with column and row number. The size of the block is normally stated as the radius from the block's center to the cells at the corners of the block. Storage of a single cell value will lead to an increase of data volume, but the volume will be reduced with squares of four or more cells (see Figure 4.27).

Forest

Water

Different
soil types

Figure 4.27
Computer memory capacity requirements can be reduced by storing homogeneous and uniform (quadratic) areas, consisting of many basis cells, as one single unit.

Figure 4.28
Quad-tree raster encoding recursively subdivides a map into quarters to define the boundary of an area. The shaded squares on this grid represent the region that is to be stored in a quad-tree. (Adapted from Green and Rapek 1990.)

4.3.5 Quad-tree model

Traditionally, the raster model is based on dividing the real world into equal-sized rectangular cells. However, in many cases, it can be more practical to use a model with varying cell size. Larger cells (lower resolution) may be used to represent larger homogeneous areas, and smaller cells (higher resolution) may be used for more finely detailed areas. This approach, known as the *quad-tree representation*, is a refinement of the block code method. In representing a given area, the aggregate amount of data involved is proportional to the square of the resolution (into cells). Because the quad-tree model is a very practical concept, it is preferable for the storage of both small and large volumes of data. The hierarchical data structuring typical of the quad-tree has other advantages, which are discussed later in this book.

The quad-tree paradigm divides a geographical area into square cells of sizes varying from relatively large to that of the smallest cell of the raster. Usually, the squares are then quartered into four smaller squares (Figures 4.28 to 4.31). The quartering may be continued to a suitable level until a square is found to be so homogeneous that it no longer needs to be divided, and the data on it can be stored as a unit.

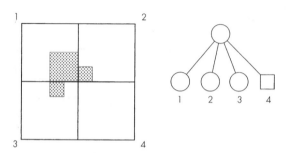

Figure 4.29
Initially, the grid is divided into four quadrants. If part of the region falls within a quadrant, it is assigned a white circle. If no part of the region falls in a quadrant, it is assigned a white square. (Adapted from Green and Rapek 1990.)

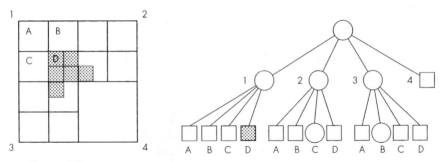

Figure 4.30
The quadrants that contain part of the region are again subdivided, which adds another level to the tree. If a quadrant is filled completely by part of the region, it is represented by a black square. (Adapted from Green and Rapek 1990.)

A larger square may therefore comprise several raster cells having the same values. However, homogeneous areas that are not square or do not coincide with the pattern of squares employed may be further divided into homogeneous squares.

The structure of the quad-tree resembles an inverted tree, whose leaves are the pointers to the attributes of homogeneous squares and whose branch forks are pointers to smaller squares—hence the name quad-tree (Figure 4.31).

The quad-tree principle can also be applied to vector data (Figure 4.32). Quad-tree models have been developed for both spaghetti and topological data. Even parts of lines/links will thus represent homogeneous units, right down to the smallest unit consisting of a single point or node. Indexing of vector data is used to give more rapid data access in specific areas. This subject is dealt with in greater detail in Chapter 12.

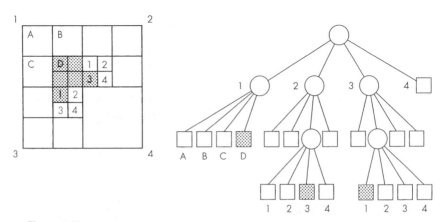

Figure 4.31
Subdivision of the grid continues until each leaf in the tree is represented by either a black or a white square. Only 19 leaves are required to store this region in a quad-tree. (Adapted from Green and Rapek 1990.)

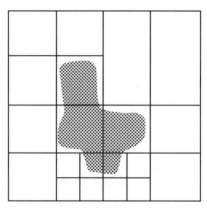

Figure 4.32
Quad-tree representation of a polygon.

The advantages of the quad-tree model are:

- Rapid data manipulation, because homogeneous areas are not divided into the smallest cells used
- Rapid search, because larger homogeneous areas are located higher up in the point structure
- Compact storage, because homogeneous squares are stored as units
- Efficient storage structure for certain operations, including searching for neighboring squares or for a square containing a specific point

The disadvantages of the quad-tree model are:

- Establishing the structure requires considerable processing time.
- Protracted processing may prolong alterations and updating.
- Data entered must be relatively homogeneous.
- Complex data may require more storage capacity than ordinary raster storage.

The compression methods are most effective with raster data having large homogenous areas. Should the raster have continuous changes in cell values, the effect would be relatively small. One should also be aware of that compressed data usually must be expanded to a complete raster when visualized and when calculations and analyses such as overlay or similar functions are to be done, and that that this expansion takes some time.

4.4 Automatic Conversion Between Vector and Raster Models

GIS applications sometimes require data in a form differing from that which is available. As a result, many GISs now have facilities for au-

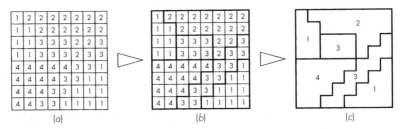

Figure 4.33
Conversion of raster data to vector data: *(a)* each raster cell is assigned an attribute value; *(b)* boundaries are set up between different attribute classes; *(c)* a polygon is created by storing x and y coordinates for the points adjacent to the boundaries.

tomatic conversion between vector and raster models. Raster data are converted to vector data through *vectorization*. The reverse process, which is just as common, is *rasterization*. In vectorization, areas containing the same cell values are converted to polygons with attribute values equivalent to the preconversion cell values (Figure 4.33). In the reverse process of converting polygons to cells, each cell falling within a polygon is assigned a value equal to the polygon attribute value (Figure 4.34).

Various routines are available for converting raster data to vector data, and vice versa. The former is the more complex and time-consuming of the two processes, and different conversion programs can yield differing results from the same set of raster data. In rasterization the following formulas can be used to place a coordinate defined pair in a raster (see also Figure 12.6):

$$\text{Column no.} = \left(\frac{x - x_{min}}{X_{max} - x_{min}}\right) N \tag{4.1a}$$

$$\text{Row no.} = \left(\frac{y - y_{min}}{Y_{max} - y_{min}}\right) M \tag{4.1b}$$

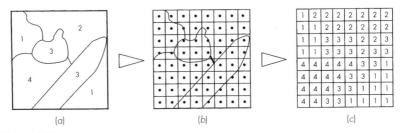

Figure 4.34
Conversion of vector data to raster data: *(a)* coded polygons; *(b)* a grid with the right cell size overlays the polygons (the polygons that contain the center of the individual cells are identified); *(c)* each cell is assigned the attribute code of the polygon to which it belongs.

N equals the total number of columns and M the total number of rows; the total should be rounded down to the nearest whole number. In connection with calculations, the origin is normally placed in the bottom of the left corner (x_{min} and y_{min}). The row number will therefore also be calculated from this origin, and must later be connected to the grid's real origin, which is the upper left corner. By rastering lines, the cells that are touched by a line will have the same value. For a polygon, the cells surrounded by the polygon are found and the cells on the inside are all given the same value.

Vectorization—converting from raster to vector data—is, as mentioned, more complicated than rasterization, especially with lines. The method, which is closely linked to scanning is discussed in more detail in Section 9.3. A great deal of aftertreatment of vectorized data is often required, for establishing topology and allocating IDs.

Normally, some information/data are lost in conversions. Consequently, converted data are less accurate than original data.

4.5 Vector Versus Raster Models

One of the basic decisions in GIS design involves the choice between vector and raster models, each of which has advantages and disadvantages. In the vector model, the observation units are end points and/or variable line or polygon magnitudes, whereas the raster model presupposes fixed observation areas in a grid. Otherwise, the models are, in principle, identical.

As discussed previously, vector and raster data have varying ability to represent reality. It is not always easy to recognize vector data's discrete objects out in the terrain. This applies especially to phenomena with diffuse borders and continuous changes, such as vegetation and population density. However, many real phenomena are related to locations. Measurements are often made at points, infrastructures are often related to lines, and administrative units are frequently described in terms of defined areas of various shapes and sizes.

Raster GIS emphasizes properties: Here, the basic units of observation are regular cells in a raster. As opposed to vector data, raster data are relatively well suited to represent continuous phenomena, provided that an appropriate cell size is selected. But not all phenomena are related directly to such grid patterns. At present, satellite data, digital ortho images, and digital elevation data account for the bulk of data available in raster form. In many countries, national elevation data models have been established based on a fixed grid (e.g., 100 $\times$ 100m). Other types of data usually have to be reworked to a greater or lesser degree to suit rasters. The accessibility of raster data may thus be a major problem and perhaps the greatest drawback of a raster GIS

	Raster	Vector
Data collection	Rapid	Slow
Data volume	Large	Small
Graphic treatment	Average	Good
Data structure	Simple	Complex
Geometrical accuracy	Low	High
Analysis in network	Poor	Good
Area analysis	Good	Average
Generalization	Simple	Complex
Continuous phenomena	Average	Poor

Figure 4.35
Some of the characteristic differences between raster and vector data.

in comparison with a vector GIS. However, as we have seen, there are methods for converting data from vector to raster.

A vector model, on the other hand, often requires the time-consuming and costly compilation of digital map data, whereas maps are integral parts of the data compiled for a raster model. Maps may be drawn for all cells as soon as they are assigned values.

Despite oversimplification from a functional viewpoint, vector data may be considered best suited for documentation, and raster data are more adept at showing the geographical variation of phenomena. Another simplification may be that vector data are preferable for line presentations and raster data are superior for area presentations. Selected characteristics of the vector and raster models are compared in Figure 4.35.

To date, the vector model has been dominant in commercial GIS implementations. The raster model, on the other hand, has been used more frequently in natural resource planning and management and in teaching because it is more easily explained and used. Many newer GISs can manipulate both vector and raster models. With dual capability, a GIS can exploit the respective advantages of both: Vector data may be converted to raster data to perform overlaying or other operations more easily performed using rasters, then converted back to vector data.

4.6 Attribute Data and Computer Registers

With the advantages of easy updating, rapid search, and the flexible superimposition of data, the computerized filing of information has become commonplace in administrative work. In many cases, inaccessible, massive quantities of traditional records and files have been replaced by workstations, from which very large amounts of information are rapidly accessible. Physical separation by rooms, buildings, national borders, or intervening distances is no longer a barrier to ready availability of information.

Computerization permits a simplification and coordination of registration efforts and can eliminate duplication and rationalize the overall filing process. In the public sector, central registers have been established as a common resource for numerous users.

Most countries have various computerized registers, including registers of:

- Customers/clients
- Motor vehicles
- Properties
- Literature
- Movables
- Buildings
- Addresses

- Citizens
- Pleasure boats
- Deeds
- Environment
- Companies
- Road and signs
- Personal data

Some of these registers are important in GIS applications. Others are of less interest. In many countries, though, work is under way to make public registers available to GIS users. Upon entry, register data are selected (structured) so that registers contain uniform and limited data. As for digital map data, register data are stored using formats. There is no general pattern for register content, but usually the items registered will have identities, locations, descriptive details, time and date notations, and sometimes references to other registers (Figure 4.36). For example, typical entries in a building register may contain all the details listed in Figure 4.37.

GIS users may be obliged to compile registers when they are unavailable in computerized form, either by using GIS techniques directly or by using other systems and subsequently exporting data to GIS. In all cases, data models should be specified.

4.6.1 Coding and entering attribute data

Attribute data may be coded for several reasons in order to:

- Establish an ID code between geometry and attributes
- Conserve computer memory

Figure 4.36
Geometric content is often limited to geometry/coordinates and topology. Attribute content often comprises location (address), various descriptions of the object, and dating (age, etc.).

BUILDING REGISTER	
Identity:	• Building number
Location:	• Address
	• Representative coordinates
Description:	• Builder/owner
	• Status (under construction, in use, demolished)
	• Type (detached house, semi-detached house, farmhouse, garage)
	• Function (residence, farm, shop)
	• Structure (concrete, frame, brick)
	• Water supply (public, cistern, stream)
	• Sewage (public, subsoil disposal, closed system)
	• Heating (electric, oil, wood)
	• Available area
Dating:	• Year built

Figure 4.37
Typical contents of a building register.

- Ease input work
- Ease verification of data entered
- Simplify subsequent searches for data in databases

The coding of geographical data is not new. Systems have been established in many fields for coding pipes, manholes, streets, properties, buildings, the names of towns, and so on. Indeed, codes have been used for many reasons, not least as file access keys or to conserve the space used on file cards.

Coding of attribute data often includes data structuring. Codes are often assigned according to a hierarchical classification system devised to ease such data operations as searching and sorting. Examples include the official codes widely used for addresses, names of towns, highways, and so on. For example, a land-use application may involve a hierarchy such as that illustrated in Figure 4.38.

Attribute data may also be tabulated, as illustrated in Figure 4.38. Tabulation and coding ease sorting, such as through finding all wooded objects or all transportation objects by using their respective land-use category codes. The column headings in the table are the names of attribute types. The type of data may be specified for each field, such as integer (land-use code), decimal (area), and text (name). Code tables may be compiled and used with the main table to produce more meaningful printouts from the system.

Attribute data may be entered relatively easily in most GISs, either manually via a keyboard or by importing data from an existing register; see Chapters 9 and 10 for further details. ID codes are usually entered together with the attribute data. They may also be registered or edited into compilations of attribute data that initially have no codes.

Level 1 Code	Attribute	Level 2 Code	Attribute	Level 3 Code	Attribute
100	Built-up	110	Industry	111	Light
				112	Heavy
				113	Other
		120	Transportation	121	Roads
				122	Railways
				123	Airports
				124	Parking
				125	Terminal
				126	Other
		130	....	131	....
200	Wooded	210	Coniferous	211	Fir
				212	Pine
				213	Ohter
		220	Deciduous	221	Oak
				222	Beech
				223	Other
		230	....	231	....

Figure 4.38
Land-use code hierarchy. Attributes are often coded in accordance with a hierarchical classification system.

4.6.2 Storing attribute data

Attribute data are usually most easily and expediently stored in tabular form. Each line in a table represents an object, each column an attribute. Attribute data are therefore often called *tabular data* and are normally stored in a relational database; see Chapter 8, especially Section 8.4.4.

Data on different types of objects are usually stored in separate tables, each dedicated to a single object type. In each table, line formats and lengths are identical throughout (Figure 4.39). The number of columns may be extended by combining several tables, either by using a common access key or by entering new attributes manually.

In principle, table design is independent of whether the geometrical data to which attributes refer are in the form of vector data or raster data. However, table content must be relevant to the objects, so each object or line must have a stable identity or access key. Data available

ID code	Land-use code	Area (ha)	Township code
1	222	22.67	0914
2	211	1.45	0916
3	121	46.80	0923

Figure 4.39
Tabulated attribute data. Attribute data can be sorted and arranged in various tabular forms.

in existing computerized registers are not always in convenient tabular form. As a result, conversions and roundabout methods must often be used to access such data for GIS uses; see Chapters 9 and 10.

4.7 Linking Digital Map and Attribute Information

4.7.1 Linking by ID

Common identifiers in map data and attribute data permit moving from map data to attribute data, and vice versa. Attribute data that basically lack georeferencing may be linked to geography. As illustrated in Figure 4.40, this is possible if the attribute data that lack georeferencing have access keys in common with attribute data that have other access keys in common with map data. The connection is then from attribute data to other attribute data to map data.

This illustrates one of the distinctive capabilities of GIS. Data that initially contain no geographical information or referencing may be given geographical dimensions and may therefore be used to enhance and present data in new ways, in maps or on screen. It is not always necessary to link geometry with attributes. In some instances, the geometry can be stored directly together, with the attribute data linked to each register object. This can occur in the case of, for example, a building register with coordinates representing each building or a register of measurement points for use in registering pollution levels. We

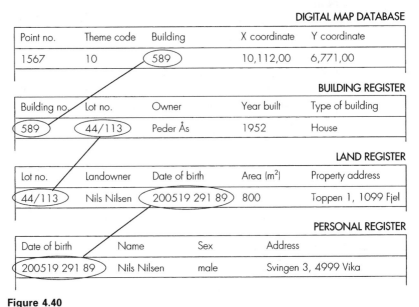

Figure 4.40
How elements of data in one register can be used as an access key to another register, thereby acting as a link between dissimilar data. A precondition is that identical information must be available in a minimum of two registers.

91

have also mentioned that identification codes and theme codes are, in principle, attribute data though stored together with geometry.

4.7.2 Linking by geometry

When attributes are stored together with geometry, it is possible to link attributes to other attributes by using overlay techniques such as point-in-polygon, line-in-polygon and polygon-in-polygon. The overlay techniques are dealt with in Chapter 14. Attributes linked to a polygon are transferred to point(s) by each point being assigned the attributes that belong to the polygon in which the point(s) are contained (see Figure 4.41). Polygons can correspondingly adopt the attributes of points. Equally, attributes linked to lines and polygons can be transferred to overlapping polygons.

Superimposing dissimilar data, such as geological data and vegetation data, is often hampered by a lack of commonality of the observations made in the field. That is, the observation areas listed in the respective attribute tables cannot be listed together because they refer to different sets of locations. In GIS this problem can be solved by using cartographic integration, in which overlay techniques are used to combine geometry from two dissimilar thematic maps into a single synthesized map. The synthesized map contains numerous new objects and areas, all of which are related to the two original thematic maps. Hence, the objects in the synthesized map comprise the least common units between the original maps and are therefore called *integrated terrain units* (ITUs; Figure 4.42). An attribute table is associated with the ITUs. In it, the ITUs are listed on the lines and the elements of the original thematic maps are in the columns. This table contains all the relevant attributes and therefore may be used in further analyses of the data.

Cartographic integration is straightforward when the areas of the original map data contain more or less homogeneous data, such as property data, land use, vegetation, and geology. Complexities arise when properties are not evenly distributed over an area. Consider, for

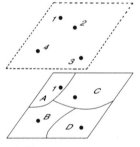

ID	Building no.	Polygon	Property
1	660	A	44/110
2	659	C	44/95
3	610	D	44/121
4	665	B	44/81

Figure 4.41
Point attributes can be linked to areas by a point-in-polygon operation.

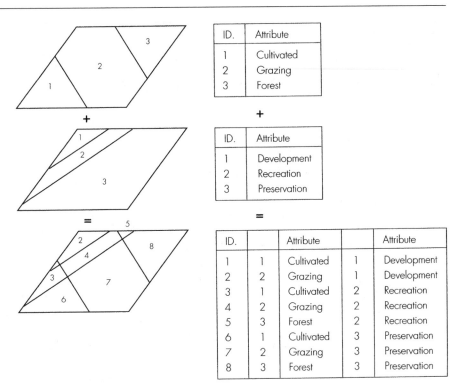

ID.	Attribute
1	Cultivated
2	Grazing
3	Forest

+

ID.	Attribute
1	Development
2	Recreation
3	Preservation

=

ID.		Attribute		Attribute
1	1	Cultivated	1	Development
2	2	Grazing	1	Development
3	1	Cultivated	2	Recreation
4	2	Grazing	2	Recreation
5	3	Forest	2	Recreation
6	1	Cultivated	3	Preservation
7	2	Grazing	3	Preservation
8	3	Forest	3	Preservation

Figure 4.42
Attribute data can be rendered comparable by superimposing geometry from dissimilar geographical units over each other to find the integrated terrain unit.

example, a typical township with an unevenly distributed population that averages 100 persons/km². An ITU may locate in an uninhabited area of the township and hence misrepresent the facts of its population. In all such cases, rules must be contrived to designate how attributes will be divided among ITUs. This can be solved by use of the fuzzy technique. The ITU is also called a *basic spatial unit* (BSU) and defined as a fundamental area unit that has homogeneous properties in the context of a particular subject.

CHAPTER 5

Advanced Data Models

5.1 Terrain Surface Representation

5.1.1 General

The models discussed in Chapter 4 describe limited parts of a two-dimensional real world. Several other data models used in GIS can extend the real world to include the terrain surface, the time factor, and movable objects. The digital representation of a terrain surface is called either a *digital terrain model* (DTM) or a *digital elevation model* (DEM). In GIS disciplines, the term DTM is often used not just for the model itself, but also for the software used to manipulate the relevant data. Only the model is discussed in this chapter; data manipulation is discussed in detail in Chapter 15.

The terrain surface can be described as comprising two basically different elements—random and systematic elements. The random (stochastic) elements are the continuous surfaces with continuously varying relief. It would take an endless number of points to describe exactly the random terrain shapes, but these can be described in practice with a network of points. It is usual to use a network that creates sloping triangles or regular quadrants.

The systematic part of the terrain surface is characterized either by sharp cracks in the terrain, such as the top or bottom of a road cut, or by characteristic points such as spot depression and spot height. The systematic part is thus best represented by lines and typical single points. Prominent terrain features can be verbally described using many terms, such as smooth slope, cliff, saddle, and so on. Geometry, however, has only three terms: point, line, and area. One cannot describe continuously varying terrain using only three discrete variables, so all descriptions are necessarily approximations of reality.

Essentially, DTMs comprise various arrangements of individual points in *x-y-z* coordinates. Often, their purpose is to compute new spot heights from the originals. A terrain model can be realized by linking height as an attribute to each point (x,y). This type of elevation model can describe only a surface and cannot handle more z values to the same point. Therefore, the term *2.5-dimensional* is often used to describe the DTM dimension (Figure 5.1). This model is best suited to visualization. In a three-dimensional (3D) elevation model, elevation is an integral part of position (x,y,z) and the model can handle several z

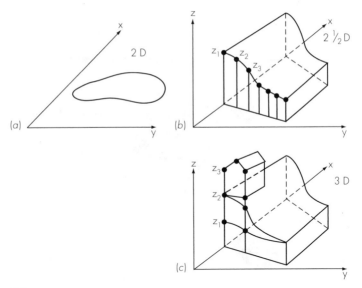

Figure 5.1
With 2.5-dimensional models *(b)*, it is possible to store only one height value for each point *(x,y)*: that is, it is only the terrain surface itself that can be described. In real three-dimensional models *(c)*, several height values can be stored on the same point *(x,y)* and the relationship above and below the terrain surface can be described in the same model. Compare with the two-dimensional model *(a)*.

values for the same x,y pair. That is, it can handle different geological layers, roof heights on buildings, or roads that cross each other, together with the terrain surface. A three-dimensional model is also suited to volume calculations.

The z value of a new point is calculated by interpolation from the z value to the closest existing points. If the points are stored in an unstructured way, *all* registered points will have to be searched to allow calculation of the z values to a new point. This can be very time-consuming even for a powerful computer. It is therefore usual to use data structures that also describe the contiguity between the points. This is achieved by using data structures based on single points in a raster (grid) or triangles covering a surface.

5.1.2 Grid model

A systematic grid, or raster, of spot heights at fixed mutual spaces is often used to describe terrain (Figure 5.2). Elevation is assumed constant within each cell of the grid; that is, the area represented by each cell is shown as a flat area in the model. Thus, small cells detail terrain more accurately than large cells. The size of cells is constant in a model, so areas with a greater variation of terrain may be described less accurately than those with less variation. The grid model is most suitable for describing random variations in the terrain, whereas the

95

Figure 5.2
Digital terrain models describe the terrain numerically in the form of *x*, *y*, and *z* coordinates. Graphic presentation can be either in the form of a grid (as in the figure) or as profiles.

systematic linear structures can easily disappear or be deformed. A possible solution is to store the data as individual points and generate grids of varying density as required. It is debatable whether the grid model represents samples on a grid and can therefore be called a point model, or represents an average across raster cells. In the United States the former seems to be the most usual. Elevation values are stored in a matrix, and the contiguity between points is thus expressed through the column and line numbers.

Different interpolation techniques are used to generate an elevation grid from source data such as points, contour lines, and break lines. In interpolation of elevation values for the cells, it is usual to assume that points located near a cell should have greater effect on the cell's value than points located at a distance. The averages of the elevations of those closest to grid points, within a given circle or square, can be assigned to the grid points with inverse weighting in proportion to the intervening distances involved. More advanced statistical methods can replace this kind of simple weighting in order to obtain a best possible model of the terrain based on available data. When the data relate to profiles or contours, grid point elevations are interpolated, in the same way, from the elevations at the intersections of the original data lines and the lines of the grid (Figure 5.3).

When elevation data are organized in a grid structure, the matrix will give direct access to neighboring cells. Thus, the interpolation of the *z* value to the new *x* and *y* points is quit simple, for example, based on linear interpolation with the nearest grid points.

The grid model may have difficulty in picking up small local variations in the terrain. This may result in one or more cells obtaining a lower elevation value than the adjacent cells, thus creating a sink (Fig-

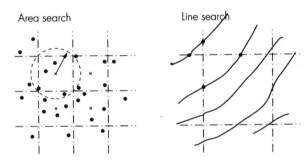

Figure 5.3
The DTM-based grid model is created either from an area search in a point cloud or from intersections with contour lines.

ure 5.4). In the model there will be no drainage from such a sink. There are functions for automatic elimination of sinks. For example, sink cells can be identified and awarded the same value as the adjacent cells. The grid model will often have a terrace effect when contour lines are used as source data, because of the high density of observation points along the contours. This effect can be eliminated in several ways, such as by thinning out the data along the contour lines prior to interpolation.

Terrain may also be described in terms of chosen or arbitrarily selected individual points (i.e., a *point cloud*; Figure 5.3). In principle, the characteristics of the terrain between points are unknown, so it follows that point densities should be greatest in areas where terrain features vary most. Only elevations are stored for the points of a regular grid, but both point coordinates and elevations must be stored for point clouds. For a given area, the amount of memory storage required for the two point arrangements differs. For describing abrupt terrain variations, such as the top and bottom of a road cut, point models are inferior.

12	11	11	11	11	11	11
12	11	10	10	10	10	11
12	11	10	9	9	10	10
12	11	10	9	9	10	10
12	11	10	10	10	10	11
12	11	11	11	11	11	11
12	12	12	12	12	12	12

Figure 5.4
A sink is shown by cells that have lower elevation values than adjacent cells.

5.1.3 TIN model

An area model is an array of triangular areas with their corners stationed at selected points of most importance, for which the elevations are known. The inclination of the terrain is assumed to be constant within each triangle. The areas of the triangles may vary, with the smallest representing those areas in which the terrain varies most. The resulting model is called the *triangulated irregular network* (TIN; Figure 5.5).

Insofar as possible, small equilateral triangles are preferable. To construct a TIN, all measured points are built and the model thus represents lines of fracture, single points, and random variations in the terrain. The points are established by triangulation and in such a way that no other points are located within each triangle's converted circle. In the TIN model, the *x-y-z* coordinates of all points, as well as the triangle attributes of inclination and direction, are stored. The triangles are stored in a topological vector data storage structure comprising polygons and nodes, thereby preserving the triangle's contiguity, which eases the calculation of *z* values for new points (Figure 5.6).

Various *interpolation algorithms* are available for selecting representative points from the basis data (grid, contours, point clouds) and for creating appropriate triangles. Should the basis data be available in grid form, it is possible to move a window (one point and its eight contiguous points) step by step over the data and remove the points that are least characteristic in relation to their contiguous points. Usually, triangles are formed by laying circles through three points and testing whether there are other points within the circle. If other points are not available, a new triangle will be formed. This method is adjusted to produce triangles with a low variance in length, which are known as *Delaunay triangles* (Figure 5.7). Each triangle is a surface, it is assumed, and has the same degree of slope and the same direction of slope over

Figure 5.5
The surface of the terrain can be described as inclined triangular areas, starting from points selected for their importance. This method of describing terrain is called the *triangular irregular network* (TIN).

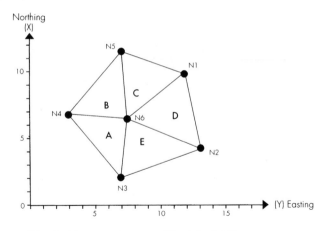

Triangle table

Triangle	Adjacent
A	B, E
B	A, C
C	B, D
D	C, E
E	D, A

Triangle/node table

Triangle	Nodes
A	N3, N4, N6
B	N4, N5, N6
C	N1, N5, N6
D	N1, N2, N6
E	N2, N3, N6

Coordinate table

Node	Coordinates
N1	X_1, Y_1, Z_1
N2	X_2, Y_2, Z_2
N3	X_3, Y_3, Z_3
N4	X_4, Y_4, Z_4
N5	X_5, Y_5, Z_5
N6	X_6, Y_6, Z_6

Figure 5.6
In the TIN model, the triangles are stored in a topological structure.

the whole surface. A local function can be fitted through the triangle for interpolation of z values:

$$z(x, y) = a_0 + a_1x + a_2y \qquad (5.1)$$

Degree of slope and slope direction are calculated from the parameters a_0, a_1, and a_2. Some GISs automatically store degree and direction of slope separately as triangle attributes.

A weakness of the triangulation method is that it often establishes a number of flat triangles when source data are in the form of contour lines and the terrain is relatively flat. This results in a great distance to neighboring lines, and points on the same line will thus be wrongly linked together in a triangle (Figure 5.8). This problem can be avoided by supplementing characteristic points and break lines in the actual areas.

As compared with the grid model, the TIN model (Figure 5.6) is

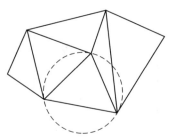

Figure 5.7
Delaunay triangulation is a method used to fit triangles in a point cloud. The circle described ensures that the triangles have good geometry with least possible variation in the page lengths.

cumbersome to establish, but more efficient to store because areas of terrain with little detail are described with fewer data than similar areas with greater variation. However, the TIN model normally requires considerably larger storage capacity than the grid model. TIN models are good for describing terrain because the sharp breaks of slope between uniform-slope facets fit certain types of terrain well. Updating with new data is relatively difficult, as it requires building a new topology for the whole model.

5.1.4 Other models

Isolines—continuous lines connecting points of the same elevation—may represent terrain in much the same way that contour lines depict terrain on conventional maps (Figure 5.9). The point densities should be greatest in those areas in which the terrain varies the most. As the intervening terrain between successive isolines is unknown, smaller elevation increments between isolines result in greater accuracy of description. An advantage of the isoline model is that it has a simple data structure, but it is difficult to interpolate good elevation values for new points. Although an isoline model may be compiled readily, amending its data is an involved undertaking. In practice, the methods used are determined by the data compilations.

Parallel profile lines connecting points of varying elevation may be used to describe terrain. The density of points along profile lines

Figure 5.8
When source data are in the form of contour lines and the terrain is relatively flat, flat triangles can wrongly be established in the TIN model.

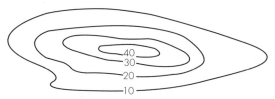

Figure 5.9
Lines that link points with the same terrain height can be used to represent the terrain surface. This corresponds to traditional elevation contours. This structure is poorly suited to the calculation of elevation values for new points.

should be increased in areas where there are major variations in the terrain. In principle, the terrain between successive profile lines is unknown, so the closer the lines, the greater the accuracy of description.

A combination of isolines and individual points may also be used to describe terrain, especially when specifying such point features as peaks and valley floors, or vital terrain lines such as the top and bottom of a fill. As mentioned previously, the grid and TIN models are best suited for calculation of the z value of new points.

5.1.5 Input

Source data are normally represented either as single points, contour lines, stream lines, or break lines, depending on the registration method. Good interpolation techniques are available to generate digital terrain models in a grid or TIN from sources of this type.

Terrain data are usually compiled from survey point elevations, measured by theodolite or GPS, from isolines digitized from existing maps, or from photogrammetric point and/or line (contour or profile) registration. Elevation data are usually collected as contour lines in vector form. Normally, photogrammetric point registration is 30% more accurate than photogrammetric line registration. Digital stereo images allow digital terrain models to be established very effectively through the pattern recognition for automatic determination of elevation points in raster or profile lines. However, we have to be aware that automation cannot differentiate between the surface of the terrain and buildings, trees, and so on; thus, the tops of buildings and trees will be part of the digital terrain model and may have to be corrected manually afterward. In the same way, digital optical satellite data (SPOT, etc.) and radar data (ERS1, etc.) can be used to generate digital terrain models.

5.1.6 Accuracy

The quality of the DTM determines how well the model describes the real world. The accuracy of terrain descriptions is governed primarily by random variations in the terrain, spreading of measured points, distance between measured points, and accuracy of points by the method of generating the model grid and triangular surfaces, and by the method used to interpolate between points in the model. We have already mentioned the problem with sinks in the grid model and flat triangles in the TIN model. In DTM, errors in the x and y coordinates also result in errors in the elevation. A number of graphic techniques have been developed to judge the quality.

In those models in which cells and profiles are recreated from a point cloud with each computation, accuracy depends on the cell or profile density. Profiles must be closer to each other to represent more rapid terrain variations, but greater profile densities naturally call for the processing of greater amounts of data.

Grid models and TIN models are always less accurate than the original data from which they are derived. In some GIS/DTMs, therefore, original data are often stored in point clouds. Models with accuracy suiting specific tasks are compiled from these as required. For example, a grid model for estimating road construction excavations may be more accurate than one intended to detail the general vegetation of the region.

5.1.7   Practical observations

Various interpolation programs compute new z coordinates for new x–y coordinates, thus facilitating specific computations such as estimating cut-and-fill volumes in road planning or assessing reservoir volumes for hydroelectric plants. Various GISs may also be implemented with functions for calculating slopes, relief shadowing, drawing in perspective in order to visualize the impact of works, computing the runoff, or perhaps draping in colors to enhance visualization; see Chapter 15 for a more detailed discussion.

The methods used to describe terrain surfaces may also be used to describe other continuously varying phenomena. Thus, population density, prevailing temperatures, or biomass production can be described quite simply by assigning the parameter involved to the z axis of all the observation points located in x–y coordinates (Figure 5.10).

5.2   Three-Dimensional Objects

All physical phenomena are located in space; thus, the world can be described as three-dimensional (Figure 5.11). A complete data model should be based on these three dimensions: ground, position, and elevation. This applies not only to terrain surfaces but also to buildings, borders, addresses, accidents, and all manner of data; a complete data model should manipulate georeferenced data in three dimensions.

The realization of three-dimensional (3D) objects in GIS still has the-

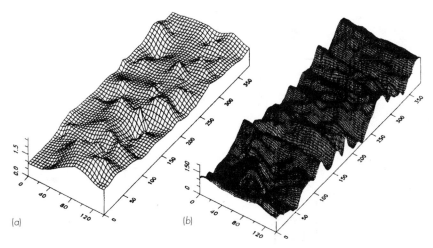

Figure 5.10
By placing various themes on the z axis, the distribution of varying amounts over the terrain can be illustrated, such as (a) differences in elevation in an area; (b) biomass production in the same area.

oretical and practical limitations (Figure 5.12). Topological data are needed for such procedures as color filling (and photo texture) of vertical areas and for data search. It is a theoretical and mathematical problem to establish topology for three-dimensional objects. The topology will be very complex and present opportunities to establish objects that are illegal (objects that cannot be oriented) (Balstrøm et al. 1994). It is also difficult to establish satisfactory routines for checking whether declared data exist in three-dimensional topology.

Figure 5.11
The world is three-dimensional, with phenomena having a location and a surface area in both elevation and ground plane.

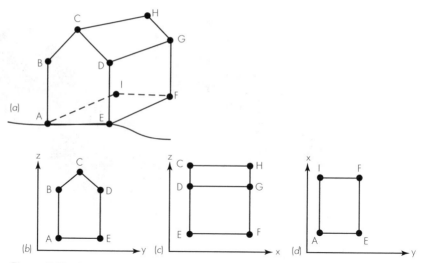

Figure 5.12
Establishing three-dimensional topology is demanding. Large volumes of data are required, which can create theoretical problems (nonorientable surfaces) that are difficult to solve. Traditional GIS operates with only two-dimensional topology, whereas in three dimensions you have to relate to three levels *(y,z; x,z; and x,y)*.

Specification of all objects in three dimensions can easily increase the amount of data collected beyond that which is needed; the amount of data will in any case be considerably larger than with the use of 2 and 2.5 dimensions. It may also influence the techniques used to collect data. The collection of photogrammetric data provides free elevation data in addition to the northing and easting data of the ground plane. When existing maps are digitized, however, elevation data must be entered manually (and sometimes inaccurately, as exact elevations may not be available for all objects). Today, we see that users have an increasing need for digital three-dimensional map data, particularly for applications connected to urban areas. At present, commercial GIS is still only capable of handling two-dimensional topology. Even though relational databases support binary large objects (BLOBs) for storage of texture (building facades or similar objects), this type of data cannot be searched for as other data can. Therefore, new database techniques have to be developed in addition to relational databases and the special databases that are generally used today. Models can be constructed, but should in this case be carried out in systems for computer-aided design (CAD).

Three-dimensional objects can be represented as cubes in a three-dimentional raster. The elements in the three-dimentional raster are known as *voxels* (volume element) and correspond to pixels in raster (Figure 5.13). Most of the raster operations on pixels can also be car-

Figure 5.13
Three-dimensional objects can be represented as cubes (voxels) in the three-dimentional raster.

ried out on voxels, although voxels require considerably larger storage volume than pixels.

5.3 Representation of Time

Cognitive science has shown that the collection of historical data is essential to understand reality and to the acquisition of knowledge in general. Time is therefore a factor that concerns us deeply. It was studied by ancient Greek philosophers such as Plato and Aristotle, and it plays an important role in Einstein's theory of relativity.

Most things change with time, and the same applies to geographical data. Consider, for example, property data. New properties are continuously under development because of the division of existing properties; therefore, new geometry occurs. However, most changes are related to title ownership, with the resulting changes in attribute values. An example of an area in which extreme changes occur is the transport sector, with vehicles continuously changing position and where frequent on- and off-loading of goods change the attribute values that are linked to a vehicle.

In Chapter 3, we classified geographical objects into continuous and discrete phenomena, pointing out that there are phenomena in the transition between the two main classes. In the same way, we can classify events according to their frequency (Peuqet 1999):

- Continuous changes
- Frequent changes
- Sporadic changes
- Unique, one-time changes

Moreover, some events are cyclical (repeated at fixed time intervals) and others are chance happenings. Classification can also be influenced by the resolution of the time reference system. There is, for example, a big difference between measuring continuous events in years or in hours.

In addition to the fact that both the geometry and attributes of objects change over time, the reference system can also be changed.

Chapter 6 discusses how this occurs when roads are changed in a distance-based reference system, but it also happens when the geometry of administrative reference units is changed (e.g., with changes in municipal borders and postal zones). Furthermore, topological changes often occur as a consequence of geometrical changes.

Time registration has to be based on a fixed reference and common method of indicating time. The global time reference system is based on a common Christian calendar (Julian calendar). Joint time reference is given through Greenwich Mean Time (UTL, universal time coordinates) and linked to different time zones. Time can be given in units such as days, month, and years (ddmmyy), and so forth.

In GIS there is often a need to report the status of a situation at a particular point in time, observe changes over time, or even to make prognoses for possible further changes. In practice, it is difficult to create a data model that is capable of incorporating all imaginable changes. The time factor is relatively often neglected in GIS, probably because we are more concerned with documenting our current situation than we are with historical changes.

However, we know that in many cases it is desirable to be able to see a pattern in changes—for example, changes in land use—and that it is important to be able to document historical data, such as land ownership. Databases are updated continuously, so unless special measures are taken, the time picture will be fleeting. A general problem in the digital GIS world is how to keep geographical data up to data without erasing old data. The use of analog technology can document time changes via different printed versions of maps. To the extent that maps have been archived, it is also possible to preserve special versions of history, although they may be unsystematic. If the time factor is not incorporated in the data model for GIS, we run the risk of losing important historical data, and society will be without history in regard to changes in land use, property ownership, and so on.

There are, however, some practical solutions possible:

1. The attributes of the objects are changed.
   a. Historical data are stored only in fixed or variable time intervals (e.g., every year, every fifth year, etc.). The attribute values between these intervals may have to be interpolated.
   b. All changes (or transactions) are registered and stored for selected types of objects (i.e., historical data have to be preserved by date stamping).
2. The geometry of objects is changed.
   a. Historical data are stored only in certain time intervals (e.g., every second or fifth year). The geometry between these intervals may have to be calculated. As in the case of attribute values, the object type will decide where it is possible to interpolate new geometry.

b. All changes are registered and stored for selected object types (i.e., all historical data have to be preserved by date stamping).

Changes in geometry can lead to changes in the relationships between objects and the resulting changes in topology, which also have to be preserved.

The most usual way of handling the time factor in GIS is to consider time as an attribute to the objects like other attributes, and observe it with fixed (cyclical) or variable time intervals. This view corresponds with the usual way of presenting geographical data as two-dimensional time overlays and can thus be realized for both vector and raster data (Figure 5.14). This model is popularly known as the "snapshot" model. Short time intervals would result in very large data volumes, and this simple approach will not necessarily create a logical connection between the various time layers. It can therefore be extremely difficult to assess what the situation might have been between two time layers.

Time is a dimension analogous to the spatial dimension. Position (instance) and distance (period) in the linear temporal reference system can be identified. Temporal topology provides information about connectivity between objects in time. The temporal reference system is defined by a calendar showing years, months, and days, and a clock indicating hours, minutes, and seconds within a day.

It can be appropriate to consider the concept of time as an extra dimension, as with elevation (Hagerstad 1970), or one can use the four-dimensional space—hyperspace—where time and space enter simultaneously (Rucker 1977) in addition to attributes (Figure 5.15). For

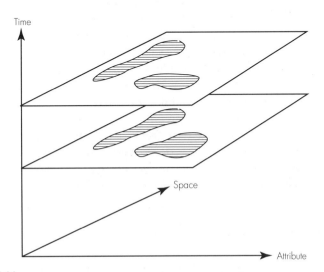

Figure 5.14
In GIS it is normal to realize time in a model with time layers. Reality is thus simplified to allow changes to occur spasmodically and not continuously.

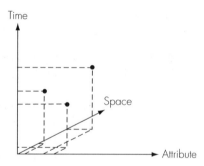

Figure 5.15
In reality the changes occur more or less by chance and are thus scattered randomly in time, attribute, and space. However, it is difficult to create suitable models that can be realized in GIS.

example, the path of a moving object could be represented as a curve described by coordinates in *x, y (z)* and *t*.

Today's GISs permit registration of time only in the form of variable or fixed time intervals. Registration in time intervals is more of a practical solution, where the principal aim is to maintain rapid access to data, simple updating, and limited data volumes rather than realization of a basic data model of which time is an integral part (Figure 5.15). Time models specify how changes in terrain over time can be preserved. Updating includes the routines to be followed for registration of changes and the speed at which changes can be loaded into the database(s) and, to a certain extent, can be viewed independently of the models.

Normal databases can handle only a single situation image of the real world. Temporary (object-oriented) databases can handle several situation images or time intervals. Such dynamic databases should be able to handle continuous changes, but are yet not in practical use.

Several prototype object-oriented systems and even a few commercial systems are available that provide some temporal support (Skjellaug 1996), but it seems that there is a fair amount of research and development work remaining before a complete data model for the time factor can be realized in GIS. Even though a data model can be created that can handle the time factor satisfactorily, we are, in practice, reliant on changes being registered and stored in the database within a reasonable time. The relevant updating and organizational procedures are dealt with later in this book (Section 11.7).

5.4 Models for Movable Objects

A considerable part of the real world consists of movable objects: vehicles on a road network that carry passengers or goods, or water running over the terrain surface. One should also be able to realize this aspect of the real world in GIS. Special models have therefore been developed to handle these conditions.

5.4.1 Network model

The network model comprises road systems, power grids, water supply and sewerage systems, and the like, all of which transport movable resources. The most usual type of networks for which GIS models are developed are road systems; the following description therefore refers to road systems. For most purposes, reality can be simplified to a model that can handle two different situations:

1. Displacement of resources or objects from one place to another
2. Allocation of resources or objects from or to a center

As with other GIS data models, this model is based on geometry and attributes (Figure 5.16). The geometry in the network is represented by lines consisting of connected lines of vector data. The geometry of a road system is represented by the center line of the road. This model assumes topological data built up of links and nodes. Every link and every node in a network must have a unique identity. The attributes are tied to links and nodes in a linear system through these IDs and are intended to describe the total accessibility of the system. We can thus state that the model is based on three basic relationships:

1. Continuous, connected networks
2. Rules for displacements in a network
3. The possibility of attribute value accumulations due to displacements

Attributes are connected to links and nodes and consist of two main categories. One sets conditions for transfer in the network, and the other specifies which resistance occurs at different locations in the network (Figure 5.17). Attributes that determine how objects can be moved in the network can have direction predetermined (one-way streets,

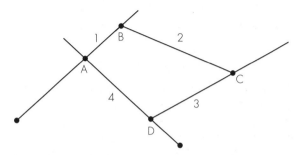

Figure 5.16
A road network can be split up into links and nodes, whereby a link is a line without logical intermediate intersection and a node is an intersection point where two or more links meet, or a start/end point.

LINK	DISTANCE	RESTRICTION		NODE	RESISTANCE
1	350 (m)	40 (km/h)		1	1.0 (min)
2	500	60		2	0.2
3	400	50		3	1.0
4	450	40		4	0.2

Figure 5.17
Attributes can be attached to links and nodes and the resulting data displayed in tables.

closed roads, weight limits, etc.). Attributes that specify resistance in the network can be speed limits, roadwork, peak traffic, traffic lights, bus stops, sharp curves, and so on. The accumulation of all resistance occurs along the different routes in the network will be the basis for choosing the best (shortest/quickest) alternative.

Once the model has been constructed, it is possible to simulate the quickest and/or the shortest route between points A and B based on the route with lowest accumulated resistance. For example, it is possible to define areas that are covered within a driving distance of 2 km and 5 km, respectively, from, say, a school and to simulate personnel transport by different means. Certain relationships must be taken into consideration when establishing network models. Links are customarily selected to carry information, which may complicate the task because data volume and complexity increase in proportion to the number of links. The initial step, how a road is divided into links, determines the nature of the nodes. If all intersections, events, and features along the road result in nodes, the number of links may be enormous, resulting in the need for large storage capacity and slow data retrieval.

The network model represents a real-world model different from those described in Chapter 3 because it is neither based on entities/layers nor is it fields/object oriented. Information in the network model is based on links and nodes, which are not found in the entity model. Nor is it object oriented, because new nodes and links will not be established wherever attributes change; they are established only where it is practical to measure the flow of resources through the network.

5.4.2 Model for movement over surfaces

In the network model, movements are limited to the network. There are, however, some situations where access is otherwise in the terrain, such as in water that flows on the surface (drainage) or in the movement of cross-country vehicles. The free flow of resources in the terrain can also be modeled by using geometry and attributes, but in this case, it is practical to use the full-coverage raster model instead of vector data. The geometry is thus represented with regular cells, and attributes are represented with coded values for each cell (Figure 5.18).

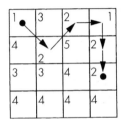

Figure 5.18
Optimizing route location on raster data.

The cells are coded with attribute values that characterize the terrain in relation to the phenomenon to be studied. In the case of drainage, direction of slope can be one theme and angle of slope another. The angle of slope indicates the resistance in the cell—large angle of slope gives little resistance, small angle of slope gives high resistance. The accumulation of cell values based on certain rules will thus give the total drainage values for different parts of the terrain.

Connective models of raster data may also be used to determine travel distances, to identify areas of given shapes and sizes, and so on. The raster model for movable objects is in many ways not unlike standard raster models. It is a question of the suitable coding of cells relating to that part of the real world which the model is intended to reflect.

5.5 Combination of Models

No models, of any degree of complexity, are perfect in relation to the real world; they are only more or less successful approximations. However, they can be better in combination than singly. This property can be utilized in the creation of hydrological data models by combining three dimensions, the network model, and the raster model (see Chapter 15 for a more detailed discussion).

The technique of multimedia integrates several types of models: vector, raster, 3D, time, and so on. The multimedia technique helps the user to develop complete mental models of spatial problems and gives the user the ability to navigate in a GIS-derived information space. However, there is still a good deal of research and development work to be carried out before the time factor and three dimensions are fully integrated in commercial GIS.

Georeferencing Systems

Before dissimilar geographical data may be used in GIS, they must be referenced in a common system. The problems associated with dissimilar geographical data are hardly new: There are numerous georeference systems that describe the real world in different ways and with varying precision. Working with GIS today, one has to relate to many dissimilar data sets of different origin in regard to georeferencing systems. Earlier, such terms as *datum, coordinate system,* and *map projections* were indicative of a field of activity for geodesy specialists. Today, GIS users have to understand what these terms involve. GPS has brought geodetic terms into the daily work of GIS, inasmuch as one must relate to global positioning, which very often differs from the traditional local georeferencing systems.

Georeferencing is defined as positioning objects in either two- or three-dimensional space. We have previously looked at several probable georeferencing systems. This chapter gives a more systematic presentation of basic methods for georeferencing; including datum, coordinate systems, map projections, and elevation systems. There are two principal methods of georeferencing:

1. Continuous georeferencing systems
2. Discrete georeferencing systems

Continuous georeferencing systems

Continuous georeferencing implies continuous measurement of the position of phenomena in relation to a reference point with no abrupt changes or breaks. It involves resolution and precision. Resolution refers to the smallest increment that a digitizer can represent; it is similar to precision, which is a measure of the dispersion (in statistical terms, the standard deviation). Accuracy, however, is the extent to which an estimated value approaches a true value (i.e., the degree to which it is free from bias).

Theoretically, there is no limit to how precise continuous measurements may be. Precision depends entirely on the measuring method. Distances may, for instance, be measured to within 1 kilometer, 1 millimeter, or 1 nanometer. Many geographical phenomena, including property boundaries, manhole locations, building details, and many map details, are measured on a continuous basis.

Continuous systems include *direct georeferencing,* which involves:

- Datum
- Coordinate systems
- Map projections

For obvious reasons, the direct method is also known as the coordinate method. The other continuous system is *relative georeferencing,* which includes:

- Offset distance
- Measurement along (road) networks

Currently, an international standard promulgated by the ISO (International Organization for Standardization), titled "Geographic Information—Spatial Referencing by Coordinates" is in preparation.

Discrete georeferencing systems

In discrete georeference systems the positions of phenomena are measured relative to fixed, limited units of the surface of the Earth. The discrete method is also known as spatial georeferencing by geographical identifiers. This method is discussed in more detail later in this chapter.

6.1 Datum

To determine an exact position on the Earth's surface, one has to know the shape and size of the Earth. As we know, the Earth is not a perfect sphere, but more like an ellipsoid with flattening (ellipticity) at the poles (Figure 6.1). The shape of the Earth is therefore, for practical purposes, often expressed by the shape of the ellipsoid. Ellipsoids of various sizes have been used, depending on the degree of accuracy with which it was possible to measure the shape of the Earth. The Greek

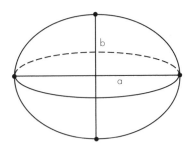

Figure 6.1
The Earth is often described using a rotation ellipsoid. The lengths of the semimajor axis *(a)* and semi-minor axis *(b)* have gradually changed through history as more exact measurements of the Earth's shape have become available.

Eratosthenes (276 -194 B.C.) is known as the first to determine the Earth's shape with satisfactory accuracy. In the 1600s and early 1700s scientists eagerly discussed the shape of the Earth, debating whether it was flattened at the poles or at the equator. These discussions ended after the French Academy of Science sent out two expeditions, one to South America and one to Northern Sweden. The results showed that the Earth was flattened at the poles. Since that time many different ellipsoid sizes have been computed and used (see Figure 6.2).

In 1924 a common international ellipsoid was accepted by the International Association of Geodesy, but it has been proved recently that the Earth's radius at the equator (semi-major axis) is 251 m less than was believed in 1924.

A datum (reference level) is a model (ellipsoid) of the Earth and gives the relationship of a coordinate system to the Earth. It is defined by:

1. Size and shape (semi-major axis a, and semi-minor axis b) of the ellipsoid
2. Positioning of the ellipsoid in relation to the physical surface of the Earth by an anchor point

The common datum for a country or area requires that there are specific coordinates for the datum origin (anchor point). For orientation of the axis of the coordinate system, a prime meridian must be defined, from which longitude values are specified. Most geographic datums use Greenwich as their prime meridian. A datum is thus a reference system in which the coordinates can be expressed, which is realized physically through monumentation of points with known coordinates.

Today most countries have their own local datums, but some countries have adopted WGS84 (World Geodetic System 1984) as a datum. WGS84 is a global datum based on the center of the Earth's mass as the anchor point. Georeferencing by GPS is based on WGS84 as the datum. Strictly speaking, WGS84 is not a datum, but a military reference

Ellipsoid	Semimajor axis (a)	Flattening $\left(f = \frac{a-b}{a}\right)$
Bessel 1841	6 377 397.155	299.1528128
Clark 1886	6 378 206.4	294.9786982
Hayford 1910	6 378 388.0	297.0
International 1921	6 378 388.0	297.0
WG584 1984	6 378 137.0	298.257223563
ETRF 1989	6 378 137.0	298.257223563

Figure 6.2
The table shows the parameters of a number of Earth ellipsoids that have frequently been used in geodetic measurements.

system. In Europe, ETRS89 (European Terrestrial Reference System 1989) is used as regional datum. ETRS89 is a special variation of WGS84 with an anchor point that was linked up to the location of Europe in 1989. The difference between WGS84 and ETRS89 will increase with time because of continental drift. The distance between Europe and America, for example, will increase by 1 to 3 cm per annum.

As mentioned, there are a number of local and regional datums around the world, which vary in both ellipsoid size and point of origin. As we shall see later, the transformation of coordinates from one datum to another is usually no simple operation.

6.2 Coordinate Systems

In mathematics there are, in principle, two different methods for designating a point's location. The location of a point may be designated:

- As an angle and distance
- As coordinates in a right-angled cartesian coordinate system

A position can therefore be indicated by distance and direction from the origin. This is known as the polar coordinate system (Figure 6.3). Polar coordinates can be transformed relatively easily to cartesian coordinates without loss of precision, provided one is operating within the same datum. In practice, the angle of the new point is measured in relation to one or more known points (see Figure 6.4).

In geodesy, the most usual right-angled coordinate systems are:

- *Geocentric coordinates,* where x, y, and z are given in a triple-axle system with the origin in the Earth's center
- *Geographic coordinates,* where latitude and longitude are stated on the Earth's sphere
- *Map projection coordinates,* where North and East, or x and y, are given on one plane

Figure 6.3
Example of unique positioning finding rectangular coordinates and polar coordinates.

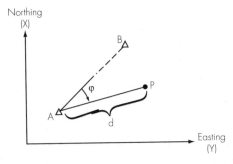

Figure 6.4
The polar location of a point is measured as the direction and distance from other fixed points in the terrain.

A coordinate system is defined in full by name, the measuring units, and the direction and the sequence of the axes. Geographic coordinate systems use degrees, minutes, and seconds as units of measure, whereas for map projection and geocentric coordinate systems, units of the metric system are normally used. A coordinate *reference system* is a coordinate system which has a reference to the Earth by a datum.

Geographic coordinates
The geographic coordinates on the surface of the Earth are *latitude,* measured in degrees north or south of the equatorial plane, and *longitude,* measured in degrees east or west of the Greenwich meridian (Figure 6.5). In the literature on the subject a number of terms are used, based on latitude and longitude, such as *geographic coordinate system, geodetic coordinate system,* and *spherical and coordinate system.* Positions

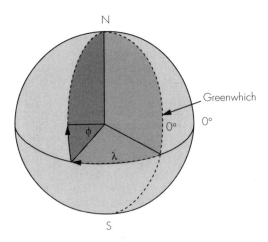

Figure 6.5
The geographical coordinate system originates from the Earth's center and defines a location in degrees in relation to the equator and the zero meridian at Greenwich.

in latitude and longitude are only relative; distances and areas must be calculated using spherical geometry and the actual radii of the Earth for the points in question. In applications, latitude and longitude are usually used in describing major land areas.

The lines that go through points with equal longitude are known as meridians, and lines that go through points with the same latitude are known as parallels of latitude. Meridians and parallels of latitude are often depicted on maps at intervals, such as at each 5 or 10 degrees (see Figure 6.9).

Map projection coordinates

Map projection coordinate systems give the position of points relative to two mutually perpendicular axes in a plane (Figure 6.3). As shown in Figure 6.3, the position of a point forms a rectangle with the axis, therefore the system is therefore also known as a rectangular coordinate system. The axes are called Northing and Easting and are often written as x and y. The orientations of coordinate systems may differ, so coordinates should always be identified unambiguously—for example, in terms of compass directions (Northing, Easting) from the origin. Many countries have national, and even local, georeference systems of rectangular coordinates.

Ordinary geometrical computations may be made in rectangular coordinate system, but deformations caused by the map projection will always arise (see Section 6.3.) This difficulty is most often overcome by constraining the axis systems to relatively small areas. Larger areas or zones are then represented by a system of several axes, displaced in relation to each other. In areas where neighboring zones meet, data must be converted to a common axis system to avoid working in two systems. Many GISs have facilities for transforming data from one coordinate system to another, based on common points in the two systems. When the common points are unknown, the parameters for the datum, project method, and coordinate system should be ascertained.

Geocentric coordinates

Geocentric coordinates are based on a rectangular coordinate system with an origin at the center of the Earth. The z axis is coaxial with the axis of the Earth's rotation and is positive in the direction of the North Pole. The x axis goes through the zero meridian in Greenwich; the y axis is orthogonal to the positive x and z axes, as shown in Figure 6.6. Geocentric coordinates are also known as cartesian coordinates.

The advantage of the geocentric coordinate system is that it covers the entire Earth, which is why it is used for GPS georeferences (see Chapter 10). Geocentric coordinates are, however, of limited practical use, and when introducing the ellipsoid for height referencing, one will only get ellipsoid height, which is not the same as for normal national elevation systems. This subject is discussed in more detail in

117

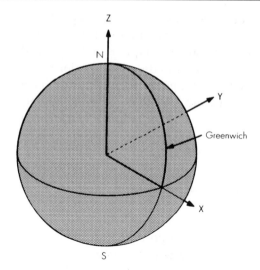

Figure 6.6
The geocentric coordinate system is rectangular, with an origin at the center of the Earth and axes as shown.

Section 6.6. Geocentric coordinates can be converted to map projection coordinates and can be used with them.

6.3 Map Projection

Georeferenced data may be drawn on maps only when referenced to a plane surface, not to the curved surface of the Earth. Various projections are used to represent the curved surface of the Earth on the plane surface of a map. Map projections can sometimes be more closely related to map drawing than to position finding. They are classified in three groups according to the underlying geometrical conversions involved (Figure 6.7):

- Cylindrical
- Conical
- Azimuthal or plane surface

As shown in Figure 6.7, the three different projection surfaces can be unfolded to one plane and thus function as a plane map. It should be remembered, however, that curved surfaces are always distorted when imaged into a plane, and the deformations multiply with the increasing size of the area represented. In the planes' touch line or the touch point with the Earth is however the scale factor equal to 1. This can be used when choosing a projection in relation to the shape of the area to be mapped.

A map projection can be defined as a mathematical flat surface representation of the geodetic ellipsoid. There are different projection for-

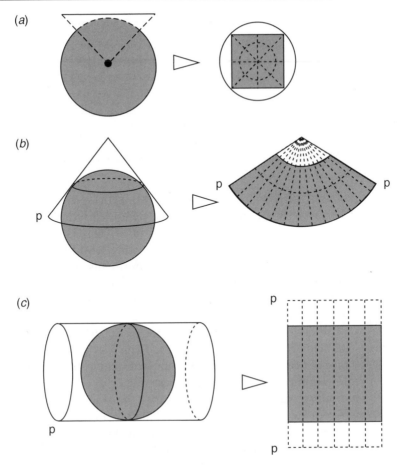

Figure 6.7
Map projections: *(a)* azimuthal projection; *(b)* conical projection; *(c)* cylindrical projection.

mulas that preserve either shape, area likeness or distance in a particular direction. Length preservation, where the scale is the same in all directions, is not possible. Cylindrical, conical and azimuthal projections can be made equidistant, equal in area, or conformal. Because fixed mathematical formulas are used, it is possible to convert to and from different projection systems without loss of accuracy.

For larger-scale mapping, cylindrical projection (Transversal Mercator) or conformal conic surfaces are normally used.

6.4 UTM

The best-known map projection coordinate system is the UTM Grid (Figure 6.8), established by the U.S. Department of Defense in 1947. UTM is based on a special type of cylindrical projection known as the

Figure 6.8
UTM is based on a cylindrical projection cutting through the globe.

Universal Transverse Mercator (UTM). UTM covers the surface of the Earth from 84 degrees North to 80 degrees South with the help of 60 zones, and each zone has a width of 6° (Figure 6.8). The zones are numbered eastward from 1 to 60, starting west of continental Alaska. The central meridian in each zone represents the zone's north–south axis, and the origo of the coordinate system lies at the meeting point of the axes and the equator. Thus, the equator represents the zone's east–west axis.

The projection is reduced by a factor of 0.9996. This corresponds to true scale at 180 km from the x axis. The result is that lengths at a distance of less than 180 km from the central meridian will have negative projection correction, and positive projection correction when the distance is greater than 180 km (Figure 6.9). At 60 km from the north axis will for example a distance of 10,000 m measured in UTM have a correct value of 10,003.60 m. To avoid negative eastern values, the origo for each zone has been given an addition of 500,000 m in the

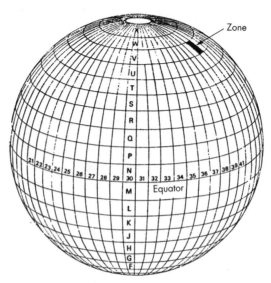

Figure 6.9
In the UTM system the Earth is also divided up into reference zones and bands.

	Regional datum	Local datum
Horizontal Datum	ETRF89	NGO 1948
Map projection	UTM	Gauss-Kruger
Zone	32	2
Northing (m)	684080.35	405032.42
Easting (m)	046364.30	−003828.82
Vertical Datum	NN1954	NN1954
Elevation (m)	2468.49	2468.49

Figure 6.10
Principle of georeferencing. The highest mountaintop (Galdhøpiggen) in Norway stated in a regional horizontal datum (ETRF89) and a local datum (NGO1948).

eastern direction. To avoid negative northern values, the origo has been given an addition of 10,000,000 m for areas south of the equator. Because there are in total 60 axis systems or zones with the same origo value, it is necessary to specify in which zone we are operating.

In addition to the zones, for rough georeferencing the Earth's surface is divided into 20 belts starting from the equator, lettered from C to X from south to north. A grid of zones and belts is thus created (Figure 6.9).

Figure 6.10 shows an example of correct georeferencing by use of a regional and a local horizontal datum.

6.5 Coordinate Conversion and Transformation

When working with GIS, we often encounter the problem of working in one system while some of the data lie in another system. In order to work with all the data simultaneously, we have to be able to transform coordinates within and between different datums, map projections, coordinate systems, and zones. When all formulas and parameter values are known, it will normally be possible to carry out such coordinate transformations without any significant loss of accuracy. Normally, GIS software has modules that support the most important transformation operation. ISO recommends the use of the term *transformation* when transferring coordinates between different datums and *conversion* when transferring coordinates between different map projections and coordinates in different zones.

Transformation from one datum to another can be a problem, because there is no fixed mathematical link between different datums. Ideally, the ellipsoid parameters, translation of the origin of coordinate system, and prime meridian should be known. If particularly high accuracy is required, then the point elevation and geoid-ellipsoid separation should also be known. Transformation of coordinates from one local datum to WGS84 is particularly difficult, but formulas and

121

parameter values are usually available from the national mapping authorities.

Within the same datum, the conversion within and between different coordinate systems and projections are clearly mathematically defined with fixed formulas. It is therefore possible to transform back and forth without loss of accuracy when all the parameters are known. This kind of one-to-one mapping of coordinate systems, often known as coordinate conversion, is widely used to convert geographic (ellipsoidal/geographic) coordinates to three-dimensional geocentric coordinates and vice versa, provided that the ellipsoid is known. Conversion from map projection coordinates to or from geographic coordinates is also possible. Most GIS, have software that can present data in different projections. In addition, conversion between different UTM zones is standard in many GISs. This function is necessary when working in an area that stretches over two or more zones.

However, problems arise when not all parameters are known. This is a situation familiar to anyone working with GIS, and the reason in many cases is that the data are inadequately documented. The usual way of solving this problem is to use common points in each system together with conversion formulas which, as experience shows, give a satisfactory result. The most commonly used formulas for two-dimensional transformation are conformal (Figure 6.11) and affine transformations. In some cases polynomials of the second order are used. The mathematical relationships are as follows:

Conformal transformation:

$$X = c_1 + ax - by \tag{6.1a}$$
$$Y = c_2 + ay + bx \tag{6.1b}$$

Affine transformation:

$$X = c_1 + a_1x + b_1y \tag{6.2a}$$
$$Y = c_2 + a_2x + b_2y \tag{6.2b}$$

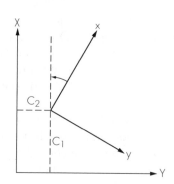

Figure 6.11
Conformal transformation.

where x and y are the coordinates to be transformed to a system with X and Y; a, b, and c are the transformation parameters.

It is also possible to carry out three-dimentional transformations of coordinates. The three-dimensional transformations formulas are:

$$X = c_1 + m(a_{11}x + a_{12}y + a_{13}z) \tag{6.3a}$$

$$Y = c_2 + m(a_{21}x + a_{22}y + a_{23}z) \tag{6.3b}$$

$$Z = c_3 + m(a_{31}x + a_{32}y + a_{33}z) \tag{6.3c}$$

where X, Y, and Z are the main coordinates system; x, y, and z are the minor coordinates entered; and m, c_1 to c_3, and a_{11} to a_{33} are the transformation parameters.

In conformal transformation the two coordinate reference systems differ only by their position and orientation in space and scale. All objects retain their original shapes. Affine transformation alters the shape of objects through unequal x and y scale alternations and unequal rotation of the x and y axes. The transformation parameters are determined by least square computation, but in any case greater or lesser inaccuracies will occur in determining the parameters. These inaccuracies also influence the accuracy of the transformed coordinates. The root mean square (RMS) of the least square computations indicates how accurately the transformation of new points can be carried out. It is usually the case that accuracy is dependent on the transformation formulas being able to compensate for the actual variations, and that accuracy of the control points are satisfactory, that the points are geometrically divided at the edge of and in the center of the area, and that an area chosen is not too large.

6.6 Elevation Referencing

We established previously that reality is three-dimensional and that elevation is an important part of position finding. The elevation of a point is defined as the distance from the chosen reference surface along the perpendicular to the surface. In geodesy two different reference surfaces are normally used: ellipsoide and geoid. Consequently, there are also two different vertical coordinate systems:

- Orthometric height system, elevation given in relation to the geoid
- Spherical height system, elevation given in relation to the ellipsoid

Most countries have established a national vertical reference system based on orthometric heights, called a *vertical datum*. Normally, the reference point for an altitude system is based on the tide gauge (Figure 6.12), corresponding to the mean sea level within a deviation of

Figure 6.12
In many countries the zero point for elevation references (vertical datum) is based on measurement of the variations in the ocean's surface and closely corresponds to the mean sea level.

maximum 1.0 m at global level. This surface, which goes through all points at zero altitude, is called the *geoid* (Figure 6.13). The geoid is affected by the mass of the Earth and therefore follows the Earth's contours—upward where there are large amounts of land mass and downward where there is less. The geoid is always at a right angle to the vertical line, and the vertical (plumb) line thus represents the direction of gravity. The geoid can be defined as the hypothetical surface of the Earth formed from mean sea level and its continuation through the continents at the same level of gravitational potential.

The fact that a geoid does not have an even and simple mathematical surface is not a problem as long as normal theodolites are used, which are adjusted with the help of levels affected by the same forces of gravity. By using the geoid and orthometric heights, one can ensure, for example, that water runs in the right direction when height measurements are carried out in water piping systems. However, problems arise as soon as GPS is used where coordinates are measured in relation to the center of the Earth. Heights on the surface of the Earth, based on geocentric coordinates, can be calculated only in relation to a mathematical surface—the reference ellipsoid—which is a mathematical approximation the surface of the Earth. Consequently, this ellipsoid differs from the geoid. The height difference between geoid and ellipsoid is known as the *geoid-ellipsoid separation* or *geoidal undu-*

Figure 6.13
Because of variations in the Earth's mass, the geoid will not coincide with the rotation ellipsoid. National elevation references are normally linked to the geoid.

lation (Figure 6.13). The difference between geoid and ellpisoid varies from place to place and has maximum values of $+/- 100$ m, when the whole Earth is considered. That is, not even the oceans of the world are mathematically even surfaces but have their high and low points just as the geoid does.

Conversion of height

The conversion of ellipsoidal height to orthometric height is impossible without knowing the geoid-ellipsoid separation at the actual point. When measurements of gravity are not carried out, the elevation of GPS points is most frequently measured by traditional surveying instruments or by finding the geoid-ellipsoid separation through measuring the ellipsoid height (using GPS) at points with a known orthometric height. If a project requires centimeter accuracy, it would not be safe to assume a constant geoid-ellipsoid separation over distances longer than about 200 m (Illife 2000), but it depends on the horizontal gradient of the geoid.

The relationship between the orthometric height *(H)* and the ellipsoid height *(h)* is given by the equation 6.1 (Figure 6.11):

$$H = h - N \qquad\qquad (6.4)$$

where N is the geoid-ellipsoid separation.

Few GISs can handle three-dimensional data satisfactorily. Most systems register height only as an attribute of the objects. Good topological models for three-dimensional objects are still to be developed.

6.7 Relative Georeferencing

Relative georeferencing includes:

- Offset distance
- Measurement along (road) networks

The relative georeferencing method is used to specify location in relation to details in the terrain.

In the offset distance method, location is specified using either direction and distance or distance(s) only from specified objects in the terrain (Figure 6.14). For example, the locations of buried pipes are often given relative to terrain features such as road edges, buildings, or poles. Where several pipelines have been laid in the same trench, their location can be specified in relation to the center line of the trench. Such measurements can be converted to existing rectangular coordinate systems where the coordinates are known for the details from which they are measured. For example, these coordinates could be taken from detail maps.

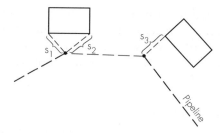

Figure 6.14
The location of a point can be expressed as the distance from physical objects in the terrain, known as the *offset distance*.

Most people will have noticed the milestones that stand along main roads in many countries. The function of these milestones is to indicate the distance (in miles or kilometers) from a given zero point and act as a visible point of reference for measurements in the road network (Figure 6.15). A network can have many zero points, often defined by different road junctions. Where the total layout of the road system is known, measurements (taken by measuring tape or odometer) along roads give good but indirect georeferencing. A road network system, and other types of continuous networks such as railway and tram, in many ways provide a schematic or topological georeferencing.

Customarily, links are selected to carry information. Because data volume and complexity increase in proportion to the number of links, how a road is divided into these links determines the nature of the nodes. If all intersections, events, and features along the road result in nodes, the number of links may be very great.

The transformation of measured distances in the road network to a Cartesian coordinate system can be beset with problems. First, there must be a sufficient number of coordinate determined points that describe the road. Second, a distance of, for example, 899 m along an in-

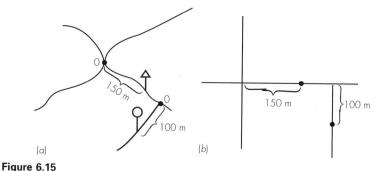

Figure 6.15
(a) In a road network the reference system is often based on measurement of distances from a given intersection. (b) This will only give in principle the topology of the network; the road network can therefore be presented only schematically.

clining road will not correspond exactly to 899 m on level ground. It is not possible to give an exact transformation without knowing a road's vertical and horizontal Cartesian coordinates.

Systems using relative georeferencing are unstable and vulnerable because the references are linked to physical objects in the terrain. Once the referencing objects are moved or removed, the reference system is also changed. For example, should a road be changed by rerouting or straightening a curve, the distance value of the reference points along the road will also be changed.

6.8 Discrete Georeferencing Systems

In discrete georeferencing systems the positions of phenomena are referenced relative to fixed, limited units of the surface of the Earth. Objects are linked to geography using the reference units as tags; this is known as the *tagging method* or georeferencing by *geographic identifiers*. With this type of system, we know that each object is located within and linked to the specified reference unit, but the location within the reference is unknown. Typical area reference units include:

- Address and street codes
- Postal codes and area name
- Property units
- Administrative zones and statistical units
- Regular grids and map sheets

The unit size determines the accuracy of registration: The smaller the units, the greater the accuracy. It is assumed that, in principle, the units are given a standard specification. In practice, however, the division between units can be relatively diffuse (e.g., it can be difficult to define exactly which geographical area is covered by an area name).

Many geographical phenomena are registered in discrete systems. Address and street codes, for example, may be used to indicate locations along roads (Figure 6.16). Reference to a whole street gives a

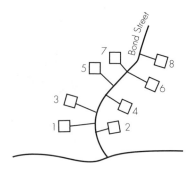

Figure 6.16
Street addresses are often used as simplified georeferencing.

clear discrete georeference. Roads and streets are often divided into selected registration units, such as city blocks, or fixed or variable distances. In addition, each house has its own number. As a road is gradually split up, georeferencing becomes almost continuous, although there will still be clear geographical gaps between units.

Postal codes, statistical and administrative units, locate phenomena within geographical districts of variable shapes and sizes. This type of division into areas can be described as irregular division, as opposed to a grid that is regular division (Figure 6.17). The relevant irregular geographical districts are often arranged in a hierarchy—for example, county, municipality, and statistical census areas.

Regular grids give positions in the same way as grids used in GIS, by row number and column number. Typical examples include the indices common on city maps or on national sheet map survey series (Figure 6.18). Most countries have an official system of division and numbering of map sheets in different scales. Reference to a map sheet number will thus provide an approximate discrete georeference. The same is true of the system with zones and belts previously described for UTM.

A so-called gazetteer provides a directory of instances of a class of phenomena, containing some information regarding position, such as coordinate references or purely descriptive references. The position can also be defined loosely by descriptive statements; for example, the position between two road exits.

Discrete reference systems are often easy to use and are therefore ex-

Figure 6.17
Discrete location registration may be based on a reference system of administrative or statistical geographical units.

Figure 6.18
A typical example of a discrete reference system is the index system common on city maps.

pedient when accuracy is not a major concern. For example, finding the address of a new house is easier than locating the exact coordinates of its four corners, but in order to find it in the field, information must be available as to how reference units are related to other objects—for example, how a beacon is related to important buildings such as city hall and the post office. Thus, finding discrete position has many similarities to finding relative position. Problems can also arise whenever boundaries are moved, unit shapes and sizes are changed, or name/title is changed. This means that if an administrative boundary or a road name is changed, all references to the old unit will be of limited value vis-à-vis the new georeference. Because data sets are not linked to the same geographical units, it can be difficult to compare data from different periods of time.

Discrete reference systems are often based on code indices in a hierarchy, which makes it easy to group objects linked to each unit and even combine them into larger units. Because georeferencing does not involve geometry, it is not possible to calculate directions, distances, and topology in a discrete reference system. So, except for regular grids, they are ill suited for presentation on maps. Nevertheless, the data registered may be linked to rectangular coordinates through reference transformations or through cartographic fixing of reference unit area limitations (e.g., by digitalization).

CHAPTER 7

Hardware and Communication Technology for GIS Applications

GIS tasks are many and varied, so no one system assembled from currently available computers, peripheral equipment, and software can be expected to meet all present and future needs. Moreover, the variety of computer hardware, software, quality, and prices on the market is broad and expanding. Thus, matching a system to defined needs remains crucial in implementing GIS facilities (Figure 7.1). As described in this chapter, the main physical parts of a GIS system collect, store, manipulate, and present data.

7.1 Computers

The heart of a computer consists of one or more integrated-circuit microprocessors, or central processing units (CPUs), which contain the circuitry necessary to interpret and execute program instructions for processing data and controlling peripheral equipment. *Reduced instruction set computer architecture* (RISC), a relatively new technological development based on high-level programming, offers a number of improvements as compared with the traditional CPU. For example, it has a wider range of basic functions and can carry out more complex operations with greater power and speed. The choice between RISC and the traditional CPU is more a matter of division of labor between the processor and a low-level operating system. Users will notice no particular difference other than the speed.

Data are stored in the working memory along with the operating system and the application program, which may be entered from secondary storage as needed. The working memory is completely solid-state electronic, and therefore access to it and retrieval from it are rapid.

Random access memory (RAM) is the most common type of working memory and is relatively expensive. Professional GIS users today normally use a PC with 128 to 164 MB of RAM. The number of operations per second is expressed in megahertz. A typical operation may be to send 8-, 16-, 32-, or 64-bit bytes through the processor, or from the

Figure 7.1
A geographical information system can comprise equipment such as that shown. It is intended to meet requirements for data collection, storage, manipulation, and presentation.

processor to RAM. In 2001, PCs were available with a processor speed of 400 MHz. Performance has doubled every two years over the last decade, and there is as yet no sign of a change in this trend. For GIS applications, RAM is normally more important than processor speed.

A byte usually comprises 8 adjacent bits, processed as a unit and capable of holding one character or symbol. *Mega-* is a prefix normally indicating a magnitude of 1 million in the decimal system. However, as shown in Figure 7.2, in the binary system used in computers it indicates a magnitude of 2 to the twentieth power.

Secondary storage is used to store permanently large quantities of data or program systems, such as GIS data that are not actively being processed. Secondary storage may consist of:

Prefix	Meaning in decimal system *Example*: frequency (hertz)	Meaning in binary system *Example*: capacity (bytes)
kilo-	$10^3 = 1,000$ Hz kilohertz (kHz)	$2^{10} = 1,024$ bytes kilobyte (kB)
mega-	$10^6 = 1,000,000$ Hz megahertz (MHz)	$2^{20} = 1,048,576$ bytes megabyte (MB)
giga-	$10^9 = 1,000,000,000$ Hz gigahertz (GHz)	$2^{30} = 1,073,741,824$ bytes gigabyte (GB)

Figure 7.2
Common prefixes in decimal and binary systems.

- Hard disks, with capacities ranging from 240 MB (in PCs) up to several hundred gigabytes
- Floppy disks, with a capacity of 360 KB to 1.44 MB
- Magnetic tapes, with a capacity ranging from a few hundred megabytes to 20+ GB (tapes are usually described as tertiary storage)
- Optical disks, with a capacity of some hundred megabytes up to 1 or 2 GB
- CD-ROM (compact disk—read-only memory) disks with a capacity of 650 MB

Solutions are being developed today that will quickly increase tenfold the storage capacity of CD-ROMs. In 2001, a typical PC for GIS work has a hard disk of 4 to 10 GB.

A secondary storage data search is mechanical and therefore slower than a working memory (RAM) search. Magnetic tape searches (and writing) are particularly slow, because the search must be carried out from the beginning of the tape. They are therefore better suited for storage of backup data than for data that are in frequent use. Search time on disks is considerably shorter, because data can be read from a given point on the disk. ROM is based on a special technology that allows the data to be written out only once onto the storage medium, thereby eliminating the possibility of making changes at a later stage. For this reason, CD-ROM is unsuitable for GIS data that will need to be changed frequently and is better suited for the distribution of large quantities of data. The storage capacity of magnetic storage media seems to more than double every three years. This is of great advantage for GIS applications that require the frequent handling of extremely large volumes of data.

The new "power PCs" can run powerful operating systems such as Microsoft's Windows NT. Windows NT can run on both PCs and workstations, thus eliminating the traditional division between the two types of machine. Among the most widespread PC systems are local area network (LAN) configurations (Figure 7.3), which facilitate multiuser access without restricting local computing ability.

Figure 7.3
Typical local area network (LAN).

7.2 Networks

Information is one of the most important strategic factors facilitating development in any area. Modern political and economic systems cannot function without a continuous interchange of reliable information. Worldwide, the communication and distribution of data is a large and rapidly growing sector.

The term *network* is used here to mean data systems linked together by electronic lines of communication and software. There has been a considerable increase in the use of networks in recent years, due to the following factors:

- Exchange of data between systems (users)
- Sharing of expensive resources (equipment)

The basis of network technology is that individual units in the system comprise common resources (server) for the rest of the network (clients). A server normally has a greater data storage capacity than the other computers in the system and is capable of handling complex program systems. Local networks are available that link local data equipment—for example, in the same building, for sharing of data storage capacity and software, as well as printers, plotters, and so on.

HUBs are used as nodes between computers, enabling different machines in the network to communicate with each other. Local networks, often called LANs (local area networks; Figure 7.3), and remote networks, often known as wide area networks (WANs), link computers together over greater distances (Figure 7.4). WANs are based on the services of national/international telecommunications companies.

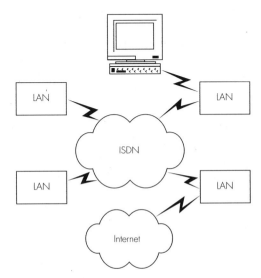

Figure 7.4
Typical wide area network (WAN).

Several standard local network systems are available, some of the commonest being Ethernet (based on bus topology) and token ring (based on ring topology) (Figure 7.5). A number of LANs can be linked together to form larger systems (metropolitan area networks [MANs]) with the help of *bridges*. A local network can be connected to the national/international network (a WAN) with the help of *gateways*. Satellite communication is also used to connect networks.

In a LAN the distance between units is normally limited to a few hundred meters. However, this can be increased to 3–4 km using amplifiers or bridges. Data transfer over greater distances has to be carried out over the public telephone network system. In this way it is possible to transfer data throughout the world. A modem (modulator/demodulator) is used to connect digital data to and from an analog telephone network.

Bandwidth is the range of transfer capacity in a network and is measured in bits per second (bps). The transfer capacity in a local network depends on how many users are using the network at the same time. However, typical maximum values are 5 to 10 megabits per second (Mbps). With the use of an analog telephone network and a modem, the transfer speed of a WAN is often limited to 300 to 14,400 bps. With the use of digital data systems, modems are not required and capacity increases immediately to 64 kilobits per second (kbps). Today many systems use a network speed of 2 Mbps. The speed on the Internet is normally 64 kbps to 1.5 Mbps. However, there are special solutions (asynchronous transfer mode [ATM]) with speeds as high as 150 Mbps. Ethernet, with a capacity of as much as 100 Mbps, is now being adopted for use. Using fiber-optic cables and fiber-distributed data interface (FDDI), it is now possible to achieve a transfer speed of 150 Mbps, which makes multimedia—the simultaneous transfer of sound, image, and data—a reality.

A network requires control mechanisms in the form of software to establish the necessary connections, known as *access protocols*. Special network operating systems such as Novell Netware and Microsoft

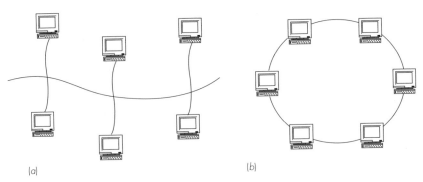

(a)   (b)

Figure 7.5
Ethernet *(a)* is based on bus topology; token ring *(b)* is based on ring topology.

LAN Manager have been developed, which can operate independently in relation to equipment and protocols. This type of operating system carries out configuration control, security control, error control, accounting functions, and so on. Two types of protocols are commonly used today:

- OSI—Open System Interconnect, developed by the International Standards Organization.
- TCP/IP—Transmission Control Protocol/Internet Protocol. TCP/IP was originally developed for the ARPAnet research network in the United States.

The TCP protocol splits information into single packages of fewer than 1500 characters before they are sent. The IP protocol acts as an envelope with an address on each package and ensures that each package arrives at the right address. On arrival at their destinations, the packages are regrouped into information with the help of the TCP protocol (Figure 7.6).

The Integrated Service Digital Network (ISDN) has a larger bandwidth (64 kb/s) and has been developed specifically for the transfer of various digital data such as text, image, and sound over greater distances. ISDN, called a *point-to-point digital carrier*, represents a physical transfer technology. The system is based on the technology and software of national and international telecommunications companies. ISDN requires a fully digital solution within a digital telephone system.

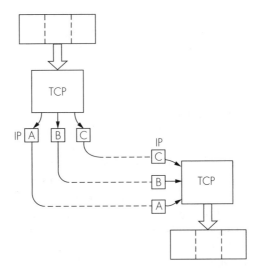

Figure 7.6
TCP divides the data into small packages for shipment. Each is then packed into IP *envelopes* and sent in different directions through the network. On arrival at the recipient, the packages are reassembled by TCP.

Figure 7.7
The Internet consists of many networks (LAN, WAN, etc.) linked together
through *paths* in the commercial telephone network.

Internet

The Internet comprises many networks (LAN, WAN, etc.), which are
linked together by lines in the telecommunications network (Figure
7.7). Internet is a service based on established physical networks such
as those of national and international telecommunications companies.
The physical elements of the network can be a modem (for home
users), ISDN, frame relay (fixed lines), Ethernet, ATM (asynchronous
transfer mode), and so on. The lowest level in the Internet is TCP/IP,
a set of protocols (language) that enables machines to communicate
with each other (Figure 7.6). All computers that are linked to the Internet have a unique address (IP number).

The Internet was started by the U.S. Defense Department toward the
end of the 1960s; it is a service that is fully available to anyone. Users
connected to the Internet can search for and retrieve information from
databases throughout the world. They can also make their own data
available to others, send electronic mail (E-mail), and send and receive
normal data files, functional data programs, and multimedia files in
the form of speech, music, images, animation, and videos. Many organizations have adopted Internet services for internal activities, in a
system known as the *Intranet*. The Intranet is separated from the open

Internet by a "fire wall," a combination of hardware and software aimed at securing the organization from external users.

Many Internet applications are useful in GIS. The Internet/World Wide Web (Chapter 8) can be used for distribution of information on geographical data in the form of *metadata bases*. User-friendly Internet solutions for data search and transfer are currently under development. In the long term, the Internet may become a "GIS trading exchange," a marketplace for the purchase and sale of geographical data. In addition, simple GIS functions are currently accessible on the Internet, and it is possible to operate GIS software available on a server from terminals.

7.3 Displays

The display (monitor) of a computer is the user's primary visual communication medium in all computer work. A display consists of a screen and its associated electronic circuitry. The screen itself is either the face of a cathode ray tube (CRT), similar to those long used in television sets and radar, or one of the newer, flat arrays of semiconductor elements, such as the liquid crystal displays (LCDs) used in laptop and notebook PCs. LCD technology is advancing, but at present the CRT remains the mainstay for almost all GIS applications because it produces more light, has higher resolution, and has a greater range of possible display colors than do LCDs. All CRTs are vacuum tubes and operate on the principle that a beam (ray) of electrons emitted by a cathode is focused and directed at an electroluminescent screen. The information displayed is refreshed periodically. Regeneration screen images are refreshed up to 60 times a second. As with TV images, the displays are based on rasters, and GIS vector map data are converted to raster data for screen presentation (Figure 7.8).

Color screens employ the three basic colors—red, green, and blue (RGB)—in varying intensities, which reproduce a wide range of hues. A high-resolution 19-inch (diagonal) screen designed for computer

Figure 7.8
Raster regeneration screen.

137

graphics may resolve 1600 points horizontally and 1280 points vertically. Screens are now available with processing functions that include vector–raster conversion, storage of vector data, and color look-up tables (LUTs) for rapid alteration of color patterns. High-definition color screens of 21 inches are used most frequently in GIS applications.

The on-screen position of the cursor—the movable, sometimes blinking, symbol that indicates the position of the next operator action—is controlled in many ways. On keyboards, this is usually accomplished by four arrow keys and various function keys. Provided that the computer involved is suitably equipped, the cursor position may also be controlled by either a joystick or a mouse, a palm-sized button-operated device that, when moved across a flat surface, produces a corresponding movement of the cursor on the screen.

Tablets are also used to control cursors and communicate with application software. A tablet is a small digitizer arranged to accept overlays containing command displays. By moving either a penlike device or a mouse, the cursor on the screen is moved and software instructions are initiated by pointing at fields in the menu.

Small penpad computers have been developed. These consist of a penlike unit, combined with a screen that is sensitive to touch and special character recognition software (Figure 7.9). These replace the keyboard as the input unit. The operator uses the pen to make selections from menus and dialogue boxes, and the software interprets handwritten input on the screen. The system can also receive data from GPS, total stations, and other field equipment. The penpad system can display an existing map and can be used in the field for inspection, data input, calculations, and simple editing.

Figure 7.9
A penpad computer makes it possible to handle data in the field as they are produced.

Some computers support the use of a light pen, which when its tip is pointed at a feature in the on-screen display, registers its coordinates and permits selective functions to be performed, such as the retrieval of information on the objects pointed out.

7.4 Quantizers

Quantizers are devices that convert analog data, such as data on ordinary maps, into digital data, as used in GIS. The quantizers used most frequently include the following:

- *Digitizer:* a combination of manual positioning and electromagnetic sensing on a plane surface; historically, the first analog-digital converter for GIS applications
- *Scanner:* an automatic electromechanical system that converts a picture into a raster of points (pixels)
- *Video:* the electronic equivalent of motion-picture photography, often followed by rasterization similar to that employed in a scanner
- On-screen digitizing and *automatic line tracer:* a function performed through software programs on raster data that have been scanned and displayed on screen

7.4.1 Digitizers

A digitizer consists of a flat nonconducting surface in which a grid of wires is embedded (Figure 7.10). Electric currents passing through the wires give rise to electromagnetic fields, which together form a pattern on the surface. A small magnifying glass fitted with crosshairs is used to locate objects on a map placed on the table. The glass is also fitted with a sensor coil, which detects the position of the crosshair center with respect to the underlying wire grid. Thus, positions are entered

Figure 7.10
Most digitizers are basically a flat surface in which a grid of thin wires is embedded for conducting electric current. The sensor coil detects the electromagnetic field around the wires, thereby registering the position of the crosshair center.

continuously or at the command of a keystroke. A keypad with up to 25 keys connected to a microprocessor allows identifiers, thematic codes, and so on, for objects that have been located to be entered, and also facilitates interaction with the program involved.

Digitizers are available in various designs and sizes. Overall size ranges from the 27 × 27 cm tablet size up to 1.0 × 1.5 m. Map digitizing requires a relatively large surface capable of accepting the common 84.1 × 118.9 cm ISO A0 format (according to the international ISO 216-1975 standard). Those used for encoding map data must feature an accuracy of at least ± 0.5 mm. A common set of characteristics includes a resolution of 0.02 mm (the smallest unit of registration), an absolute accuracy of ± 0.15 mm, and a repetitive accuracy of ± 0.025 mm. The best digitizing tables have an accuracy of 0.075 mm. Accuracy is dependent on the density on the grid of wires and on the equipment picking up signals from the wires.

Some digitizers are fitted with background illumination (like ordinary light tables), an advantage in carrying out many digitizing tasks. A digitizing table is ideal for accurate digitization of all types of map information.

7.4.2 Scanners

Scanning is a means of encoding through conversion to a digital image comprising a regular grid of pixels rather like the array of dots that make up a newspaper photograph. Scanners are classified into three main categories according to their basic mechanism and physical size: flat-bed, drum, and continuous-feed.

Flat-bed scanners resemble automatic drafting machines. Map image information is registered via a photosensitive pickup that rides on a track, which is mounted on a beam that itself moves in the transverse direction (Figure 7.11). Scanners can be extremely accurate and capa-

Figure 7.11
Scanners are operated by means of a photosensitive pickup being moved with great precision over the surface of a map. Photosensitive cells register the different gray tones on the map, thereby creating a raster image showing all the lines and white areas on the map.

ble of high resolution. Resolution is normally measured in dots per inch (dpi). Flat-bed scanners usually have very good resolution, normally 500 to 1000 dpi. Pixel size is then as small as 25 μm (0.025 mm) and 50 μm (0.050 mm), which corresponds to a resolution of 2.5 cm or 5.0 cm in detailing terrain from a map with a scale of 1:1000. Accuracy is very good, ordinarily ± 0.10 mm.

Drum scanners employ the same process as flat-bed scanners, except that the map to be encoded is mounted on a drum that rotates beneath a fixed track on which the pickup rides (Figure 7.12). Normally, the pickup digitizes one grid line for each drum rotation. Most drum scanners have a resolution of 400 to 800 dpi, some even as much as 1600 or 2000 dpi. Drum scanners are usually less accurate than flat-bed scanners but adequate for all normal GIS use. Degree of accuracy is normally expressed as a percentage, such as ± 0.1%, which would correspond to ± 0.6 mm for a map width of 60 cm. Drum scanners can scan an entire map in the course of two minutes.

With *continuous-feed scanners*, documents to be scanned are fed past a row of sensors as wide as or wider than the maximum width of the documents scanned. The basic structure is simpler and less expensive than that of table or drum scanners, so accuracy and resolution are limited accordingly. Therefore, continuous-feed scanners are best suited to ordinary document scanning, which is why they often are called *document scanners*.

Scanners can register color, usually in a three-pass process involving transillumination of the scanned map and successive magenta, yellow, and cyan blue filtering. Some scanners convert raster data to vector data in the scanning process itself, simultaneously presenting the results on screen.

There is a draw-back to using scanners, in that scanning senses everything on a map, including labels, coffee stains, wrinkles, and other unwanted information. Therefore, extensive editing is almost always necessary unless a map is drawn as needed, which is rarely the

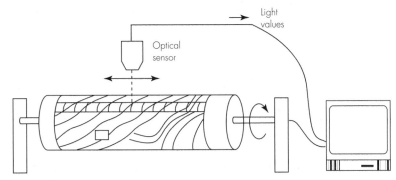

Figure 7.12
Drum scanner principle.

case. Because editing is expensive, many organizations now routinely redraw maps to be scanned. Scanners are ideal for the quick and cheap production of digital background maps for later improvement by vectorization and pattern recognition. With more efficient software for pattern recognition, scanners will probably be able to handle many more demanding data conversion tasks of the future.

7.4.3 Screen digitalization

Digital raster data obtained from a scanning process can be digitalized directly on a computer screen by using the computer's pointing device (mouse, tracball, etc.) to navigate in the map image. Software programs have also been developed that can perform an automatic line-tracing function on raster data on screen. To a certain extent, an operator has to control the process for following special lines (such as water pipes), for solving conflicts between crossing lines, and so on. In addition, the operator can also choose between manual single-point registration or manual stream registration. On-screen digitalization has the same degree of accuracy as table digitalization and also has the advantage of being more user-friendly, in that the operator sits comfortably and can create a panorama and enlarge complex areas as needed.

Raster data are transferred to vector data automatically through coordinates being computed continuously for all points on a line, as they are identified by the software. Automatic line tracing is slower than pure raster scanning but more rapid than manual tablet digitizing, and the results are lines in continuous form (vector data). On-screen digitalization is increasingly replacing much of the work previously carried out on the digitalization table.

7.4.4 Video

Video cameras are also used to digitize maps, documents, and pictures. The resolution and accuracy attainable are relatively poor (typically, 512×512 pixels), although adequate for some applications. The use of video images in GIS is very limited at present.

7.5 Plotters and Other Output Devices

7.5.1 Ink-jet printers

Ink-jet printers form images by projecting droplets of various colored ink onto paper. The technique has long been used in the printing and labeling industries and was first used in computer printout devices in the mid-1970s. One variety, the pulsed ink-jet printer, in which one or more columns of ink nozzles are mounted on a head that traverses the medium, is capable of high printing speeds, up to 400 characters per second. From a cartographic viewpoint, most of the early ink-jet printers had relatively poor accuracy. They were therefore used mainly for such low-accuracy applications as replicating screen displays. However, in the early 1990s ink-jet printers with high resolution, up to 400 dpi, became available, both in letter sizes and in larger sizes suitable for use as a plotter for GIS applications. Today, the color ink-jet plotter is the most popular plotter for GIS applications, being relatively economical, fast, and having satisfactory accuracy ($\pm$ 0.25 to 0.5 mm) and cartographic quality (400 to 600 dpi). Ink-jet plotters can be used for both vector and raster data.

When plotting on paper the complementary colors cyan, magenta, and yellow are used. These are known as subtractive colors. Black is also used to deepen the dark areas, and the entire range is abbreviated as CMYK. Some of the new plotters use, in addition, lighter color variations of cyan, magenta, and yellow to bring out lighter nuances.

7.5.2 Electrographic plotters

Electrographic plotters include electrostatic, electrophotographic, and electrosensitive plotters. The electrostatic and electrophotographic processes, which are used most frequently in GIS applications, involve an intermediate electrostatic image. The electrosensitive process, which involves direct interaction between an electric current and the surface of a special paper, is used in various high-resolution applications such as photo-typesetting.

In an electrostatic plotter, the image is first written as a pattern of electrostatic charge, either directly onto specially treated paper or onto a drum coated with aluminum oxide. The pattern is then made visible by washing the charge with a colloidal suspension of particles of pigment carrying the opposite charge. In the drum design, the visible image is then transferred to plain paper by the application of pressure and an electric field.

Electrostatic plotters, which are frequently used in GIS applications, have an advantage over pen plotters in that they can plot from both vector and raster data. The drawback, however, is that most models require special paper, which means that existing maps cannot be amended. Electrostatic plotters require climatized rooms, as the paper is affected by humidity. The resolution of a good electrostatic plotter is usually on the order of 200 to 400 dpi.

In an electrophotographic plotter, a beam of light writes the image onto the electrically charged surface of a drum. The light discharges the charged surface to form the image, which becomes visible once the drum is dusted with a dielectric powder (toner). The image is then transferred to paper by direct contact and the application of an electric field. The toner particles are finally affixed to the paper by heat and/or pressure. Electrophotographic plotters plot in large formats and normally have a resolution of 400 dpi.

The electrophotographic process, also known as xerography, is the same as that used in most office copying machines. It is also used in laser printers, in which lasers are the light sources. A wide range of laser printers are now available for use with text and graphics, including the printing of map data in monochrome and colors. Most handle paper formats up to standard letterhead size—ISO A4 (21.0 × 29.7 cm) or the U.S. 8½ × 11 inch standard—but printers for ISO A0 (84.1 × 118.9 cm) are also available. Today many laser printers have a resolution of 600 dpi and can reproduce 21 shades of gray.

The light-emitting diode (LED) is a relatively new technology, similar to a laser. The main difference is that instead of using a laser beam, an array of LEDs is used to transfer digital data to the paper through several processes.

All electrographic plotters can produce final images in black and white or color, depending on the pigment color (magenta, yellow, cyan blue, and black, in successive passes). Wherever particularly high resolution is needed for GIS applications, electrostatic plotters are the devices of choice.

7.5.3 Thermal plotters

Thermal plotters and thermal transfer plotters use heat to produce images. Thermal plotters use local heating to warm thermo-sensitive paper, which is coated with two separate components. Once heated, these combine to produce a wide range of colors of a reasonable quality. Thermal transfer plotters transfer colored wax to produce high-quality images. Both types are available to handle sheet sizes up to ISO A3 (29.7 × 42.0 cm). Thermal plotters have limited color coverage and the colors soon fade. Direct thermal plotters are a new kind of thermal plotters, which yield very good cartographic products and are used in GIS.

7.5.4 Film printers

In a film printer, a laser or LED draws a raster image on a photosensitive medium. The image resolution is very fine, ranging from 400 to 2500 dpi, and the accuracy is good. The technique is normally used to produce film originals for printing, as used for printing maps in color and for production of photographic films from satellite data. Digital map data may be entered directly into devices that produce films for printing. Film printers can also plot on photo paper, and large-format film printers are also used today for high-quality plotting of satellite images and ortho images.

7.5.5 Comments

Output speed for all types of plotters is determined by the time needed to transfer data to the plotter and plotting speed. Plotters for GIS applications should have a relatively large memory capacity, because it is often necessary to transfer 20 to 30 Mb at a time for plotting. Plotter speed is regulated by the complexity of the data and the plotting technique. The fastest plotting is achieved by electrophotographic, laser, and direct thermal plotters, whereas ink-jet plotters are slower. Black-and-white plotters are normally faster and cheaper than color plotters. (In addition to machine costs, one should also consider operating costs, such as paper, toner, and maintenance. Experience shows that such costs can be quite high and quickly approach the purchase price of the equipment. For a single A0 plot, running costs today can vary between $1 for an ink-jet plotter and $.50 for a laser device (Ireland 1996). Today, PostScript and HPGL are the normal formats for color plotters.

One should be aware of the fact that plotters seldom replicate on-screen colors exactly. This is due to the fact that computer monitors have a greater spectral purity than can be achieved by plotting. This problem can be solved through extensive calibration.

GRID

Global environmental problems are both large and complex. Until the mid-1980s, most environmental problems were addressed individually and locally. However, as the underlying problems are global, there is a need for the global approach that is now possible using computer technologies. In 1985, the United Nations Environmental Program (UNEP) established the Global Resource Information Database (GRID) as part of an overall effort to collect, manipulate, and promulgate global environmental and natural resource data. By using GIS, remote sensing, telecommunications, and other modern technologies, these data may be made readily available to professionals, government officials, decision makers, managers, and the general public. GRID's mandate is to acquire data and information relevant to prudent management of the global environment.

GRID finds it increasingly important to make maps, data sets, and other environmental information products available on the Internet/WWW. *The State of the Environment of Norway* is the first comprehensive national State of the Environment report made accessible on the WWW. The GRID-Arendal home page also gives hyperlinks to related institutions worldwide and will be an important tool for any Internet user searching for environmental information.

CHAPTER 8

Basic Software and Databases for GIS

8.1 The Foundation Stones of GIS Software

GIS software is under continuous development, with respect not only to the specific GIS program functions, but also to operating systems and other general program tools. For this reason it is not possible to give a comprehensive description of a GIS software program that is relevant in the long term. In the following section, therefore, we present just a few of the basic relationships. More detailed and up-to-date information can be obtained from suppliers' manuals and brochures.

The GIS application software, the database management systems, and the operating systems represent the foundation stones of GIS (Figure 8.1).

GIS application software may be divided into four main categories:

1. Functions for registering, entering, and storage of data
2. Functions for correcting and adapting data for further use
3. Functions for processing and analyzing data
4. Functions for data presentation

These four individual GIS application functions are explained in Chapters 13, 14, and 15, and operating systems and database management systems are described in this chapter. In addition, GIS-related software such as the human–machine interface, CAD, WWW, and multimedia are also discussed briefly in this chapter.

Geographical information systems often comprise complex program systems for various applications. This is because many systems were originally devised for special, limited tasks, such as mapping the network of a particular energy supplier. However, the current trend is toward more general systems (tool kits) from which users may select components suited to specific applications. The tasks involved can be demanding, so general systems often have relatively high entry levels for new users. Some GIS software suppliers can supply their own development tools for the programming of special applications.

Moving a software system from one computer to another can be extremely difficult because of the differences in the respective operating systems. However, the process is becoming easier now that standards

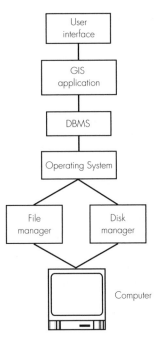

Figure 8.1
A DBMS runs on a computer just like an application program. The operating system, with file manager and disk manager, lies between the DBMS and the physical computer.

have been established in the market. Modern software systems are flexible and, to a large extent, based on official or de facto standards. One such standard is the graphics kernel system (GKS). This set of computer graphics routines is intended for use by application programmers and is standardized by the International Organization for Standardization (in the ISO 7942 and ISO 8651 standards). GKS ensures consistency; a screen can be identified as a screen regardless of its maker. Among the de facto industry standards that have been established are HPGL, a plotter language developed by Hewlett-Packard, PostScript, a page description language developed by Adobe Systems for word processors and desktop publishing packages, and SQL for relational databases. Considerable standardization is now being completed, of which perhaps standards for data models and handling spatial data in data bases will be among those of greatest immediate importance to GIS users.

Many software systems boast libraries containing a wide range of types and models of digitizers, screens, plotters, and other peripherals, which the software supports through having the relevant drivers. A driver is a software routine that handles the specific details and characteristics of a single peripheral device so that it can work correctly with the system. The user may select peripherals from the

147

library listings in order to customize the software to a particular set of peripheral devices.

8.2 Operating Systems

An operating system is a basic program that administers the internal data processing in a computer. Important elements of an operating system are the file manager and disk manager, which deal with the file structure and page structure of the computer, respectively (Figure 8.1). Different operating systems, such as Windows 98 and 2000, Windows NT, OS/2 for PCs and UNIX (Linux, BSD) for workstations, are used in different types and makes of computer and for different GIS software. As application programs cannot be moved from one operating system to another without extensive modification, the choice of application program is often determined by the choice of computer.

Windows NT is platform independent by taking advantage of the 32-bit addressing technique of 486-, Pentium-, and RISC-based processors. There are no technical reasons today that Windows NT cannot be ported to run on any computer platform; the system is capable of carrying out more than one task at a time (multitasking).

Current trends are providing more options, and many suppliers of GIS programs now offer versions matched to different operating systems. Computer manufacturers are increasingly offering wider ranges of equipment, such as UNIX in workstations and Windows NT in PCs. There appears to be a clear trend toward increased adoption of Windows NT as the standard operative system for GIS software.

8.3 Communications Between Users and Computers

Users and computers communicate at the human–machine interface (MMI). The features of MMI are vital in GIS applications, where users need to control communications with their computers by deciding what is to be done at what time and in what order—always with the option of canceling an action midway through a communication.

In particular, MMI communications require input/output devices, such as ordinary keyboards, special function keys, mice, joysticks, and so on, along with their supporting software. Tablets are also used in interactive human–machine communications. A tablet is a graphical input device in which positions can be selected by using crosshairs instead of pointing on screen.

The dialogue between a user and a computer is *interactive* when user commands result in immediate system responses in the form of messages, error messages, new choices, and so on. On-screen dialogue may be controlled by:

- User commands to the system
- User responses to system messages or queries
- Menus of options that elicit user input
- Icons (pictorial symbols used in menus) to which the user points the cursor

Commands

The first software systems were controlled by relatively cumbersome sets of commands and had no interactive features. Many GISs contain remnants of these earlier complexities, although choices between commands and menus are often available. To use commands, users must enter relatively long instructions, which in turn require a knowledge of numerous basic commands (Figure 8.2).

Modern systems are often structured so that users may compose and store sequences of composite commands compiled from large registers of basic commands. The programming involves system macro instructions and makes it possible to develop customized applications. Composite commands are best suited to repetitive operations for a specific product or application. Although software systems are now menu driven, many experienced operators prefer commands to menus because commands permit greater flexibility and often achieve final results more rapidly.

Queries and messages

The dialogue may involve the user's either answering system queries or following instructions given in system messages; examples are given in Figures 8.3 and 8.4. Dialogues of the type illustrated in Figure 8.4 are well suited to the needs of many end users, such as those who use PCs or workstations while serving the public. Standardized query routines, known as *Structured Query Language* (SQL), have been developed for relational databases. These routines are discussed further later in this chapter.

COMMANDS (ArcInfo, ESRI)

– RESEL road ARCS route > 0 AND veh CN 'T'
  (logical selection among road and vehicles)

– Lineset last.lin
  (activating symbol table)

– ARCLINES road route route.lut
  (plotting instructions for selected roads using
   route no. to select symbol through look-up table)

Figure 8.2
System commands can be given in the form of long strings of command instructions.

```
┌─────────────────────────────────────────────────────────────┐
│  PROMTS                                                       │
│                                                               │
│  Give number for selected route of roads          : <   >     │
│                                                               │
│  Select roads maintained by lorry (L) or truck (T)  : <   >    │
│                                                               │
│  Plot the roads selected, Y / N                   : <   >     │
│                                                               │
└─────────────────────────────────────────────────────────────┘
```

Figure 8.3
Command activation can be in the form of questions to the operator. These
questions can often be adapted by the user for special functions.

Menus

Communication between users and computers by pointing and clicking on menus is now commonplace. A menu is a program that elicits
user input through an on-screen display of options, which may include commands or keys to submenus (Figure 8.5). The software system is then said to be menu driven.

Menu functions are available in a variety of "windows" programs
(e.g., MS-Windows, X-Windows, Motif). Windows techniques permit
simultaneous operation of different programs by displaying information on them in separate rectangular areas on-screen (Figure 8.6). Windows, menus, and dialogue boxes may be moved around on screen as
needed, for example, when they obscure part of an active application
on screen.

Many GISs allow the user to create his or her own menus, composed
of a series of commands. Thus, the user interface can be tailored to fit
an individual application. A special macroprogramming language is
used for this purpose.

Icons

Icons are selected by pointing and clicking. Each pictorial icon on
screen represents a command. The icons may be simple, such as a set

Tone Table Picture	Standard Table Picture	Align Edges	Align Corners	Align Centers	Align Reference Points		
	Get Mis	Create Mis	Undelete Mp	Undelete map	Current Mp	Current Visible Picture	
		Delete Mp	Delete Visible Picture	Position Picture	Move Any Picture	Move Visible Picture	List Mp

Figure 8.4
Plotter menu from an image-processing system. Commands are activated by
positioning the menu table on the digitizer and pointing and clicking the mouse
in the appropriate fields.

Figure 8.5
In modern GISs, dialogue with the user is generally based on screen menus
where one can point and click in the command fields.

of boxes for color selection; or they may be more complex, such as a
trash can to represent discarding (Figure 8.7).

Ancillary features

Many systems contain help functions that users may activate for further information on commands. Once a command is activated, an explanation of it can be requested using a contextual search. Users may
often select various function parameters. If a parameter is not specified by the user, the system will revert to its own standard (default)
value.

Response time is crucial in all interactive work, because response
times that are longer than $\frac{1}{10}$ second may lead the user to doubt that
the system has understood the last command entered. However, batch
processing may be advantageous in some cases, as with final drawing.
In batch processing, commands are collected in a file and executed
when computer resources become available.

Figure 8.6
The screen can be split into several fields by means of the windows technique.
This allows several different commands to be activated simultaneously.

151

Figure 8.7
Command fields in screen menus can be presented in the form of easily comprehensible symbols called *icons*.

8.4 Database Management Systems

It is doubtful that there would be any GIS without database software. There are a number of different systems for organizing and handling of data in a data base. It is not possible to give a complete view of all database management systems here, so we will mention a few main points.

8.4.1 Files

Files are the foundation stones of all database software and are therefore discussed first. Files comprise *records*, each of which contains *fields* (Figure 8.8). Each record contains data concerning a single topic or affiliation; each field contains an item of data consisting of one or more characters, words, or codes that are processed together. *Keys*, which are codes used to access information, help to retrieve records from files. Keys are associated with one or more fields of a record.

Data in a file may be stored sequentially, as in a line (Figure 8.9). New records are simply added at the end of the file. This simple file structure is called a *simple list*. Records may be of variable length, in which case the beginning of each record must contain information about its length; such information is not necessary, however, when the length of the record is fixed.

Searching for data in a sequential simple list is time-consuming. For example, if the search is for points stored in a random sequential file, all records in the file must be queried to ascertain whether the desired points have been located. Therefore, storage in unorganized sequences is recommended only for relatively small quantities of data.

	Field 1	Field 2	Field 3
Record 1	110	Oslo	10.61
Record 2	115	Stockholm	9.15
Record 3	116	London	18.33
–			
–			
Record n			

Figure 8.8
A file is divided into records and fields.

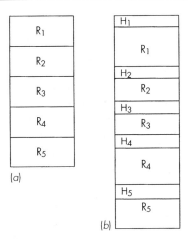

Figure 8.9
There are two categories of sequential files: *(a)* fixed record length (R_1, R_2) and *(b)* variable record length. The headers (H_1, H_2, etc.) specify the length of the record.

The search time may be reduced if the data in a file are structured—for example, when they are sorted by ascending coordinate magnitudes. The search may then be limited by the largest magnitudes entered. This file structure is called *ordered sequential files*. A search could also start in the middle of a file, thereby allowing the search to progress successively, by dividing intervals into two, deciding each time which half contains the point sought, and carrying on until the point is located. This process, known as *binary search*, is considerably faster than sequential searching. Even though this method considerably increases search speed, it has a disadvantage in that new records have to be continuously put in their correct order, otherwise one would end up with a simple list file structure.

In GIS, users want to be able to use flexible search criteria, not just to find a particular point, but also, for example, to find objects with a particular attribute. In this case an ordered file would be of little help. An index to the data store is needed to increase search speed. The address, or position of an item in a file, may be stored in a separate file. Known as an *index file* or *pointer file*, it contains a set of links that can be used to locate records in the data file. Searching for items in the data file may be compared to searching for entries in a well-indexed book, by first finding page numbers in its index. There are two types of index files: *direct files* and *inverted files*. In direct indexed files a direct connection is established between a particular code (attribute) and the record number of all objects having this code. Inverted files are more flexible than direct files and are to a greater extent adapted to unspecified search.

An index file is an additional structure that the user has to establish prior to starting a search. Suitable indexing requires that the user

153

know the current search criteria. Organizing index files is time-consuming, but once the files are established, a search will be rapid and can be constantly repeated.

The largest practical disadvantage to files is that each program application has to be specially tailored to the structure of the files. When changes in the file structure are made, application software also has to be changed. Usually, unstructured files are not suitable for handling large quantities of data or data that change continually. Considerable double storage can easily occur, and extremely complex situations have to be handled by supplementing the data. Simultaneous use of the data by several users is very difficult to handle when the data are in the form of separate files. In addition, it is not just one single file being used simultaneously but many files. When files have to be shared simultaneously among several users, a central system must be established to control who has access to files and which operations can be carried out by each user. Despite these drawbacks, the file-based systems have been popular within GIS. Many of the leading systems on the market use file systems and tailor-made management systems for storage and management of the geometric data. However, special indexing routines have been developed to increase efficiency in pulling out data (Chapter 12). Attribute data are usually stored and handled by standard relation DBMS. See the following section for further discussion.

8.4.2   Databases and database management systems

The difference between a file and a database is semantic and varies somewhat from one discipline to another. For GIS purposes, a *file* is regarded as a single collection of information that can be stored, and a collection of files is regarded as a *database*. A database comprises one or more files that are structured in a particular way by a *database management system* (DBMS) and accessed through it (Figure 8.10). A DBMS can therefore be defined as a software package for storage, manipulation, and retrieval of data from a database (Figure 8.11). DBMSs have

Figure 8.10
A DBMS is located between the physical storage of data and the user.

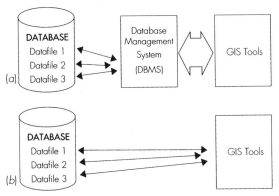

Figure 8.11
A DBMS *(a)* ensures easier communication between a database and GIS tools than through direct manipulation of files *(b)*.

been developed to handle the complex tasks of multiple files in a database and data amendment. The database management system has been developed from files' management systems.

Database systems consist of:

- Data
- Hardware
- Software

The data in a database system are stored on disks with associated input and output devices; thus, a database is a physical collection of files structured by DBMS software. We shall now look at the software aspects. The starting point is that one has geographical phenomena described in a *data model* (Figure 8.12). This is the data model which will be realized in the data base. The work of structuring data to fit in a data model is further described in Chapter 12.

All database systems aim for ease of search and linking of tabular data. A DBMS must be able to manipulate several types of objects and various relations between objects. DBMS has a question compiler that checks the questions and sends them on to the database processor, if the questions are correct (Figure 8.13). To take out data from the database, a connection has to be established between "the object" in the question and physical location of the object on the disk. This is done by the data manager, who uses the operative system to gain physical access to the data.

Figure 8.12
The data fed into the database have to be structured so as to produce useful applications and products.

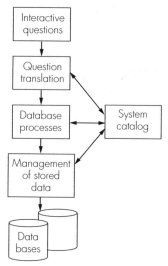

Figure 8.13
The figure shows the principles for handling of questions to a database.

Most users will be little interested in how a database functions, but they are quite interested in the query and search functions, report functions, transaction handling (changes, add and delete data), backup copying, availability control, and the like. This type of functionality is available today from general database management systems. Yet databases are no easy option. They can be both misused and overused. The homeowner who uses a database on a PC to keep an updated record of the contents of the home freezer is solving the daily "What's for dinner?" problem with technological overkill; keeping a pencil and paper next to the freezer is more efficient. It is important to choose the right database system so that data model(s) can be realized and tasks solved efficiently. This presupposes a knowledge of certain basic principles of DBMS.

8.4.3 Database structures

In GIS the data in a database system can be structured according to different models.

The first database solutions for GIS were based on databases called *flat-file databases,* which comprise files only, with very limited supporting hierarchy or other interrelationships. Traditional model structures are relation structure, hierarchical structure, network structure, and object-oriented structure. The hierarchical and network structures were already available at the end of the 1960s, but it was soon apparent that they were less suitable for GIS, partially because all queries had to be known in advance and accounted for in structuring and entering data, which is not necessarily natural or suitable in GIS applications. Today those structures have been more or less out-competed by relations-based structures.

The choice of a DBMS ill suited to an application may escalate both work and costs. Database structure is perhaps most important because it affects the response time, storage capacity, flexibility, and other parameters that together determine the user-friendliness of a database. Structures differ, and no single structure is best for all applications. The various methods of data storage differ primarily according to the facilities or operations supported by the DBMS. In general, we can differentiate between *record-based structures* and *object-based structures*.

8.4.4 Record-based structures

Relational database systems

A record-based structure is based on a collection of files in which each geographical object is represented in a record and attributes are represented in a fixed set of fields. A relational database system is one in which the DBMS supports a relational model of the data stored. That is, it is more of a guideline of how *to represent* data than a structure *for* data. In many ways it is comparable to a table, which is why *tabular database* is an alternative term for *relational database* (Figure 8.14).

In the relational data structure, there is no hierarchy and no specific keys point to the various records. All fields in a row may be regarded as being connected to each other permanently, yet individual tables contain no information on other tables. A relational database is thus a collection of tabular relations, each of which has a set of attributes. Each table (permanent or temporary) in a relational database is known as a relation. The tables have a name and a certain number of named attributes, which are placed in a column. Each row lists attribute values for one object. Any column (field) in a table may be a key for accessing data from another table; the system permits all objects and attributes to be related to each other. Thus, a collection of data on many objects with complex relationships may result in a large number of tables—a weakness of the relational approach. A mathematical relation

Building ID	Property	Owner	Year	Type
589	44 110			
610	44 50			
955	44 99			

Property ID	Owner	AREA	ADDRESS
44 99			
44 50			
44 110	Nils Nilsen	6,51	9999 Toppen

Figure 8.14

Relational database. Each field in a table can be the key to locating data in another table.

157

is, in theory, a quantity. This is what has given the name *relational database*.

Relational databases have three basic attributes or functions: primary key, relational join, and normal forms. Values in a single column, or a combination of values in multiple columns, can be used to define a *primary key* for the table and act as the addressing mechanism. A primary key is thus an item in a table that contains a value identifying rows in another table; it represents a relationship between two elements of a relational database. When columns and keys are defined, the table can be filled with data. Data that lie in different tables are linked by using a function called a *relational join*. Relation join is based on matching a key to its primary instance. A join operation associates an object to another table using a key. Attributes of that table can be added to the objects as a new column. This is very similar to overlay operations, which are discussed later (Section 14.11). *Normal form* is the identification of interdependence between the data. There are three normal forms that should be satisfied in relational databases. The first criterion is that there should be no duplication or recurring key columns, so that the same data or attributes cannot appear under different names. Thus, it should be possible to identify a row in the table solely by stating the key value. A breach of the second normal form occurs when the function of an attribute is dependent on only part of a compound prime key. The third normal form is achieved when a relation fulfills both the first and second normal forms and, in addition, has no attribute dependent on another attribute, except when the second attribute is a prime key. Data have to be split up into a series of tables in order to fulfill the normal forms.

There are no pointers in the permanent tables that are defined and established. Internal temporally indexed tables are used to direct intertable communications, which means that the system must open at least one index table for each connection between data tables. This can result in relatively large databases and slow access. Moreover, as there are no pointers, searches for records or fields must be sequential through the tables. The result is that relational database systems are inherently slower than hierarchical or network database systems. However, a field may be located quickly in a given table, and the data accessed may be manipulated.

Relational DBMSs are the type used most frequently for attribute data in GIS applications, primarily because of their simple, flexible structures. Another reason is that they support the complex relationships common among real-world geographical objects. Multiple entries are less frequent than in hierarchical or network database systems, although search times tend to be longer. This is particularly noticeable in specific operations on digital map data—for example, when forming polygons and with automatic text placement.

Standard relational databases require that only single values (text, numerical integer, numerical real) be stored in each cell. They are therefore basically unsuitable for vector data, which often consist of lines with many and varying numbers of points. For a long time this has limited the use of pure relational models for GIS. However, lines can be stored if each line segment (vector) and each point is stored separately (Figure 8.15).

This has been the reason that many GISs have developed their own effective structures for handling geometry. For a long time, attributes have been satisfactorily handled in relational databases. However, relational databases can now also handle so-called abstract data types (ADT), which also include geometrical data. Therefore, a number of newer GIS solutions are based on the storage of both geometry and attributes in one and the same database, realized as a BLOB (binary large object) and ADT. The different GIS suppliers today have their own way of handling this, such as with API (Application Program Interface), an interface between database and GIS software. In recent years computer capacity has been so good that it is considerably easier to handle large and complex amounts of data in relational databases. But the most usual database solution for GIS is still a combination of file-based systems for geometrical data and standard relational DBMS for attribute data.

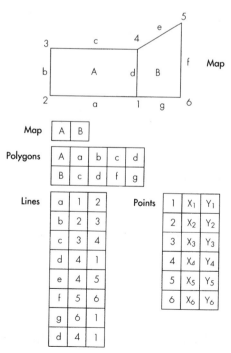

Figure 8.15
Map data stored in a relational database.

Structured Query Language

The simple structures of relational database systems have permitted the development of standard query languages, one of which is Structured Query Language (SQL). SQL gives users access to data in relational DBMSs by describing the data they may wish to see. SQL also allows users to define data in a database and to manipulate those data. The additional functions SQL supplies to relational databases are very useful for many GIS applications. Efficient use of SQL presupposes that the data satisfy the three normal forms.

A new table is set up in SQL through the function CREATE TABLE. With this function, the names of the attributes are given, as well as the primary key. For the SQL search and retrieval of data three basic functions are used:

SELECT—state what the result should contain, often a given number of column names.

FROM—State from where the data shall be retrieved, often a given number of table names.

WHERE—State which rows are to be included in the result. These rows are those that fulfill a given logical condition and can be retrieved from different tables (relational join).

Relational algebra may be performed using two classes of storage and retrieval operations. The *set operations* include union, intersection, difference, and product. The *relational operations* include selection (accessing rows), projection (accessing columns), joining, and dividing. Relational joining links tables and creates a new table from data retrieved from the various tables. The new table need not be stored physically in the database.

There are six logical operations in SQL:

1. $=$ Equal
2. $<>$ Not equal
3. $<$ Less than
4. $>$ Greater than
5. $\leq$ Less than or equal
6. $\geq$ Greater than or equal

Logical search in relational databases employ these six types of logical operations, also called set algebra or Boolean algebra. The operators AND, OR, and NOT are used to test whether a statement is true or false.

There are five aggregate functions:

1. The total of all the rows, satisfying any conditions of the given column, where the given column is numerical
2. The average of the given column
3. The largest figure in the given column

4. The smallest figure in the given column
5. The number of rows satisfying the conditions

Changing of data content in the tables is carried out by:

INSERT—Add one or more rows.
DELETE—Remove rows that fulfil a given condition.
UPDATE—Change the column value in a given row.

SQL is used most frequently in searching, although other query procedures are also used. Complex GIS functions such as data search within specified rectangles or circles, creation of buffer zones, and overlay require operations that are not implemented in standard SQL. However, several suppliers of GIS software have developed special spatial SQL dialects. These are applicable in particular to systems that use relational databases for storage of both geometry and attributes.

Standards are now being developed by ISO and CEN (Céomité Européen de Normalisation) in the database area, particularly in regard to spatial queries, which are made available through SQL3. The major database suppliers have also become deeply engaged in the development of spatial data solutions in relational databases in order to be able to handle spatial operations as overlay—in this connection often called *spatial join*.

Other record-based structures

In addition to relational databases, there are a number of other record-based structures. The best known are *hierarchical data structures* and network structures. A hierarchical database system is one in which the DBMS supports a hierarchical structure of records organized in files at different levels, with connection between the levels (Figure 8.16). Hierarchical data structures are easily expanded and updated. However, they require large index files and are susceptible to multiple entries. Searches are rapid, but search routines are fixed and constrained by the structure.

A *network structure* is an expansion of the hierarchical structure. Each element, or collection of like records, has connection to several different-level elements (Figure 8.17). The purpose of the network structure is to improve the flexibility and reduce the multiple entries of the hierarchical structure. A search need not pass through all levels, but can take shortcuts. However, the volume of index data is greater than that of the hierarchical structure. Both hierarchical and network structures are used infrequently in GIS applications, but some elements may be used for tailor-made systems.

8.4.6 Object-oriented database systems

Hierarchical, network, and relational database systems are intended for administrative tasks and are not particularly suitable for

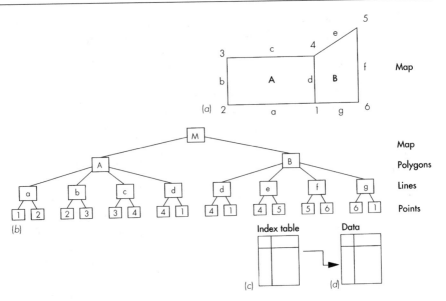

Figure 8.16
Map data stored in a hierarchical database system. Note the double storage of lines and points.

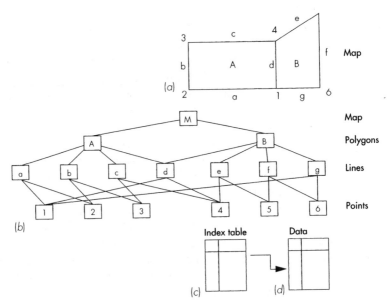

Figure 8.17
The storage structure in a network-based database system: *(a)* a map section; *(b)* polygons stored in the network system; *(c)* a pointer table that specifies where the map data *(d)* are stored.

representing a conceptual data model of geographical reality. All these systems are record oriented; that is, the data they contain are filed record by record, rather like cards in a conventional card index file. For example, in a relational database, all the elements that together constitute a particular map object may reside in several records in various tables. Furthermore, geometrical data and attribute data are often separated, each residing in a separate database.

Object-oriented database systems, however, attempt to overcome these difficulties by more faithfully representing the real world, which comprises complex homogeneous objects with varying internal and external relations (Figure 8.18). In object-oriented systems, data modeling and data manipulation are in many ways synonymous. Objects are manipulated as independent homogeneous entities, yet may comprise several other objects in a hierarchy. The flexibility of an object-oriented system also makes it easier to integrate the time factor in the data model.

Object-oriented database systems employ simple, albeit "intelligent," databases. The requirements for user expertise are, however, correspondingly higher because users must be able to define all relationships between objects in the database system. The chief advantage of object-oriented database systems is that they are easier to update on a consistent basis. For example, a line (which is an object) may be a property border, a road edge, or a zoning boundary. In an object-oriented database, it needs to be updated only once, whereas in other database systems the data pertaining to it may be in various locations, requiring multiple updating operations.

At present, object-oriented database systems are not used widely in GISs; a few GISs do use them in the storage and manipulation of data because of their ability to handle lengthy transactions in a multiuser distributed environment. Object-oriented systems do not appear to

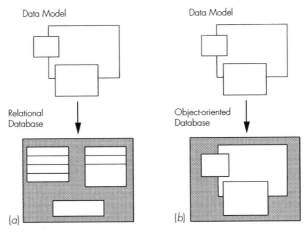

Figure 8.18
Object-oriented database *(b)* compared with a relational database *(a)*.

have the widespread use anticipated when they were launched in the 1980s.

The implementation of database for GIS use is discussed at greater length in Chapter 12.

8.5 Computer-Aided Design

Computer-aided design (CAD) has many features in common with GIS, but these systems are intended to perform completely different tasks. CAD contains data for construction of a future reality or construction of technical devices and is therefore well suited to graphical construction of artificial objects. GIS normally contains data observed in the field for an existing or previous reality, suited to the analysis and simulation of a reality. Such analysis functions (e.g., overlay) that are based on, among other things, topological data structure, are not available in CAD.

Many of the artificial objects in CAD have identical geometry and the same nongeometrical values. General descriptions common to many objects are therefore often stored only once. One of the differences between CAD and GIS is that the presentation parameters in CAD, such as type of symbol, color, and scale, are stored together with the objects, whereas in GIS these parameters are normally defined in connection with the presentation.

We have already seen that GIS is based on a two-dimensional model of reality and, in addition, 2.5 dimensions in digital terrain models. CAD allows access to a three-dimensional tool (i.e., the same x,y pair may have several z values). CAD may therefore be better suited to handle a digital terrain model (DTM) than GIS. With CAD it is possible, as in GIS, to produce maps and all kinds of terrain perspectives, but CAD can also be used to control machines for computer-aided manufacturing (CAM).

8.6 Multimedia

Multimedia has a user interface comprising text, sound, and image and is making inroads in the GIS world (Figure 8.19). Multimedia program systems and equipment comprise:

- Processing, storage, and exchange of data
- Processing of all types of data: numbers, text, graphics, images, animation, sound, and video
- Integration of all kinds of data
- Interactive use of data under machine control

In GIS, multimedia includes such elements as hypermaps, carto-

Figure 8.19
Multimedia comprise software and hardware that include integrated data in the form of sound, images, numbers, text, and video that can be treated interactively.

graphic visualization and animation, digital images and satellite data, sound, hypertext, virtual reality, and hypermedia spatial databases. Multimedia techniques enable the establishment of information from images, maps, and so on, by direct manipulation. The use of animation techniques allows multimedia to show spatial changes over time. The most extreme virtual reality experience is obtained through turning on a computer and submerging oneself in its reality. Equipped with two screens for eyes, three-dimensional sound, and data gloves, one can be transported into a world where almost all exterior sensations are artificial.

Sound, image, and video are long streams of binary data that can be of random length. In relation to databases, therefore, we can talk of binary large objects (BLOB). Database systems must be capable of handling BLOBs of random length and of deciding how they will be stored, depending on size, access time, and transfer speed. Many GISs are already able to store and display a BLOB quite easily.

165

8.7   World Wide Web

World Wide Web (WWW) is software developed for use on the Internet. WWW (originally developed by CERN—Conseil Européen pour la Recherche Nucléaire—in Switzerland) is a freely available program system whereby users can design screen images on the Internet to display their own data in the form of text, pictures, video, and sound. WWW includes a set of standards for the treatment of communication protocols and for structural hypertext. Marked text and objects interconnect the client machine and data lying on the server(s) at other locations on the network. A special program called Hypertext Transfer Protocol (HTTP) is used for linking and transfer of data (Figure 8.20).

Web home pages are created by using a marking language known as Hypertext Mark-up Language (HTML). With the help of Hypertext, home pages can be linked to other home pages and to other Internet resources. Hypertext is a nonsequential text where special references are created between different parts of the text (links), which can be traced via a computer. The hyperlink technique is also used to create a multimedia system based on special references between different segments of information (e.g., from picture to text and from text to sound). WWW has a simple technology using a fixed user interface and has therefore become very popular with public and private institutions/users alike. The Internet can be compared to a national highway system, whereas WWW is a delivery service for packages and letters to customers/subscribers who use the telecommunication network as a transport system.

WWW acts as a client/server model in which the navigation software runs on a local machine and the server software on a host machine. Each page (WWW site) lies on a server and has a unique address—Universal Resource Locator (URL). The URL address consists of several parts that tell which Internet protocol is to be used, which

Figure 8.20
Using the hypertext technique, it is possible to point and click on selected text segments (or other objects) and obtain more information via the Internet than was clicked on.

Internet resources are to be contacted, and the identification of the server. This could, for example, contain the country code and information on a particular catalog at the server.

In principle, there are three different techniques that can be used for geographical data on the Internet (Figure 8.21):

1. Transfer on the net from server to client of application software and data, for example Java (a portable and object-oriented language)
2. Preinstalled application software on the client. Only data are transferred on the net from server to client, the so-called "plug-in" solution
3. The client functions as a Windows terminal. Only screen images are transferred on the net from server to client

The solution can, in practice, be based on a combination of these three techniques. The most common GIS functions available on the net are "selection" and "download" of data, interactive selection by point and click, and zooming and moving in map images, as well as a few simple analysis functions. The ultimate goal from the user's point of view should be to have access, through the Internet, to the original databases distributed on different servers in different institutions all around the country and, ideally, all around the world. By a simple selection menu, each user should be able to compose a data profile and have the data transferred automatically to his or her PC. A solution of this kind

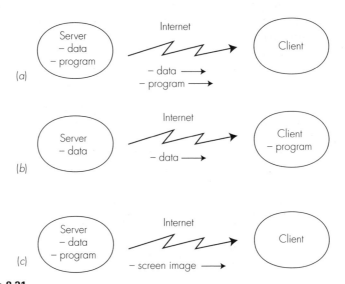

Figure 8.21
Geographic data (and other data) can, in principle, be realized on the Internet by three different methods: *(a)* transfer of application software and data from server to client, *(b)* preinstalled application software on the client and transfer only of data to the client, *(c)* the client acts as a Window terminal to the server—only screen images are transferred from server to client.

requires a considerable degree of standardization. The organization "Open GIS Consortium" have already developed a test version (Web Mapping Test Bed) in the United States.

8.8   User Requirements

GIS users may be divided into groups according to their functions and needs:

- End users, who seek problem solutions and see final products only in the form of maps and reports
- GIS specialists, who advise and solve problems for end users, GIS operators, and data compilers
- GIS operators, who understand the functions of a specific system so as to manipulate the data
- Data compilers, who understand the data but not the system

The value of GIS depends on how well users state the problems and how well GIS specialists provide solutions. End users frequently need only a limited knowledge of GIS. However, GIS specialists must be familiar with the general principles of GIS, data and data compilation, cartographic facilities, and, not least, with the potentials and limitations of the system employed. In other words, a specialist must be able to sketch alternative production routes.

A GIS specialist has to be expert in several fields, although not everyone in a GIS organization will need to have the same capabilities. Outside consultants may also be engaged. Normally, it will take about a year for an operator to become thoroughly familiar with a larger system. However, he or she can be productive long before that. In many cases, it is possible to complete a task without specific knowledge of all the available functions. The initial training of operators preparing to work on larger systems usually lasts 14 days and is followed by additional training for specific parts of the system. Furthermore, whenever new computers are installed, training courses are held to familiarize staff with the operating system, backup routines, and other facets of their operation.

The need for computer knowledge depends on the size of the system and the task confronting the operator. A single-user PC is easier to operate than a multiuser workstation system. Professional staff members need only moderate computer knowledge to learn how a system works. However, extensive computer knowledge is a prerequisite for staff who will be responsible for operating the system, and they will require an in-depth understanding of the geographical information system, the operative system, and the peripheral equipment.

Most organizations that have implemented GIS have been obliged to conduct some development for matching older data systems (such as those containing older registers), improving existing programs, interfacing new peripheral equipment, and so on. Developments such as these require computer expertise, which may be either acquired in-house or contracted externally. Digitizing personnel and other data compilers should be familiar with the attributes with which they are dealing, but need only a limited knowledge of the GIS involved.

8.9 Working Environment

Work at a GIS station can be demanding; therefore, the working environment involved in any GIS implementation should always be taken into consideration. Digitizing maps can be physically tiring and mentally tedious. Clearly, such work cannot be performed continuously for an entire working day, but should be split up into shifts through rotation or with longer breaks. Spring-loaded armrests are available to ease arm loading in digitizing work. The same precautions apply to the prolonged use of a mouse for pointing and clicking. A good working position and sufficient breaks are also necessary for good work at a screen.

The working environment should be relatively compact, so that manuals, maps, manuscripts, notes, and the like are all within reach. A U- or L-shaped layout is often best. Screen height above the floor, distance from the user, and keyboard and tablet positions should all be freely adjustable. Lighting should be indirect to minimize screen reflections.

As concentration is vital in GIS work, workplaces should be shielded from disturbance. Printers, plotters, and other noisy equipment should be located away from the operator. In some cases where the equipment produces heat, air-conditioning or fans may be required. Staff training is essential to the success of the system, and system handbooks and other documentation should be in the operator's native language, if at all possible.

CHAPTER 9

Data Collection I

9.1   Introduction

At the outset, sources of data used in GISs may be (Figure 9.1):

- In various digital forms: vector, raster, databases, spreadsheet tables, satellite data, and so on
- Nondigital graphics, such as conventional maps, photographs, sketches, schematic diagrams, and the like
- Conventional documents in registers and files
- Compilations in scientific reports
- Collections of survey measurements expressed in coordinates or other units

The data sources for a comprehensive GIS are probably more numerous and of greater variety than for most other information systems. GIS is a true mixed-data system. Most often, the data mix makes for a lively system. Although most of the problems associated with collecting data for GIS uses were identified by the mid-1970s, data collection remains the most expensive and time-consuming aspect of setting up a major GIS facility. Experience indicates that data collection accounts for 60 to 80% of the total cost (time and money) of a fully operational GIS, whereas the purchase of equipment accounts for not more than 10 to 30%. Other costs include training, development, and administration. As shown in Figure 9.2, the costs of data collection are dominant even when spread over an implementation period of several years.

There are four methods of producing digital map data:

1. Digitizing existing maps with a digitizer
2. Scanning existing maps
3. Digital photogrammetric map production
4. Entry of computed coordinates from field measurements

The first three methods, which produce digital data from ordinary maps, may all be divided into three phases:

1. Preparation
2. Digitization/scanning
3. Editing and improvement of data quality

Figure 9.1
Sources of data can be both numerous and varied.

9.2 Digitizing Maps

Preparation

Data must be prepared at frequent intervals and carefully organized to facilitate the work of digitizing. Manuscript maps (Figure 9.3) may be used to group and highlight information relevant to the various themes involved. For example, roads shown on a base map may simply need highlighting in various colors, according to the type of road. Or the operation may be more comprehensive, as when collecting and grouping network information from various maps and sketches onto a single common map. In practice, these operations invariably demand some redrawing of maps prior to digitization. As a rule, changes and amendments to graphics are achieved more easily on paper maps prior to digitization than they are on digital data. Experience has shown (Aronoff 1989) that even with extensive redrawing, manuscript maps save as much as 50% of digitization costs. They are therefore considered vital in digitization. The identifier data necessary for relating

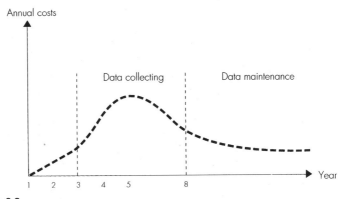

Figure 9.2
Data collection costs versus implementation time: a typical cost curve resulting from the introduction of GIS into a larger organization.

171

Figure 9.3
Typical manuscript map showing game registration in Norway.

digital map data to computerized attribute data are usually organized on manuscript maps.

Digitizing

As illustrated in Figure 9.4, digitizers are used in manual digitizing. Digitizers have their own internal coordinate systems, which may be related to terrain coordinates by cross-registering checkpoints with known terrain coordinates. These cross-registrations permit the computation of transformation parameters, which may then be used to compute terrain coordinates from digitizer coordinates. Some pro-

Figure 9.4
Map digitizing: a digitizing table complete with map and with the internal coordinate system defined.

grams convert coordinates continuously during digitization, whereas others convert them in the editing phase.

Conformal transformations involve transferring digitizer coordinates to terrain coordinates through parallel translation, rotation, and scale alteration. All objects then retain their original shapes. Affine transformations alter the shapes of objects through unequal x and y scale alterations and unequal rotations of the x and y axes. This method compensates for variable shrinkage in the map. See equations 6.1a and b and 6.2a and b.

The checkpoints are selected in respect to number, location, and quality. Because the formulas for conformal transformation have four unknown parameters, at least two checkpoints (with x and y) have to be registered; affine transformations have six parameters and require at least three checkpoints. In practice, the parameters have to be determined by redundant observations, and the number of checkpoints should normally be 3 to 5 and 4 to 6, respectively, for conformal and affine transformation. The points on the map are located so that the whole of the digitizing area is surrounded, as well as providing one or more points in the center of the area to capture possible variations (see Figure 9.4). Points are selected that can be uniquely identified and measured with high precision, such as grid intersections, polygon points, border markers, house corners, and so forth.

Once the known points are registered and transformation parameters calculated, digitizing may begin. Two methods are used. In *point digitizing*, a crosshair is placed over the position to be recorded and a key is pressed to enter the datum. In *stream digitizing*, a crosshair is drawn along a linear symbol and data on positions are automatically entered at preset intervals of either time (such as five per second) or distance (such as five per centimeter). Data entry may also be curve fitted automatically, with more entries as the curvature radius decreases (Figure 9.5).

Figure 9.5
Principle of stream digitizing. Various methods can be applied in digitizing to select a point on the line based on, among other factors, a given arrow height value.

173

Point digitizing requires that the operator continuously select points and key in data entries. This type of digitizing is most suitable, therefore, for individual points and for lines comprising straight segments, such as property boundaries and building outlines, which need not be followed between successive data point entries.

Stream digitizing requires more sophisticated computer software (partly because data-entry speeds may be high) and greater operator skill. Once these requirements are met, stream digitizing is suitable for digitizing sinuous lines, such as contours and the borders between different soil types. The drawback is that unless its parameters for point selection are selected carefully, stream digitizing may generate unnecessarily large quantities of data. Point digitizing is preferable to stream digitizing for high-accuracy entry. Lines of fixed curvature, such as property borders on curving roads, may be described by radii of curvature rather than by their constituent points. During digitizing, each object is automatically assigned a unique system-ID (SID). Digitizing also includes manual entry of thematic codes for object types and ID codes that link object types to attribute data (if SID is not to be used). For example, digitizing a building includes entry of the thematic code for buildings and an ID number for the specific type of building. A new ID number is entered for the next building, and so on. Codes are keyed in or entered via a menu.

In some systems the ID codes of map objects may be entered directly, provided that stored attribute data are available interactively on screen.

Digitizing accuracy and progress should be verified as work proceeds. Verification may be either manual or semiautomatic. The former requires objects on a manuscript map to be crossed off as they are digitized and that partially completed maps be printed out as the work progresses. In some systems, results can be displayed continuously on screen—in effect, a semiautomatic, human–machine variety of echo check (a comparison of received/entered data with original data). The screen display enables the operator to verify the work performed and to avoid double entries.

Skilled operators can normally achieve crosshair placement accuracy of $\pm$ 0.05 to $\pm$ 0.06 mm.

The time required for digitizing depends on the complexity of the map in question and the extent of coding. A rule of thumb holds that digitizing may progress at a speed of one polygon per minute. Operator skill is decisive, in terms of the time required for digitizing and its quality. Most GIS facilities have found that operators experienced in cartography are preferable to operators experienced in computers or other disciplines. Operators must be trained to understand the logical connections between property borders, pipe networks, borders between soil types, and other themes that are digitized, and to master the software tools they use to the extent that they can correct their own digitizing errors.

As previously mentioned, scanned data are digitized directly on the screen. Screen digitizing has the same functions as table digitizing; in addition, systems are available for screen digitizing that have functions for automatic line tracing and vectorization of raster structures. A line can have "intelligence" built in, so that small gaps in the line do not result in a stop, the marking of objects that are digitized, automatic thinning algorithms, or other undesirable effects. It is also possible to join the scanned data together so that digitalization can be carried out independent of map sheet borders. In addition, one can zoom in to study details. All in all, screen digitizing is therefore often more efficient than table digitizing.

Data editing and quality improvement

Normally, the product of manual digitization must be edited and the quality improved. Tasks include:

- Error correction
- Entering missing data
- Forming topology

Digitizing often introduces error. Typical errors include gaps between lines that should be contiguous but dangle (overshoot) or fail to meet (undershoot), intersecting lines that should not overlay yet do and create small "meaningless" polygons, lines meeting at the wrong point or intersecting each other, objects with improper or missing codes, contour lines that cross each other, and missing details (Figure 9.6). In most systems, line lengths and polygon areas are calculated continuously during digitalization. Polygons with very small areas are probably caused by inaccuracy and can be removed. Overshoot, undershoot, and small meaningless polygons can be removed automatically; however, one can also risk removing real data. Automatic methods should therefore be used with caution, and the parameters for removal selected with care.

Figure 9.6
Typical errors that can occur during digitizing. Three types of error often occur, and these may require extensive correction work. The type of error illustrated in part *(b)*, where dangles occur, is easy to correct automatically. It is therefore generally recommended that during digitizing, situation *(b)* be preferred to *(a)*. The error shown in part *(c)* creates a new, meaningless polygon.

175

Some errors are corrected more easily than others. For example, it is better that the digitization of two lines contiguous at a point result in their crossing rather than in their not meeting at all. Crossed lines are more easily processed, particularly as the original contiguous lines (and their incorrect crossings) are normally parts of the border of a closed polygon, and a polygon of incorrect geometry is more easily corrected than a polygon that remains open. Some systems support "snapping" on a program for recognizing and connecting to nearby points or lines, so that nodes, closed polygons, and contiguous lines may be created quickly.

Forming topology also reveals many errors. Typically, the same polygon number will be assigned to both sides of an unconnected line in a polygon structure. This is illogical, as all lines in a polygon structure must mark the borders between adjoining polygons. Several other control functions are also incorporated in the formation of topology. Error correction, editing, topology structuring, and other tasks may be performed using suitable programs.

Digitizing will often result in unnecessarily large data volumes, and an important part of the finishing work will be to remove data that are of little importance for the description of the geometry. In Section 4.2.4 several current methods are discussed for filtering out data. Among these methods are the use of the Douglas-Peucker algorithm, tunnel filtering, and the spline function.

Data are verified visually on screen and/or on test printouts. Geometry, ID codes, and any thematic codes may be plotted and verified. A test printout may be laid on top of an original on a light table and compared with it, piece by piece. Where several themes are digitized, this comparison is made easier if each theme is printed out and compared with the original separately. Errors may be corrected and missing data may be entered either in the digitizing program or in the GIS programs. However, editing data is often time-consuming, so that when the original data are poor, editing may take more time than the original digitizing.

9.3 Scanning

Maps are scanned to:

- Use digital image data as bases for other (vector) map information
- Convert scanned data to vector data for use in vector GIS

Preparation
Scanning requires that a map scanned:

- Be of high cartographic quality, with clearly defined lines, text, and symbols

- Be clean without extraneous stains
- Have lines that are not too thin (preferably of a width of 0.1 mm or greater)

In scanning tasks, work is often moved from the preparatory phase to the editing phase. However, if the data encoded are to be used directly as digital image data, the quality requirements for the scanned map may be relaxed and preparation may be negligible.

Scanning

Scanning consists of two operations, scanning and binary encoding. Scanning a map produces a regular grid (pixels) of 256 gray-scale levels, which range numerically from 0 (complete black) to 255 (pure white) (Figure 9.7). Normally, map scanners use a pixel size of 25, 50, or 100 μm (1 μm = 0.001 mm). As a black-and-white line map comprises only black lines against a white background, scanning must clearly differentiate black from white. Accordingly, the scanner must be precisely adjusted for a gray-scale level corresponding to the map demarcation between black and white. If the level is set too high, vital information represented by thinner lines may be lost; if too low, extraneous information, such as dirt, stains, and dust, may be included (Figure 9.8).

Finding the most suitable gray-scale level can be time-consuming, but once it is found, a map may be scanned in minutes. With a threshold set to differentiate black from white, a gray-level array may be binary encoded. As shown in Figure 9.9, each pixel is then binary coded as either one or zero, depending on whether the gray level is over or under the threshold set. Scanning produces voluminous data. With a pixel size of 50 μm, there are 400 pixels per square millimeter and six pixels in a line 0.3 mm wide. A 60 × 80 cm map encodes to about 200 million pixels.

Pixels are stored in rows and columns, and many adjacent pixels have identical values. Therefore, programs are available to compact the aggregate amount of data stored. Binary data are particularly suited for compression: Run-length encoding compresses data by about 90%, and some advanced compression techniques can attain 98%.

Each element of an image contains basically only one gray-scale level, a color value or possibly one binary value, but no information as to what this value represents in the real world or whether it has any logical connection with the adjoining element. A scanned map image thus contains no information on inner linear or area structures.

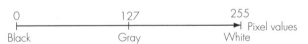

Figure 9.7
Brightness value range.

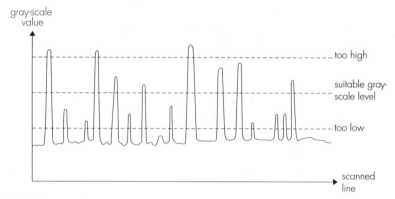

Figure 9.8
Gray-scale values along a scanned line. The peaks appear as lines on the map, whereas the bases are shown as white areas. Tests are carried out to establish the correct gray-scale value for each map being scanned.

Therefore, there is no automatic link of attributes to structures in order to define whether they are roads or rivers, properties, or buildings. From a visual standpoint, it is possible to see structures in the data, but the data themselves supply no information on cohesive lines or areas. In popular terms, a scanned image can be described as "unintelligent."

Color scanning is done in a three-pass process with successive magenta, yellow and cyan filtering. This results in several million color variations, which are normally reduced to 256 variations or converted to 256 gray tones.

140	150	67	155	152	149	147
141	139	59	158	160	162	154
150	144	66	151	161	155	160
160	162	158	72	162	153	154
151	159	158	157	70	167	167
149	153	145	156	75	166	165
147	146	156	155	68	151	152
150	155	157	156	161	65	160
170	171	160	161	162	150	61

0	0	1	0	0	0	0
0	0	1	0	0	0	0
0	0	1	0	0	0	0
0	0	0	1	0	0	0
0	0	0	0	1	0	0
0	0	0	0	1	0	0
0	0	0	0	1	0	0
0	0	0	0	0	1	0
0	0	0	0	0	0	1

(a)   (b)

Figure 9.9
Digital raster image: gray level array/binary (black, < 80; white, > 80). On the left is a raster image with gray-scale values. The corresponding binary-encoded raster image is shown on the right. All gray-scale values equal to 80 or more are lines on the map and have been given binary representation 1; the remaining gray-scale values are white areas on the map and have been given binary representation 0.

Data editing and quality improvement

Where raster data are intended to be used only as a "dead" background image, the data will require minimal or no further treatment and can be used either as gray-scale images or binary data. There are several ways of carrying out quality enhancement of scanned raster data:

- Upgrading raster data through division into three categories: shapes, potential symbols, and noise
- Noise removal and pattern recognition of symbol candidates in raster data
- Thinning and vectorization
- Separating vectorized data into shapes, potential symbols, and noise
- Vector data pattern recognition of symbol candidates
- Georeferencing
- Error correction
- Coding
- Supplementing missing data
- Forming topology

The last five methods are identical to those for quality improvement of data from a digitizing table. *Noise,* a term borrowed from acoustics, designates such information as stains and extraneous lines that are registered but not required. Data may be improved by structuring binary information in three categories: shapes, potential symbols, and noise. The noise category can then be filtered out, together with the potential symbols. If the data are to be used solely to compile a digital map image, only the noise need be filtered out.

Raster data contain no codes for map objects, but whenever map information is divided among various overlays (topography, inventories of facilities, land use, etc.), each overlay may be scanned individually to produce data relevant to its single theme. Each thematic overlay can then be converted to vector data with a common thematic code.

Raster to vector: vectorization

Some scanners vectorize as they scan. Otherwise, vectorization is usually part of the subsequent editing process, or part of the process of automatic line tracing. The vectorization of structures (lines, symbols, etc.) may be summarized in six steps:

1. A number of pixels forming a structure, such as a line, are registered.
2. All pixels transverse to the line, except those in its center, are stripped off (skeleton plotting).
3. Starting at one end, the pixels are connected one by one along the line (linearization).
4. Line curvatures are checked against set maxima which, if

exceeded, indicate that the line is no longer straight, so linearization terminates.

5. Coordinates are determined for the start and end points of the terminated straight-line segment, and a vector along the segment is formed accordingly.
6. Little by little, lines and structures are assigned coordinates and vectorization continues.

The result of vectorization is a continuous structure delineated by a sequence of coordinates designating the midline of the structure. A magnified illustration of a line of pixels is shown in Figure 9.10.

Vectorization may be subject to errors and defects, including:

- Deformations or interruptions of lines intersecting at nodes
- Vectorization of extraneous stains and particles on the original map
- Vectorization of alphanumeric information and text
- Unintentional line breaks, resulting in divided vectors
- Dotted-line symbols (trails, soil-type boundaries, etc.), resulting in many small vectors
- Smooth curves that become jagged (i.e., introduction of unwanted inflection points)

Many systems include error correction routines to circumvent these problems and ensure scanning quality.

During automatic line tracing the operator leads the system to start in a line structure of raster data, and the system follows the structure more or less automatically on screen. The process includes continuous vectorization of raster data. All operations are on screen, so the operator need intervene only in conflicts where there is a break in the line or the line is crossed by other lines. The operator can steer the system into special line structures, such as tracing water pipelines on a scanned utility network map, thus allowing the data content to be separated

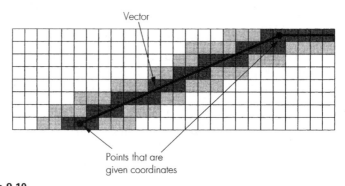

Figure 9.10
Vectorization of a scanned line on a map. The darkest pixels are the skeleton that is used in vectorization; the light rastered pixels have been stripped off; the pure white pixels appear as a white area.

into themes. No other map information, other than that which is being traced, will be vectorized.

Pattern recognition

In scanned data, numbers, letters, symbols, and other items of map information cannot be differentiated from one another directly; all information comprises pixels of gray levels or colors. However, *pattern recognition programs* are available to analyze raster and vector data and to elicit features of a known pattern. The recognition of shapes and symbols is best performed on raster data, as vectorization invariably involves a loss of information. Nonetheless, raster data are usually vectorized prior to recognition.

To ease recognition, vectorized data are usually divided into data according to shapes and relevant to potential symbols (after noise data have been filtered out). Recognition of letters and numbers is particularly useful, as is recognition of lines and specific map symbols (Figure 9.11).

Pattern recognition usually involves a system that is "trained" to detect specified patterns by assigning probabilities to their possible presence. Patterns are usually specified analytically or with the use of templates (model patterns) for comparison. For example, pattern recognition is used in the scanning of nautical charts in which the original registrations often comprise a multiplicity of depth numbers and letters. Recognition rates as high as 95 to 98% have been attained. However, results of pattern recognition are highly dependent on the quality of the original.

Coding and other editing of vectorized raster data

Scanning and vectorization invariably introduce errors and obliterate some data. Consequently, data must be verified and corrected. Vectorized raster data have no thematic codes, so the various map themes cannot be distinguished (unless pattern recognition has been employed) and map objects have no ID codes that link them to attributes. The data must therefore be coded by assigning codes to individual objects; this is usually achieved manually on screen.

Alternatively, a reference table can be established. To be able to link an object type to the cell values in a raster layer, one first has to

Figure 9.11
Pattern recognition. The figures in the binary raster image *(a)* are thinned, vectorized, and pattern recognized *(b)* and then translated to digital numerical values *(c)*.

181

identify which type of object is represented by a certain gray tone, color, or binary value. A table can thus be established that will give the link between each object type and the corresponding cell value. The same object type can have cells with varying gray-scale or color value, even within the same raster structure. Without accurate calibration, uncertainty will be linked to these reference tables. It will often be impractical to establish such tables for binary data, inasmuch as all cells in raster structures (contour lines, road lines, buildings, etc.) will have the same binary value.

Once vectorized data have been coded, the subsequent manipulations are the same as for vector data. Where a map is scanned from several thematic layers or where line tracing is carried out on one or more themes, each group of vectorized data can be assigned a theme code.

Integrating image data in GISs

Some GISs can manipulate and display both vector data and raster (image) data on screen. In principle, raster data may be:

- Scanned maps
- Scanned documents, photographs, etc.
- Satellite data
- Video images of maps, photographs, aerial photographs, etc.

Vector data and raster data may be displayed simultaneously on a screen, supported by storage split between raster and vector data and accompanied by an editing program for both types of data. Raster data are normally represented as gray-tone images or color. To be meaningful, the two sets of data entered must employ the same coordinate system for the images displayed.

The orientation of the raster image in relation to direction North in the terrain is dependent on how the map happens to be placed during scanning. The scale of the raster image will also be random, and the map paper can be affected by shrinkage. To obtain correct georeferencing, raster data have to be transformed to the terrain's coordinate system (Figure 9.12 and Figure 9.21). The relative positions of pixels, stated in row and column numbers, are transformed to terrain coordinates using selected points whose coordinates are known (e.g., the corners of houses, grid intersections, and road intersections), and values for the transformed cells are interpolated from the original raster image (see Section 9.6). The transformations may be conformal, affine, or of another type, depending on the systematic errors in the scanned data, corresponding to those in table digitizing. Usually a raster file has a prefix header, which contains such details as the transformation parameters, terrain coordinates of the lower left and upper right corners, the pixel size, and so on.

Scanned map data may be used to compile a base image whenever the map generated is for orientation only. An example is the pipe data of a network to be documented in raster form, but for which the pipe

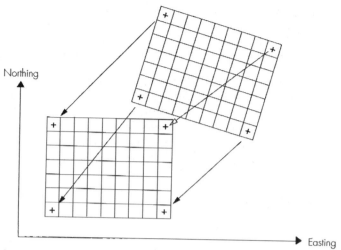

Figure 9.12
Transformation of raster registrations to a coordinate system.

routes are in vector form. Other examples include satellite data and images integrated in vector GIS. Applications of raster images include forming the bases for interactive correction and the amendment of vector data.

Scanned documents, video images, and other inputs may be integrated in GIS. Raster data may be stored on a video disk or a hard

Figure 9.13
The integration of document and scanned images within GIS makes it possible to point at a building and bring up an image of the corresponding title deed.

183

disk. A raster image and the GIS into which it is entered are linked via a table of identifiers of the map objects and pointers that indicate their relevant storage locations (Figure 9.13). Video images from a video disk can usually be screened more rapidly than scanned map data from a hard disk, which must be unpacked from storage.

Packed and stored raster data must be unpacked prior to being displayed on screen. Unpacking may be time-consuming, but once it is completed and the unpacked data are stored in the working memory, the manipulation response time is satisfactory. Another disadvantage is that the lines on enlarged raster maps may be annoyingly thick.

9.4   Manual Digitizing or Scanning

The advantages of manual digitizing are that it:

- Can be performed on inexpensive equipment
- Requires little training
- Does not need particularly high map quality

The disadvantages are that it:

- Is tedious
- Is time-consuming

The advantages of scanning are that it:

- Is easily performed
- Provides rapid compilation of digital map images

The disadvantages are that:

- Raster data files are large and consume memory space.
- Attribute data cannot be linked directly to line and polygon structures in raster data.
- Raster data lack "intelligence" and hence can be used only as bases for drawings.
- Improvement of data quality can be labor intensive.
- Clean maps with well-defined lines are required.
- Relatively expensive equipment is required where standards of accuracy are high.
- Expert personnel are required.

Although scanning produces large quantities of data, compressed raster data need not require more memory space than vector data of customary format (i.e., ASCII). One reason is that raster data lack the voluminous drawing instructions that are essential in vector data. During the 1990s, manual digitizing has been the technique used most frequently, but more recently scanning has taken over a lot of the market, particularly because of the opportunities for on-screen digitizing, efficient packing routines, and powerful computers. Further developments in pattern recognition will undoubtedly contribute to more widespread use of scanning.

9.5 Aerial Photographs and Photo Interpretation

The concept of aerial photography was patented in 1855 by French writer and photographer Gaspard-Felix Tournachon (1820–1910), who wrote under the pen name of Nadar. In 1856, Tournachon daringly ascended in a balloon from a village near Paris and took the world's first aerial photograph. Since 1909, where airplanes were first employed, aerial photography has become a discipline with its own theory and practice. Because they are taken from above, all objects in aerial photographs appear as viewed from various vertical angles. So aerial photographs almost always require interpretation in order to obtain information for mapping or depicting details as seen from the ground. Of course, almost all photographs used for their information content are interpreted in some way, but the interpretation of aerial photographs for mapping purposes is most common.

The combination of aerial photography and photographic interpretation provides information on relatively large areas without necessitating survey inspection on the ground. Roads, lakes and rivers, buildings, farmland, and forests are clearly visible in aerial photographs (Figure 9.14). Other characteristics, such as various types of vegetation, soil, and geological formations, are more difficult to interpret. Accurate interpretation depends on both experience and verification by ground control. Consequently, photographic interpretation has become a skilled profession. Interpretation is based on extracting information from the shape of objects, as well as size, pattern, shadows, gray tone or color, and texture (by comparison with larger contiguous areas). Indirect indicators, such as surface shape, runoff patterns, locations, and seasons, may also be used.

Aerial photographs may be black and white on film sensitive to wavelengths of 0.4 to 0.7 μm; color on film with three layers sensitive within the 0.4 to 0.5 μm (blue), 0.5 to 0.6 μm (green), and 0.6 and 0.7 μm (red) ranges; or infrared on film sensitive within the range of 0.7 to 1.1 μm. Black-and-white film is most commonly used. Infrared film, however, may assist interpretation, particularly of vegetation, because plants emit varying radiation patterns throughout the infrared spectrum. Most aerial photographs have a resolution, or smallest discernible detail, of about 0.025 mm.

Experienced photographic interpreters can map the details of vegetation, geology, and other terrain features from aerial photographs and a few ground controls. Almost all natural resource mapping is based on aerial photographs and photographic interpretation, along with topographic maps. Borders and characteristics of the various objects

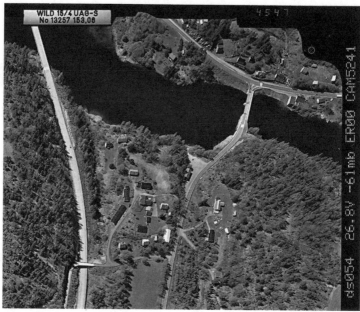

Figure 9.14
Aerial photographs have a resolution of 0.02 to 0.03 mm and therefore contain a considerable volume of information. Panchromatic black-and-white photos are often used, as shown. Information from the photo can be interpreted by means of gray tones, shadows, patterns, textures, and so on.

depicted in a photograph are drawn both in the field and in the survey office.

Like most photographs, aerial photographs are taken through a lens that views a relatively large area and replicates it, inverted, in a smaller area on a negative. The geometry involved in this process is termed *central projection* (Figure 9.15). A point on one plane (the ground surface photographed) is projected onto a second plane (the negative), so that the original point and its image on the second plane lie on a straight line through a fixed point (focal point of the lens) not on either plane. Aerial photographs always involve some distortion, because their vantage point is not exactly vertical and because terrain is seldom either completely horizontal or completely flat. Consequently, the information in aerial photographs must be transferred to a ground coordinate system for computation of lengths and areas or before they can be entered in GIS with other map data. Where elevations in the terrain are known, aerial photographs can be transformed to true map images (ortho images); see Figure 9.15.

Aerial photography is still mainly based on photographic recording methods, but digital cameras are now being introduced into the market. Registration in digital cameras is, in principle, carried out in the same way as in remote sensing instruments. Aerial photographs can be made digital by scanning, using scanners with high resolution (15

Figure 9.15
Central projection and orthogonal projection of terrain.

μm, etc.). Digital aerial photographs in combination with digital terrain models and control points form the basis for production of scale correct images—ortho images. Digital ortho image maps (both black-and-white and color) can be quickly established over large areas and have become very popular in recent years, especially in connection with the identification of property boundaries and as a means of communicating with decision makers. There is still the disadvantage that plotting of ortho images and digital images in general on ordinary plotters (ink-jet, etc.) gives relatively poor quality as compared with the screen image and the analog photo.

Photographic data are used to create and revise maps using a variety of photogrammetric instruments (described further in Chapter 10).

9.6   Remote Sensing

Satellite remote sensing of the Earth began in the 1960s, when the technical capabilities of satellites operating in Earth orbit converged with the increasing ability of computers to manipulate large quantities of data. Two observation modalities evolved: *passive optical,* which deals with reflected sunlight and emitted thermal radiation, and *microwave,* which, like radar, deals with the transmission and reflection of energy in the microwave portion of the radio-frequency spectrum. Apart from the nature of the phenomena observed, the difference between the two modalities is technical. The optical spectrum is considered to extend from the very short wavelength (shorter than 0.4 μm) in the ultraviolet region to a wavelength of 1000 μm in the longer infrared region. This is the short-wavelength limit to radiation which can be generated and detected conveniently by microwave electronic devices.

Satellites carrying optical instruments (radiometers) can take continuous pictures of the Earth's surface. In principle, the sensors may perceive energy at wavelengths anywhere in the optical region of the electromagnetic spectrum, from 0.4 to 1000 μm. Traditionally, the optical region is divided into subregions, each of which is termed a *spectrum* (Figure 9.16). Sensors are available for all spectra, from ultraviolet to far infrared, but not all may be used in optical remote sensing. This is because not all portions of the total optical spectrum are transmitted equally through the atmosphere; many are absorbed, some almost completely (Figure 9.17). So the variations of atmospheric transmission, and absorption with wavelength, limit the optical spectrum that is useful in remote sensing to transmission bands, or *windows,* in the range of 0.4 to 15 μm.

Satellites' instruments are quipped with a number of detectors that are located singly, in a line, or in an array. Different sets of detectors can register simultaneously in several multispectral limited areas of the electromagnetic spectrum, giving so-called *multispectral registration.* One can thus obtain pictures of several spectral areas over the same area of the Earth and at the same time. The spectral areas (channels) selected are those areas where the most interesting radiation slips out through the atmosphere.

Each detector registers radiation from small square areas on the Earth's surface. The size of the area gives the system's ground resolution. The registration of each area is stored in the satellite in the form of a picture element (pixel), with a digital value corresponding to radiation intensity. The images are compiled successively along a track

Spectrum name	Wavelength range
ultraviolet	shorter than 0.4 μm
visible	0.4 μm to 0.7 μm
near infra-red	0.7 μm to 1.0 μm
solar reflected infra-red	1.0 μm to 3.0 μm
mid infra-red	3.0 μm to 15.0 μm
far infra-red	longer than 15.0 μm

Figure 9.16
Spectrum classes.

Figure 9.17
Electromagnetic spectra.

(path) of satellite motion as projected on the surface of the Earth. The sensor perceives a cross-track swath, with the motion of the satellite along the track generating the raster images of the surface scanned. Some instruments have a mirror that sweeps across the track and reflects radiation to the detector. Data are stored in the satellite and are transmitted via microwave link to Earth stations. Each pixel value in existing satellites is normally stored as 8 bits (a byte) with values that can vary from 0 to 255. Satellites launched today normally use 11 bits for data storage, which gives a broader value range.

Water, forests, pastures, and other features reflect light from the sun differently and yield characteristic patterns in the relation between wavelengths and amount of reflected energy (Figure 9.18). The patterns can be recognized in the data registered by the satellite. This

Figure 9.18
Typical spectral reflectance curves for vegetation, soil, and water.

forms the basis for automatic and visual interpreting and classification of data.

The so-called Earth resource satellites are placed in fixed solar-synchronous orbits, which ensures that the satellites pass the equator at the same local time on the side of the Earth illuminated by the sun. This results in the satellites covering the same areas at the same local time. To achieve this, the satellites are placed in orbits that form a slight angle to the equator. However, this results in the areas adjacent to the poles not being covered.

The four major Earth resource scanning satellite systems in operation, which are relevant to GIS, are of the passive optical type. These are LANDSAT, SPOT, IRS, and IKONOS. LANDSAT is the generic name for a series of six satellites launched between 1972 and 1999 by the United States; SPOT, designated by the acronym for *Systeme Probatoire d'Observation de la Terre*, uses four satellites launched by France between 1986 and 1998 (Figure 9.19). Both systems are used primarily for mapping vegetation, geological features, and soil types, and to some extent for water parameters and linear structures such as roads and rivers.

The SPOT satellite's orbit is polar (passing nearly over the pole—inclination, 98°) at an altitude of 832 km. The satellite is solar-synchronous; that is, once every 26 days it passes over the same point on the equator at the same time of day. In other words, it scans the en-

Figure 9.19
Satellite in orbit. Mapping satellites trace a continuous track over the surface while circling the Earth. The reflected electromagnetic radiation from each swath is registered as digital values from which images can be constructed. *(a)* SPOT; *(b)* LANDSAT (NOU 1983:24 Satellitfjernmåling).

tire Earth every 26 days. Every circuit of the Earth takes 101 minutes. SPOT has a two-sensor system. One sensor registers the visual spectrum with a resolution of 10 × 10m. The other is a multispectral sensor that registers three channels (green, red, and infrared) with a resolution of 20 × 20m. A single satellite image, or scene, covers an area of 60 × 60 km. SPOT can be programmed from the ground to change the recording angle so that the same geographical area can be covered from different positions in space. This allows for much more frequent coverage than every twenty-sixth day. The same technique can be used to give stereoscopic pairs of images. All regular recording from SPOT over particular areas has to be ordered and programmed ahead of time.

The orbits of the Landsat satellites are both polar (inclination 98.2°) and solar synchronous, at altitudes of 705 km. LANDSAT 4 and 5 were launched in the 1980s, and the last one, Landsat 7, was launched in 1999. The Landsat 4 and 5 satellites have a repetition frequency of 18 days. They have two types of sensors. One, the Thematic Mapper (TM) sensor, registers seven channels in the visible and infrared spectra with a ground resolution of 30 × 30m, except for the thermal infrared channel, which has a ground resolution of 120 × 120m. The other, the Multi Spectral Scanner (MSS) sensor, registers four channels with a resolution of 80 × 80m. Landsat 7 scans the entire Earth every 16 days and uses 98.8 minutes per circuit. Landsat 7 has an enhanced thematic mapper (ETM), which is an improved version of the Landsat 4/5 TM. The panchromatic band has a nominal ground resolution of 15 m, and the seven additional channels have a resolution of 30 m, with the exception of the thermal infrared (IR) which has a resolution of 60 m. A single Landsat scene covers 185 × 185 km. Landsat performs continuous recording, and it is therefore unnecessary to order recording over special areas in advance.

Because they perceive reflected solar energy, almost all SPOT and Landsat sensors fail to "see" through cloud cover or in darkness. The single exception is the Landsat mid-infrared spectral band sensor, which perceives emitted energy; it is often referred to as a *thermal infrared sensor*. See Figure 9.20 for a comparison of the characteristics of different satellites.

In recent years several new satellites have been launched that supplement and partially replace data from Landsat and SPOT. The Russian Resource satellite was launched in 1994. This has the same channels as LANDSAT TM, but a geometric resolution that is no better than 170 m. An Indian satellite, the IRS-1C, has been operative since September 1996. The IRS-1C has a sensor (LISS—Linear Imaging Self-Scanning System) with a spatial resolution of 23.5 m in four multispectral bands and a panchromatic sensor with a resolution of 5.8 m. It also has a sensor called WiFs, with a resolution of 189 m.

Satellite			Sensor		Number of channels	Resolution (m)
ERS1	Europe	1991	AMi	Radar	1	26
ERS2		1995	ATSR	Multispectral	4	1000
Radarset	Canada	1995	SAR	Radar	1	9-100
SRS-1C	India	1995	WiFS	Multispectral	2	23
SRS-1D		1995	LISS	Multispectral	3	23
LANDSAT 5	United States	1984	MSS	Multispectral	4(6)	82
LANDSAT 7		1999	TM	Multispectral	6	30
					1	60
				Pancrom.	1	15
SPOT 2	France	1970	HRV	Multispectral	3(4)	20
SPOT 4		1998		Pancrom.	1	10
IKONOS	United States	1999		Multispectral	4	4
				Pancrom.	1	1

Figure 9.20
Characteristics of some Earth resource satellites.

The ground-swath width is 142 km. The altitude is 816 km, and the orbit is sun synchronous with a repeat coverage of 22 days at the equator. IRS-1D, launched in 1997, is a clone of IRS-1C with a few minor improvements.

Data from high-resolution satellite instruments are now available on the market. IKONOS was launched in the fall of 1999 and offers a panchromatic image of the Earth's surface at 1.0 m resolution and a color image (blue, green, red, and near infrared) of 4.0 m resolution; the ground-swath width is as small as 11 km. Like SPOT, IKONOS can be programmed to make oblique recordings so that the swath width is extended to 33 km. This forms the basis for stereoscopic viewing. Repetition frequency is 14 days, but this can be improved to 1 to 2 days with the help of programming for oblique recording. IKONOS does not carry out continuous registration, and recordings therefore have to be ordered and programmed ahead of time. There are plans to launch more satellites with high-resolution instruments.

There are also satellites with radar instruments that are possible resources for GIS use. ESA (European Space Agency) has launched ERS-1 (1991) and ERS-2 (1995) in orbit around the Earth. ERS-1 and 2 have a resolution of 26 m. Radarsat was launched by Canada in 1995. The Japanese JERS-1, sent up in 1992, has both radar-based and optical instruments. The signals sent from the radar instruments and reflected by the Earth are not hindered by clouds (see Figure 9.17). Radar images can thus be created independent of cloud conditions and are thus particularly suitable for areas with extensive cloud cover. Data from radar that "looks" to the side, known as SLAR (Side-looking Airborne Radar), form the basis for stereoscopic coverage and can be used for automatic production of digital terrain models. This is based on a technique known as interferomertry and the "matching" of picture elements from different recording positions in space. The accuracy of such terrain models will normally be $+/- 5$ to 20 m in height.

9.6.1 Correction

Radiometric correction

The measured magnitude, or radiometric value, of each pixel may be degraded by noise caused by:

- Sensor inaccuracy
- Atmospheric attenuation and scattering of reflected energy
- Variation of reflected energy as solar incidence varies with the curvature of the Earth, such as shadow effects
- Variation of reflected energy due to changes of surface slope

As most of these causes of pixel noise cannot be quantified directly, radiometric corrections are based on assumptions and models. The goal is to apply corrections so that the pixel values representing any one phenomenon will be the same, regardless of the pixel's locations in the overall image.

Geometric correction and resampling

Geometric correction is necessary if the data generated are to be used in mapping or for performing various analyses. The geometrical accuracy of data are influenced by:

- Rotation of the Earth during sensing
- Curvature of the Earth
- Variations in surface elevation
- Satellite instability in orbit
- Instability of recording devices
- Inaccuracies in the sensor projection system
- Atmospheric aberrations

Geometric correction of satellite data may be performed in various ways at two levels of precision:

- *Level 1:* systematic errors due to:
  - Rotation of the Earth
  - Instabilities of the satellite and its instruments (panorama effect, striping, angle of observation)

Correction of these systematic error sources is based on mathematical models.

- *Level 2*
  - Internal image improvement
  - Transfer to chosen cartographic projection

- Reduction to mean terrain elevation
- Transformation based on map coordinates
- Corrections for terrain variations

Correction of Level 2 errors is based on ground control points and possibly digital terrain models. For transformations based on map coordinates, a new grid is first established that has the correct geographical location, but with empty cells (Figure 9.21). Each new cell/pixel in the transformed grid will cover a slightly different geography than the old pixels and, through an operation known as *resampling*, the new grid will be filled up with values. This happens first with the help of ground control points, the grid transforming itself back to the original grid by use of conformal or affine transformation. It is thus possible to find where the new pixels meet the old grid. Values for the new pixels have to be interpolated from the old ones with help of various methods (Figure 15.11). The value of a new pixel can be established as the same as that of the old pixel it lands inside. This is the fastest resampling method. Or it can be determined by linear interpolation from the 2×2 closest pixels, or by polynomial interpolation from the 4×4 closest pixels. See Chapter 15 for further details on interpolation.

Note that the approach used here is different from the method used for transformation of vector data. With vector data the transformation is from old system to the new system, whereas with raster data the transformation is from the new to the old. This transformation and resampling technique is also used in geometric correction of scanned map data, scanned aerial photographs, and other types of digital raster data. The resampling technique is also used when changing the reso-

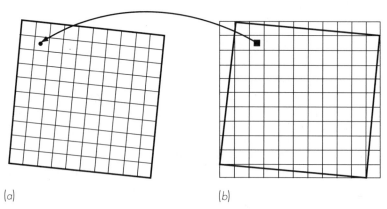

(a)   (b)

Figure 9.21
Principle for geometric correction and resampling of satellite data and other types of digital raster data; *(a)* original raster grid, *(b)* new and correct raster grid. The new pixels are assigned values interpolated from the old pixel values. ■ = new center cell point which cell value is to be determined , ● = corresponding point with known cell value in the original grid.

lution of satellite data—for example, to change the resolution of IRS data from 5.8 m to 5.0 m, or when Landsat TM 30m is adapted to SPOT Pan 10 m. However, one should be aware that even though the resolution is increased, the pixel values will not be improved. They will still be dependent on the original registrations.

After all corrections have been made, geometrical accuracy may actually be better than resolution. For example, data with a resolution of 10×10m may be geometrically accurate to ± 6 to ± 7 m, and data with a resolution of 80×80 m may be geometrically accurate to ± 50 to ± 70 m. This implies that these data are best used for overviews of areas that were poorly mapped previously, such as in developing countries.

9.6.2 Analytic methods

As a rule, satellite images must be analyzed and interpreted to yield the information desired. Satellite data are digital, so analyses and interpretations may be partially computerized. Three analytic methods are now common:

1. Color enhancement
   a. RGB color adapting
   b. HLS/IHS color adapting
2. Classification
   a. Supervised classification
   b. Unsupervised classification
3. Segmenting

In practice, these three methods are often used together. The analytical products are often used in combination with other sources, such as existing maps, and all sources are normally interpreted visually by skilled persons.

Color enhancement

Satellite image data have no colors in the sense that the eye perceives color in the visible spectrum. All coloring of satellite data relayed to Earth is done after the data are compiled into images. Normally, the data are processed interactively on screen in order to compile color images that enhance the themes of interest in mapping, such as coniferous and deciduous forest, cultivated fields, and so on. As the colors do not need to be the same as those seen by the unaided eye in daylight, the images created need not resemble conventional color photographs. In fact, in most cases they do not.

Colors are assigned to the individual spectral bands perceived by a satellite sensor assemblage in order to yield a *false-color-enhanced image*. For instance, red, green, and blue (RGB) colors may be assigned to three LANDSAT spectral bands to produce an image in which deciduous trees appear orange, coniferous trees brownish, earth and rock blue, and water black.

The basic colors may be adjusted individually, although the disadvantage of the simplest approach—basic color adjustment—is that adjusting any one color affects the aggregate image. For instance, amplifying the red basic color in an image not only makes the red pixels redder, but also adds yellow to the green pixels. Often, a more suitable method of improving false-color-enhanced images is to adjust the color hue, lightness, and saturation (HLS/IHS). The final color-enhanced images may be interpreted in the same way as aerial photographs.

Classification

Classification is based on the assumption that like phenomena in an image belong to the same class on the ground and have similar characteristic spectra; that is, that they have a *spectral fingerprint*. There are two approaches to classification that are distinguished primarily by their initial assumptions. In *supervised classification,* ground truth data from direct in-field observations are used to identify the initial parameters used in classification. *Unsupervised classification* needs no ground truth in its initial stages, but the final images and maps produced must almost always be verified in the field.

SUPERVISED CLASSIFICATION

Supervised classification begins with a compilation of *training data.* First, each class to be mapped, such as water, cultivated fields, pine trees, fir trees, and so on, is named and assigned a color. Next, a skilled analyst works on screen, circling areas known to contain the classes of objects named. Spectral band means and deviations are calculated for each encircled area. In other words, ground truth data are used to "supervise" subsequent manipulations.

Once the training data are compiled, the analyst initiates automatic classification of all classes with ground truth in the entire image. Each pixel is classified according to the greatest probability of its belonging to a particular class. Hence, the entire image may be classified using relatively few in-field observations. Various statistical methods are used to assign pixels to classes, including contextual classification, which takes account of the values of neighboring pixels. The pixels of the classified image displayed on screen are colored according to the class colors assigned in the training data.

UNSUPERVISED CLASSIFICATION

Supervised classification requires ground truth data when starting the classification. The classification is also limited to the classes with ground truth. Unsupervised classification is used to overcome these obstacles. Unsupervised classification starts with an *unsupervised cluster training procedure.* The first step is *clustering,* a way of ordering data by sorting pixels into classes according to their spectral values (Figure 9.22). In other words, clustering involves classification by spectral

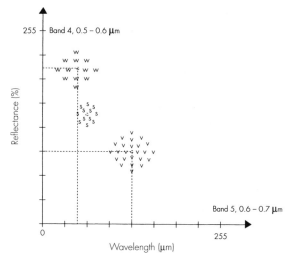

Figure 9.22
Clustering is a process whereby pixels that are adjacent in a spectrum are grouped together. Several parts of the spectral bands can be used simultaneously; w, water; s, soil; v, vegetation.

characteristics rather than by relation to ground truth observations. The analytical procedure employed may be interactive and includes four main steps:

1. Each cluster is assigned a value corresponding to its center.
2. Every pixel in an image is placed in the nearest cluster most closely matching its value, according to a method of distance measuring.
3. When all pixels of an image have been assigned to clusters, new centers are calculated to form the means of the constituent pixels.
4. Steps 2 and 3 are repeated.

The process stops when the clusters no longer change between successive iterations. The clusters are then labeled by assigning colors, and in-field observations are used to liken the colored clusters to ground phenomena.

Segmenting

Segmenting is usually performed automatically by a computer program. It starts with the comparison of a pixel value with the values of its eight immediately neighboring pixels. Of these eight, those with values—within set limits—that are closest to that of the start pixel are selected. These pixels are then compared with their neighbors, which are also selected if their values are within the set limits. The pixels selected form a segment. The process is repeated until no more neighboring pixels fulfill the criterion for selection. At this stage, a new start pixel in a new area is chosen, and a new segment is built up successively. Segmenting thus takes account of both geometry and pixel values in grouping pixels.

9.6.3 Integrating satellite data in GIS

The data of a geometrically corrected satellite image that were acquired via a satellite spectral band may be regarded as a thematic layer of pixels. Multispectral data can therefore constitute a GIS overlay. Images taken at other times or by other satellites may be adapted to the same map projection, coordinate system, and pixel size, and thereby form a new GIS overlay. Map overlays in raster form (geology, vegetation, etc.) may be integrated in the same system, provided that their respective geometries and map references are the same as those currently used in the system.

The classified raster data may be converted to vector data by using suitable conversion functions. The data may then be entered in vector GIS. It is also possible to use data from different sources simultaneously. For example, SPOT PAN can be used together with LANDSAT TM to take advantage of SPOT's good geometric resolution and Landsat's many channels. To combine the two, data from LANDSAT with 30 m resolution is resampled to data with 10 m resolution. This results in better quality of the end products but is, of course, a more expensive solution.

CHAPTER 10

Data Collection II

10.1   Surveying

In surveying, measured angles at, and distances from, known points in a coordinate system are used to determine the positions of other points (Figure 10.1). Surveying field data are almost always set out in polar coordinates, which may be transformed to rectangular coordinates using the relationships

$$\Delta x = D \cdot \cos (\alpha + \alpha) \qquad \Delta y + D \cdot \sin (\alpha + \alpha) \qquad (10.1a)$$

$$x_p = \Delta y + Y_1 \qquad x_p = \Delta x + X_1 \qquad (10.1b)$$

Heights are computed from horizontal or inclined distances (Figure 10.2). Vertical angles are computed from known points to new points, or vice versa. The corresponding transformations between polar and rectangular coordinates are

$$\Delta e = D \cdot \cos Z \qquad (10.2a)$$

$$E_p = \Delta e + E_1 + i - s \qquad (10.2b)$$

Surveying has traditionally produced numerical data in the form of angles measured by transits and theodolites and distances measured by tapes and chains. Starting in the 1950s, however, the traditional optical and mechanical tools of the surveying trade were gradually supplanted by more accurate electronic and electro-optical devices for measuring distances. Modern surveying is computerized: A *total station theodolite,* as illustrated in Figure 10.3, can now store and process digital measurement data either in the field or, subsequently, through an interface to a computer. Data may be coded or assigned themes in subsequent computer editing.

Surveying point location accuracy usually varies from ±1 to ±10 cm, depending on how measurements are made. Surveying is usually employed whenever up-to-date and accurate information is needed and for ground control points and measurement of property lines, buildings, manhole locations, and the like, as well as for smaller areas and for other details that have to be measured accurately. Measurement speed depends on topographical variations and the degree of modeling desired. As a guideline, surveying of complete terrain

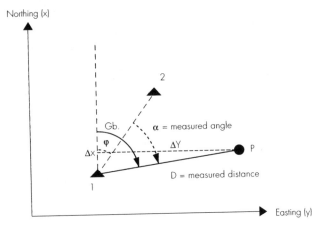

Figure 10.1
The geometrical principles of surveying (polar coordinates). Polar coordinates with direction and distance can be converted to right-angle coordinates.

details can cover about 0.1 to 0.2 hectares (ha) per hour. Photogrammetry is usually employed in mapping larger areas.

Computed coordinates are most often used to supplement and update existing maps, and coordinate lists are compiled for data on survey control stations, properties, manholes, and other fixed objects. Normally, surveying requires two people, one to operate the instrument and one to walk the terrain and select survey points with a measuring prism. Other systems allow the instrument to be remote-controlled, and the same person can operate both instrument and prism. Together with a penpad, they form a complete one-person measuring station. The choice of survey point is decisive in how reality can be reproduced in GIS. In situations where a high degree of accuracy is required, special instructions and measuring procedures are often cre-

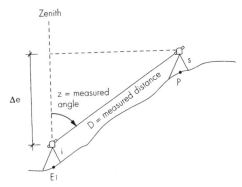

Figure 10.2
Surveying measurement of height. Height differences can be calculated from the measured zenith distance and the given or measured distance.

Figure 10.3
Most surveying is performed with the use of modern equipment known as a
total station. With this equipment it is possible to take measurements in the field
and register observations through an interface with larger computer equipment
in the office. (Courtesy of Geodimeter.)

ated. Measured data are coded directly under measuring, in relation
to both connectivity and object type.

Entry of coordinates
Survey data may be entered in GIS whenever accurate map data are
required—for example, when determining property borders for legal
purposes. New data that are intended for computers are transformed
in digital form to GIS. However, larger amounts of older analog data
are available where coordinates have to be keyed in manually, either
directly to GIS or via a simple registration program. Data for each
property should be verified, by identifying nodes and closing polygons, as entry progresses.

   This process is time-consuming—two or three times longer than the
corresponding task using a digitizer, inclusive of all preparatory and
editing tasks. However, the data produced are easier to apply than
digitizer data.

10.2   Satellite Positioning System

Global positioning system
The global positioning system (GPS) is a military satellite-based navigational system developed by the U.S. Department of Defense. The

GPS and electronic chart display and information system

The global positioning system (GPS) may be integrated with an electronic chart display and information system (ECDIS). As discussed in this section, the normal accuracy of GPS is ±25 m. Accuracy better than ±5 m, which may be needed in narrow channels, is available in differential GPS, which requires coastal reference stations. The accuracy provided by GPS and differential GPS is better than that of many older nautical charts. This means that accurate nautical charts are a prerequisite for the sensible use of GPS.

Updating is essential to the successful use of ECDIS, as navigational conditions continually change. National cartographic and hydrographic agencies regularly issue updating information, which may be entered into an ECDIS. Tests are now being conducted on transmitting updating information to ships via the INMARSAT satellite communications system.

An ECDIS incorporates three information subsystems:

1. An electronic navigation chart (ENC) that combines nautical information, such as lighthouses and markers, and hydrographic information, such as coastlines and sea-bottom topography, for display on screen. The basic information is usually digitized from traditional nautical charts.
2. Position information from the ship's radar and positioning system.
3. Steering information from the ship's steering system.

ECDIS provides a complete electronic navigation and warning system that displays the ship's position and other information vital to navigation. The system also provides audible and visible warning signals in the event that:

- The course of a sailing or a planned sailing is charted through nonnavigable waters (anti-aground).
- The ship deviates from its planned course.
- A planned change of course is not made.
- The ship does not deviate from a course of potential collision with other moving objects (anticollision).

If radar images are entered into the ECDIS, other vessels and their maneuvering in nearby waters may be displayed. If an autopilot is connected, the ECDIS can follow a planned course automatically. Changes of course are made after the navigator enters an acknowledgment of a warning of course change. As is the case in GIS, an ECDIS user can build up a database containing descriptions (attributes) of objects shown on digital maps.

system comprises 29 satellites, with 24 operational in six orbit levels and at an orbit height of 2020 km out in space. In principle, 4 to 6 satellites are available—15° over the horizon—at any time, everywhere on the Earth. Each satellite is equipped with four very accurate atomic clocks. The satellites' clocks and orbits are controlled by five ground stations that transfer data on changes in the orbit parameters and time to the satellites. Continuous signals are transmitted from the satellites. The signals contain data on the updated orbit parameters and the points in time for the signals leaving the satellite.

Determination of location on the ground is based on the transmitted radio signals, with exact orbit parameters and exact times being captured by antenna on small mobile receivers on the ground. Based on the signals, one can calculate:

1. The satellite's position in space (x, y, z) in relation to the Earth's center—based on the orbit parameters
2. The distance between the satellite and the receiver on Earth—based on two different calculation methods (code-based and carrier phase)

When the position and distance to the satellites are known, the position of the receiver is calculated in two or three dimensions as:

3. The point of intersection between the rotation circles around the satellites (Figure 10.4)

The following mathematical formula (for code-based receivers) can be used for calculating the position of the receiver:

$$(x - x_i)^2 + (y - y_i)^2 + (z - z_i)^2 = c^2(\Delta t_i - b)^2 \qquad (10.3)$$

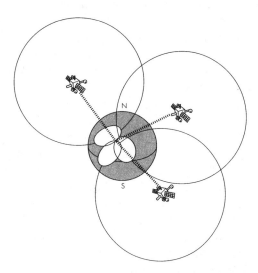

Figure 10.4
Global positioning system (GPS). GPS positions are determined from the intersection of cones.

203

where:

x, y, and z are the unknown coordinates of the receiver

x_i, y_i and z_i is the position of each satellite observed

c is the speed of light

Δt_i is the time difference between sent and received signals from each satellite

b is the unknown clock error (lack of synchronization between the clock in the satellite and the receiver)

There are four unknowns here (x, y, z, and b), and therefore signals from a minimum of four satellites are needed to calculate the x, y, and z coordinates of the receiver. However, signals from only three satellites give x and y coordinates. To improve accuracy (primarily with regard to determination of the time difference between satellite and receiver), there is, in practice, a need to register signals from more than the minimum required.

Because in using the formulas we operate with differences in distance in space, it is correct to say that with GPS, vectors (linear sections in a given orientation) are measured instead of distances. Treatment of the measurement readings is thus based on calculation of the vector in a coordinate system with the Earth's center as origin. GPS is based on the WGS84 as datum and on geocentric coordinate system, which may be transformed to other coordinate systems. GPS allows the position of points throughout the world to be determined 24 hours a day in all kinds of weather conditions.

Normally, GPS yields an accuracy of approximately ±25 m with no further corrections. Accuracy can be increased further by combining data from two receivers simultaneously, provided that one receiver (the reference station) is placed at a ground control point (Figure 10.5). The difference between the given coordinates at the reference station and the calculated GPS coordinates provides a basis for correcting new GPS-measured points in the area. This is called *differential* GPS measurement. This kind of differential measurement can determine the accuracy of new points to a few centimeters.

There are two different receiver systems, the so-called code-based receivers and the carrier phase receivers. *Code-based receivers* calculate the distance from satellite to receiver by measuring the time it takes the electromagnetic (or radio) signal to go from the satellite to the receiver (see equation 10.3). Under normal conditions this method will give an accuracy of +/− 25 m without the use of differential GPS, and with the use of differential GPS an accuracy of +/− 5 m. The code-based receivers are simple handheld instruments that are relatively inexpensive and have limited accuracy. *Carrier phase receivers* calculate the distance, as in electronic distance measurement, by calculating the number of whole wavelengths as well as the partial wave between satellite and receiver. The carrier phase instruments require longer ob-

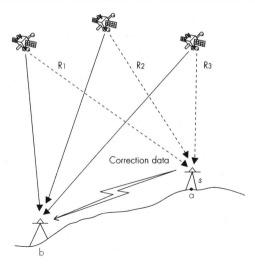

Figure 10.5
Differential GPS. GPS values are measured at two stations, one of which has
known coordinates *(a)*. Correction values can then be used to calculate the exact
position of the unknown point *(b)*, making it possible to calculate the exact
position of survey control stations on land or of mobile objects such as boats
and cars.

servation time and more elaborate calculations than the code-based instruments. They are also larger and more expensive than the code-
based, but give greater accuracy, $+/-$ 1 to 3 cm with differential GPS.
This type of receiver is therefore used for geodetic measurements.

In addition to measuring methods, accuracy is dependent on the
distance between stations, measuring time, and the satellite's position
in relation to the point to be determined. Normally, accuracy will be
reduced by a distance of about 0.5 cm per kilometer from the reference
station. The satellite geometry, *dilution of precision* (DOP), can be controlled by choosing an observation period where the geometry is good.
With a distance between the two receivers of less than 10 km, a measuring time of 10 to 120 minutes, and good geometry, phase measurements can, as mentioned, result in position accuracy of ± 1 to ± 3 cm.
Once a particular point has been determined, the neighboring points
can also be determined with a few seconds of measurement time,
called the *stop-and-go function* or *kinematic technique*. This means that
GPS can also be used for collection of terrain details. In addition, GPS
can be connected in the field to special data screens for sketching and
other operations.

Accuracy in centimeters is attainable by virtue of extensive, rapid
computations of error corrections. In the field, relatively straightforward real-time differential measurement and computations may result in a position being determined to an accuracy of ± 5 to 10 m, as
for cars and ships in motion. In some countries a service has been

established that broadcasts GPS corrections over the radio and phone network. This is particularly useful in boat and vehicle navigation.

As discussed in Chapter 6, the height reference in WGS84 is based on the ellipsoid and thus does not coincide with normal national height systems, which use the geoid as a reference surface (Figure 10.6). The height difference between ellipsoid and geoid is normally not known. Thus, one possibility is to calculate the height difference by measuring with GPS in points with a known height. The height of new GPS points can, in limited areas, be corrected by the calculated difference. Alternatively, the height of a GPS point has to be determined by traditional leveling or by trigonometrical height measurement.

One of the drawbacks of GPS is that it requires a direct line of sight between the receiver and each satellite accessed. In many instances the terrain, buildings, trees, and other objects may block the satellites from view or may disturb the signals. This can hinder the measurement of the required points and objects. This problem is also noticeable by mobile users.

GPS is most useful for:

- Locating new survey control stations and upgrading the accuracy of old station positions
- Determining points in property boundaries
- Measuring terrain features that are difficult to measure by conventional means
- Positioning offshore oil platforms
- Updating road data with a GPS receiver in a car
- Marine navigation, including integration with electronic charts
- Car navigation
- Determining camera-carrying aircraft positions to reduce reliance on fixed marks in aerial photography
- Determining differences in elevation

GPS has been integrated directly with several GISs. In general, GPS data may be regarded in the same way as other digital map data. Cod-

Figure 10.6
In GPS, the coordinates are expressed in relation to the Earth's center. Elevations are expressed in relation to the rotation ellipsoid, but these will not coincide with elevations stated in the national systems because these are based on the winding geoid.

ing of GPS data in GIS may be performed in the field to determine accurately the positions of intersections, road signs, the ends of bridges, road surfaces, and the like.

The cost of a GPS receiver depends in part on the accuracy it can achieve. High-precision receivers like those used in geodesy are the most expensive. At present, they are slightly more expensive than electronic theodolites. However, GPS instruments are becoming less expensive and are so efficient that they are taking over most traditional surveying areas.

Low microprocessor costs and large-series production will certainly boost the market for GPS receivers and spawn new applications, especially those systems that combine GPS positioning with GIS route finding for car navigation.

GLONASS

GLONASS is the Russian equivalent of GPS. GLONASS also consists of 24 satellites and is based on the same principles as GPS. With GLONASS there is no military interference with civilian signals, and the system has an accuracy of ± 25 m without use of reference stations. Receivers are available that can use both GPS and GLONASS simultaneously and thus obtain a better geometry in the measurements.

10.3 Photogrammetric Mapping

Photogrammetry means "picture measurement." The word describes a basic process first used in 1851 by A. Laussedat in France. Laussedat's original work had little practical application, because the lenses and photographic printing techniques of his time were too inaccurate for precision measurements. Toward the end of the nineteenth century, however, E. Deville, a Canadian, devised a stereoscopic machine to process photographs taken from theodolites. Subsequently, phototheodolites were developed specifically for that purpose. Phototheodolite techniques were eventually used widely, particularly for mapping the Alps of Central Europe. In 1909 the first instrument for mapmaking was built. As in so many other fields, it was military necessity that accelerated development. During World War I, balloons and airplanes were found to be excellent platforms for photography. The airplane eventually proved to be the more successful, because a single aircraft could fly along a prescribed course and take numerous overlapping photographs (Figure 10.7).

An overlapping pair of two photos forms the basis of stereoscopic viewing of the photographed terrain. Such a pair of photos can be placed in different instruments to recreate the ray path from the photography (Figure 10.8) and, thus, the three-dimensional terrain. These instruments are called stereo plotters.

Figure 10.7
With the use of aerial photography for mapping purposes, the entire area is covered by aerial photographs. There is usually a 60% overlap along each flight line and a 20% overlap between flight lines.

Figure 10.8
A stereo plotter; the system shown is the Wild autograph. *Left:* aerial photography. *Right:* a three-dimensional reconstruction of the terrain based on aerial photographs. The right image can be seen only with the right eye, the left image only with the left eye. When the images are positioned correctly in relation to each other, the brain will merge both image impressions into a three-dimensional model. The light beams in the model can be reconstructed in both analog and computerized forms (Andersen 1980).

A stereo plotter is the instrument most commonly used to transfer aerial photographic information to planimetric and topographic maps which, in turn, may be the bases for vector data. In a stereo plotter, a photographic interpreter views two overlapping aerial photographs, taken from different positions, to form a three-dimensional image. By moving a floating point around in the three-dimensional image, the operator can draw in roads, rivers, contours, and other details. The plotter has a digital transmitter, and all movements used in drawing in the photogrammetric model can be numerically encoded and entered in computer storage The traditional analog instruments are now being replaced by more or less digital systems, such as analytical plotters and digital photogrammetric workstations.

In an analytic plotter, software replaces the old analog rods. All movements of the instrument are digitally controlled, but the viewing optics are traditional and the images analog. Digital photogrammetric workstations use a powerful PC and scanned digital images. The two overlapping images are shown on the same screen, with rapid exchanges of their polarized light. The operator uses glasses with synchronized polarized glass so that the right eye sees only the right image, the left eye only the left image. Computer programs now replace all the movable parts of previous instruments. A mouse, tracer ball, or similar device is used to move a pointer around in the model. The digital images allow digital terrain models to be established very effectively through pattern recognition for automatic determination of elevation points and profile lines. It is thus possible to produce ortho images very efficiently with this type of instrument.

However, we have to be aware that automation cannot differentiate between the surface of the terrain and buildings, trees, and so on; thus the tops of buildings and trees will be part of the digital terrain model and may have to be corrected manually afterward. Automatic routines have yet to be developed for the recognition of buildings, roads, rivers, and other such details in the photos. If a solution can be realized, it will mean a true revolution in all map production.

Either stream encoding or individual point encoding may be used, as in the corresponding digitizer process. In digital instruments, transformation parameters are used to transform the internal system coordinates continuously to terrain coordinates. The transformation parameters are calculated from the photographic coordinates of points known from field survey control or aerial triangulation. Unlike the digitization of existing maps, processing photogrammetric data directly encodes elevation information at any given point within the stereo model (see equations 6.3 a–c).

In mapping, new thematic codes and ID codes are entered whenever the category of an object changes—for example, from residential to industrial use. Photogrammetric encoding requires both trained operators and their concentration. Experience indicates that ID codes are more efficiently entered on screen in the final editing phase than during map construction.

The coordinates of such individual points as manholes, drains, and property boundaries may be registered in large-scale aerial photographs. Thus, on the ground such points are often marked with white paint or white paper prior to photographing to ease their identification in the resultant photographs. Aerial triangulation, a special technique for point location based on measurements in the stereoscopic model, may be used to register marked points with an accuracy of ± 5 to ± 10 μm on photographs, corresponding to 3 to 6 cm on the ground in an image scale of 1:6000. For other details, photogrammetric accuracy is 10 to 20 μm.

The product of photogrammetric mapping must be edited to ensure that map data correspond to the map image desired. Map editing programs are used either during the mapping itself or during the subsequent editing phase. Again, experience indicates that editing can be as time-consuming as traditional manual redrawing. A 60 × 80 cm map in a scale of 1:1000 can involve 1 to 3 MB of data, depending on the complexity of the map.

When producing base maps with details and topography, map data can normally be entered at a rate corresponding to the registration of 10 to 12 decares an hour in 1:6000 scale images, whereas editing may proceed at a rate of 20 decares an hour. For 1:15,000 scale images, the corresponding rates are 20 and 40 decares an hour. GIS is unable to distinguish between a digitized map and a digitized stereo model. However, a map is inherently generalized, whereas a photo is not, especially in respect to elevation data.

10.4 Collection of Attribute Data

Just as digitizing work may be divided into three phases, so the acquisition of attribute data may be divided into:

- Preparation
- Entering
- Editing and quality enhancement

Preparation

Data of interest are often recorded on paper and then scattered among a multitude of different conventional registers and files. Consequently, the initial acquisition of data may require a major organizational effort involving many different agencies. For instance, registering all the undeveloped land in a county may require data from all the agencies in all the towns in the area.

In some cases, registration has been organized in accordance with statutes and regulations. For example, many countries have their own regulations governing the registration of fixed property. In other cases, it may be necessary to supplement new field registration with local knowledge. Attribute data may be collected effectively only if a suitable collection structure is available or is specified as an initial step. A common approach is to complete various forms, either on paper for manual registration and subsequent entry via a PC, or to hold them in files for direct keying-in on a laptop or PC. Field registration of attribute data based on penpad in combination with GPS has become quite usual and popular in recent years.

The data fields of the forms should, of course, correspond to the data fields of the database. Registration should be unambiguous and provide for all data variations to ensure that data always locate in the correct data fields. Consider, for example, setting up a GIS for a water supply system. Such a system will require a systematic structuring of data.

Piping
1. Water mains
2. Supply mains
3. Service pipes
4. Sewer pipes
5. Venting
6. Manholes
7. Drains
8. Inoperative piping
9. Inoperative piping not to be removed
10. Basins
11. Booster stations
12. Pump stations
13. Pressure zones
14. Stop valves
15. Joints and joining methods
16. Nonreturn valves
17. Pressure reduction valves
18. Separators
19. Vent valves
20. Fire hydrants, above ground
21. Fire hydrants, below ground

Service pipes
1. Type
2. Tapping
3. Sprinklers
4. Stop valves
5. Water meters

Manholes
1. Type
2. Piping accessed

3. Status (operational, planned, abandoned)
4. Location
5. Construction

Miscellaneous
1. Units of length
2. Dimensions
3. Model numbers
4. Valve numbers
5. Supply areas
6. Soil conditions
7. Road traffic conditions
8. Damage
9. Repairs
10. Service interruptions

These data may be divided roughly into three categories: locations, components, and network parameters.

The acquisition of attribute data often requires verification in the field. Verification may entail going down a manhole (Figure 10.9) to sketch or photograph details, rodding pipes to ascertain direction and junctions, measuring and leveling, poring over old files and drawings to locate details of dimensions and materials, interviewing retired plumbers, and so on. It is thus apparent that it requires a great deal of work to produce attribute data of satisfactory quality. As far as piping data are concerned, a rule of thumb is that 30 minutes are used per manhole and additional time is required to measure the route of the service pipes. It is claimed that 800 worker-years would be needed to convert the sewerage network in England to digital format, including

Figure 10.9
A manhole. A large number of the details in this picture will have to be registered for entering in a GIS.

in-field registrations. The establishment of a computer-based national property register (3 million properties) in Norway required 150 worker-years. Once these data have been collected and structured in lists, on forms, and/or on maps, entering data in GIS may begin.

Thus, experience shows that the collection of attribute data can be very labor intensive. As discussed in Chapter 12, it is therefore important to carry out a thorough data structuring prior to starting collection. When carrying out data structuring, great emphasis must be placed on determining priorities as to what data should be collected—data with most use in relation to collection cost should, in principle, have high priority, and data of least use in relation to cost should have low priority.

Entering data

A convenient way to enter data in GIS is to use a database application program on a laptop or PC for initial entry, subsequently exporting the database file to GIS. PC database application programs support forms that can be designed and filled out on screen, allowing a specific form to be designed for each type of object.

There should be space on the form to include verification of formats in all data fields to ensure, for instance, that text is entered into a text data field and only figures are entered in a numerical data field, and that values entered are restricted within set extremes. It is occasionally advantageous to call up a graphic display of map data on screen as the attribute data are entered. Whenever data *are* available in computerized form, they may be entered directly.

Editing

Automatic verification functions reveal only formal errors. Incorrect information, such as a wrong name entered in a name field, can be spotted only by manual proofreading. Major errors and meaningless text are usually easy to spot, but incorrect spellings, inversions, omissions, and other less obvious errors are more difficult to find. Consequently, it is better to verify attribute data by printing out sorted lists, which may then be checked line by line by a copy reader.

It is important to eliminate errors in entered data, although it is invariably time-consuming because it requires mostly manual work.

Once the digital map data and attribute data are verified, corrected, and entered in GIS, the identifiers linking the databases should be checked and all missing identifiers marked. Checking in GIS will not identify incorrect links due to map data code errors or mistakes (in legal fields) in the attribute data entered. Experience shows that the usefulness of attribute registers is greatly dependent on the quality of the data. If users cannot rely on the data, then the data will not be used. Active use often reveals errors, which again emphasizes the importance of data quality.

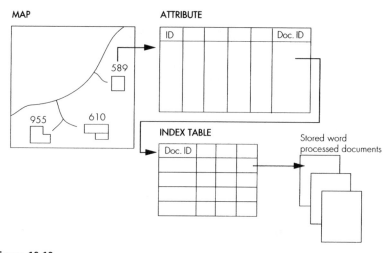

Figure 10.10
Documents in a word processing system can be integrated into GIS via an index table between the attribute table and the word processing system.

10.5 Text Data

Public administration agencies, public utilities, and other organizations dealing with georeferenced data may frequently need to integrate text data in GIS. There is substantial demand for the integration of procedural systems and GIS by public service officials. Such information may, for example, concern tasks related to questions as to which cases are being handled in respect of particular property or where a property is located for which a building application is being handled. All kinds of documentation may be involved—from letters and property deeds to zoning directives and service overviews. Text data from word processing systems and spreadsheets may be integrated into GIS via a table of object identifiers and pointers to text or spreadsheet file addresses (Figure 10.10). A GIS may also conduct searches of text documents. In practice, this requires careful organization in the naming and filing of text data. GIS graphic map images may be imported in word processing and incorporated in texts, using programs such as Microsoft Word or PostScript. Text with maps may then be printed out on a laser printer.

CAD drawings can be integrated into GIS in the same way as text. This is done by placing pointers in the GIS tables (see Figure 10.10). In addition, digital drawings produced in CAD can be transferred as normal map data through common exchange formats (e.g., DXF).

CHAPTER 11

Data Quality

The use of analog maps leads us to assume that all objects on a map are of similar quality. This is not the case with GIS, where data of differing origin and accuracy may well be mixed. In addition, the introduction of GIS into mapping has produced a completely new (and perhaps less professional) user group and producer group. In such a situation, an understanding of the concept of quality will be very important, and the marking of data according to its degree of quality a necessity.

The applications of digital data are indeed many and varied, but digital data systems are not inherently more accurate than the analog systems they supplant. In general, digital systems are capable of processing data more precisely than analog systems, but their overall accuracy still depends on the accuracy of their source data, which in most cases remain analog.

To see why this is so, let us consider time, perhaps the most accurately measured quantity in everyday life. In the 1950s, extremely accurate ships' chronometers and other timepieces became available, at a high price; ordinary wristwatches and clocks were accurate to only a few minutes a day. In most applications, clock accuracy was a question of how much a user was willing to spend for a finely made mechanical timepiece. National standards of time were kept by using large electronic clocks that derived their signals from the oscillations of quartz crystals housed in temperature-controlled ovens. These clocks were completely analog and accurate to the order of 1 part in 10^9, or about 1 second in 10 years. Thanks to microchips and digital techniques, similar levels of accuracy are now common in everyday wristwatches and clocks. Accuracy is still due to quartz crystals, although it is the digital techniques (and their ultracompact realizations using microchips) that have made such accuracy available in small timepieces. In terms of overall accuracy, the quality of a timepiece now depends both on the quality of the crystal oscillator and on the precision with which its signals are processed to give the final display. In technical terms (marketing oversell of expensive timepieces notwithstanding), user needs determine accuracy, and accuracy follows price (Figure 11.1), because greater accuracy requires a higher-quality crystal (or atomic) source and more complex processing of its signals.

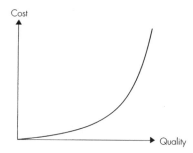

Figure 11.1
Relationship between quality and cost. There is a strong correlation between data quality and cost—costs increase with quality. Quality is therefore often determined on the basis of cost.

In GIS, the situation is similar: The final accuracy of data depends on the qualities of the original input data and on the precision with which input data are processed. In principle, the user dictates the need for greater accuracy. This is provided to some degree by improved technology, which is purchased or developed in response to the original need. Although technical achievements are virtually limitless, from a scientific viewpoint, the actual level of achievement, and thus the available technology, is determined—and consequently limited—by costs. Higher accuracy entails higher initial data quality and more precise processing, both of which increase system costs. Furthermore, the requirements for up-to-dateness affect the measure of data quality and hence system cost (Figure 11.2). In general, the costs of updating may

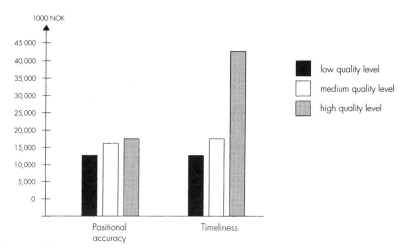

Figure 11.2
Simulation of cost sensitivity in relation to various quality standards for the Norwegian national road database. The bar chart shows clearly that timeliness is here much more susceptible to quality choice than to positional accuracy (see Figure 17.16).

216

strongly influence the determination of data quality. In theory, therefore, GIS data quality is a compromise between needs and costs. In practice, however, the choice is often a question of what is currently available or can be acquired within a reasonable amount of time.

Delays may be due to a variety of causes. Where snow falls in winter, users of aerial photographs must wait until the snow melts before they can proceed; similarly, users of satellite images may have to wait for cloud cover to disperse over an area of interest; users of map digitization produced by national mapping agencies may be assigned to backorder queues because the speed of agency work is often constrained by budget allocations. Finally, users may suffer delays in resolving copyright or other legal difficulties concerning the ownership of data. Any of these factors may cause data to be of lower quality than originally envisioned.

In summary, four aspects of data acquisition comprise the criteria for selecting data accuracy:

1. Needs
2. Costs
3. Accessibility
4. Time frame

Quality can be defined as one or more characteristics of geographical data that describe the extent to which it is fit for use. The data quality components are intended to document the quality of the data sets. The International Organization for Standardization (ISO, 2001) has proposed that the following elements should be used to describe how well a data set meets the criteria set forth in its product specification:

1. Positional accuracy
2. Attribute accuracy
3. Temporal accuracy
4. Logical consistency
5. Data completeness

These five elements describe the quantitative quality of a data set.

The following overview elements should be used to describe the non-quantitative quality of a data set:

1. Purpose
2. Usage
3. Lineage

These three elements are generally considered to be the indirect goals of accuracy.

The Federal Geographic Data Committee (FGDC) has outlined, through its metadata standards, a set of characteristics defining data quality. This list of elements is almost identical to the ISO's, but temporal accuracy is not listed as a main element. However, the FGCD

Figure 11.3
Data set quality should always be evaluated on the basis of the original specifications.

recommends, for example, the use of cloud cover percentage as an additional quality element for satellite data.

The quality characteristics are important because they encourage multiple use of data both within an organization and between organizations. The users can judge whether they can use particular data and whether they have received what they expected, and the producer can judge whether its products meet market expectations. The quality of a data set should always be assessed according to the specifications for that particular data set (Figure 11.3).

First of all we will look at the basic terms accuracy, precision, and resolution, and later at the special quality parameters used in GIS.

11.2 Measuring Accuracy and Precision

Accuracy versus precision
Although commonly considered to be identical, accuracy and precision express different aspects of measurements and should be carefully distinguished. *Accuracy* expresses how closely a measurement represents the quantity measured, or closeness of agreement between a test result and the accepted reference value. *Precision* expresses the repeatability of the measurement. For the measurements normally of interest in GIS, precision is limited only by the instruments and methods used. In physical measurements, precision is customarily indicated by the number of significant figures carried in a quantity expressed. For instance, a figure of 101.23 cm indicates that the distance is measured with a precision of a fraction of a millimeter. The last decimal place may be uncertain. It is thus quite possible to measure with high precision and produce a result that is inaccurate. One can envision a situation where a measurement is made with high precision in inches, but a wrong result is given in centimeters.

Accuracy should be stated in terms of an interval in which a true value is assumed to lie. For instance, 101.21 ± 0.04 cm indicates that the true value is assumed to lie between 101.17 and 101.25 cm. However, some textbook authors may use a notation with estimated value ± standard deviation. Care should be taken to avoid misunderstandings.

The terms *accuracy* and *precision* are particularly relevant in relation to geometry and quantitative attributes, but in GIS one must also relate to temporal precision and accuracy.

Measurements of accuracy

Inaccuracy in the data basis can be caused by any of the following, singly or in combination: random error, systematic error, and gross errors. Statistical methods are generally used to specify accuracy requirements and total accuracy. As in GIS, statistics is a tool used to describe reality; because reality cannot be described exactly, the methods used will always contain an element of uncertainty.

Standard deviation (σ) is the square root of the variance. For a series of n observations x_j, where j takes the integer values 1, 2, up to n, with known expectation μ, the variance Var(x) and the standard deviation σ are given by

$$\text{Var}(x) = \sigma^2 = \lim_{n \to \infty} \sum_{j=1}^{n} \frac{(X_j - \mu)^2}{n} \tag{11.1}$$

The estimate of σ from a limited number of observations (n is finite) is often denoted s. Hence

$$s^2 = \sum_{j+1}^{n} \frac{(X_j - \mu)^2}{n} \tag{11.2}$$

$$s = \pm \sqrt{s^2} \quad \text{(standard deviation)} \tag{11.3}$$

If the expectation for X_j (denoted μ) is not known, the expectation may be substituted by the mean of all observations and the number of redundant observations reduced by one:

$$s^2 = \sum_{j=1}^{n} \frac{(X_j - X)^2}{n - 1} \tag{11.4}$$

$$\overline{X} = \frac{\sum_{j=1}^{n} X_j}{n} \tag{11.5}$$

The standard deviation takes the number of redundant observations into account (equation 11.4). The root mean square is always divided by the number of observations (equation 11.2). Thus, if the difference between the number of redundant observations and the total number of observations is relatively small (e.g., the number of observations is large), the difference between the root mean square (RMS) and the

219

standard deviation may be negligible. The standard deviation is the measure of how close a measurement/observation lies to the value expected.

In surveying, error ellipse (two-dimensional) and error ellipsoid (three-dimensional) are utilized to visualize the uncertainty of single points (Figure 11.4). The standard error ellipse has semimajor and semiminor axes equal to standard deviation in each direction. The probability (approximately 60%) of the point's being outside the standard error ellipse is much greater than the probability of its being inside. Therefore, it is better to visualize with confidence error ellipses or ellipsoids scaled to give, for instance, 95% probability of the point's falling inside.

The precision of observations may be visualized with colors or with error ellipses or ellipsoids, depending on their dimensions. For one-dimensional observations, different colors are used to indicate different classes of precision when the observations are shown on the screen. In two dimensions, error ellipses are used, and for a vector in three dimensions, error ellipsoids. The likelihood of observations/measurements lying within given boundary limits is known as *confidence* and is expressed as a percentage. For example, the requirement could be that observations should have a confidence of 95%.

Deviations may be random or systematic. The systematic errors may be computed according to the equation

$$m = \sum_{j=1}^{} \frac{X_j - \mu}{n} \tag{11.6}$$

The systematic component may be relatively unimportant when data originating from the same source are used, but it can be disruptive in data originating from dissimilar sources.

Serious errors (outliers, gross blunders) can always occur in the data and have to be removed. The definition of what constitutes a gross blunder is a matter of opinion, but the limit is often set as three times the standard deviation for the data in question. It is difficult to spot se-

Figure 11.4
The error ellipses visualize the uncertainty in the points with regard to both size and direction.

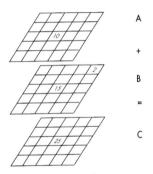

Figure 11.5
Errors in addition of the attribute values in a raster layer are transmitted to the total.

rious errors with only a few redundant observations. There is a certain likelihood that geometrical data will contain serious errors that are not detected and removed. For example, these can arise in connection with in-field measurements, calculations, or manual entry of coordinates into the database.

Redundancy is used to remove blunders. *Outliers*, which are probable blunders, are detected through statistical testing. A statistical test will reveal larger outliers, whereas smaller outliers may remain. *Internal reliability* is, for each observation, the largest possible remaining outlier after the test. The impact of this largest remaining outlier on the coordinates calculated and similar factors is called *external reliability*. External reliability is a worst-case scenario.

Data accuracy may be verified through reference to other, more accurate data that have been designated as error-free for the purpose. These reference data may be compiled from survey measurements in the field, from maps of considerably larger scale, and from more precise instruments for the measurement of attributes such as wind and temperature.

Errors may accumulate in data processing. For independent error sources, the standard deviation of a multiplication $C = A + B$ is as given by

$$s_C = \sqrt{s_A^2 + s_B^2} \tag{11.7}$$

Errors can accumulate even as data are being processed (Figure 11.5). For example, when two overlays of a raster GIS, $A = 10$ and $B = 15$, with standard deviations $s_A = \pm 1$ and $s_B = \pm 2$, are added up to a resultant C, the standard deviation is as given in equation 11.7, provided that A and B are uncorrelated.

$$C = 10 + 15 = 25$$

$$s_C = \pm\sqrt{1^2 + 2^2} = \pm 2.2$$

11.3   Resolution and Sampling Rate

Spatial resolution refers to the minimum size of objects that can be registered. In practice, spatial resolution has been closely linked to map scale—large map scale gives high resolution and small map scale gives low resolution. *Sampling rate* refers to the pattern of density in space or time of the observations. Normally, resolution and spatial sampling rate will coincide, but in principle they are different. For example, in raster data one can have small pixels (high resolution), but with greater distance (low sampling rate) (Figure 11.6). Vector data can have small objects that are observed/described with one or few points. In raster data the observation unit (cell size) selected can be too large to register all the objects desired (Figure 11.7).

Normally, low resolution gives poorer accuracy than high resolution, because observations with low resolution usually do not have the same closeness to the "true" values as observations with high resolution. The importance of resolution is particularly visible in relation to areas: Low resolution gives generalized and greater areas, whereas high resolution gives smaller areas. In raster data resolution normally refers to pixel size—for example Landsat TM with a resolution of 30 m, and SPOT Pan 10 m.

In GIS attributes can also be described as resolution. For quantitative attribute data, resolution is determined by the precision of the measuring instrument, and for qualitative data it is determined by how many object types or classes are defined. It is generally the type of usage that determines the resolution of qualitative data.

Temporal resolution is determined by the size of the timepiece (hour, day, week, etc.) that describes an event. The time collection interval for each measurement in GIS can vary from seconds to hours to days, depending on the measuring instrument and type of event. Once again, it will be the type of usage that determines temporal resolution. Temporal resolution should not be confused with sampling rate, which gives the frequency of repeated observations.

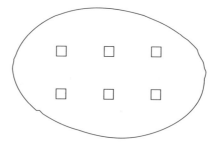

Figure 11.6
The figure shows a situation with high resolution—small grid cells, but with a low sampling rate; large distance between the observations/grid cells.

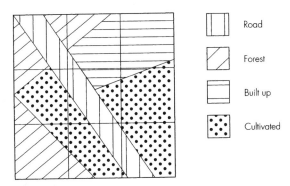

Figure 11.7
The observation unit (cell size) selected here is too large to register all the objects desired.

11.4 Data Storage Precision

Data storage of both geometry and attributes can also affect data accuracy whenever entries comprise many significant figures. Storage with floating point *single precision,* in which a single computer word (string of bits) is used to represent a number, results in precision to seven significant figures. Storage with *double precision,* in which double the usual number of bits are used to represent a number, results in precision of up to 15 significant figures. If data are transferred from a single-precision system to a double-precision system, digits after the sixth or seventh are unreliable. Single-precision systems may also introduce noticeable errors in rounding off when decimal numbers are repeatedly multiplied, as, for example, when computing nodes.

11.5 Positional Accuracy

In ordinary mapping, accuracy is inversely proportional to map scale: a map to scale 1:1000 is more accurate than one to scale 1:100,000. The corresponding case for digital map data is rather more complex. Dissimilar data are stored in the form of terrain coordinates, regardless of their respective sources and accuracy. In other words, all data are stored to a scale of 1:1 and may then be used to generate maps of any scale.

The use of insufficiently accurate data may eventually prove more expensive than the original acquisition of the data. A GIS can plot smooth lines on a map and perform convincing computations and analyses, all on the basis of its stored data and regardless of the accuracy of the original data entered. Moreover, a GIS can perform calculations of cut-and-fill works from data accurate to ±2 m as comprehensively as it can from data accurate to ±0.2 m, and plot equally fine

223

profiles from both. However, the potential consequences of using the less accurate set of data are severe: The results may be totally useless or, at best, cause serious miscalculation of construction costs.

Positional accuracy is proposed by ISO (2001) to cover the following elements:

1. Absolute or external accuracy—closeness of reported coordinate values accepted as or being true
2. Relative or internal accuracy—closeness of the relative position of features in a data set to their respective relative position accepted as or being true
3. Gridded data position accuracy—Closeness of gridded data position values to values accepted as or being true

The goal of absolute georeferencing is to locate objects exactly as they are located on the ground, as related to a common coordinate system. Accuracy is checked by comparing the actual position with a value that is accepted as being true. "True" position usually means an independent source of higher accuracy. Correspondingly, the relative position is checked by comparing the actual position with a relative position that is accepted as being true. Relative position means, for example, measuring a distance in a road network from fixed points (milestones) along the road.

Gridded data are represented geometrically by a matrix. The position of cells in the matrix can be checked against positions that are accepted as being true. Gridded data can also be transformed to a terrain coordinate system. When transformed, the position of the gridded data can be compared with terrain values that are accepted as being true.

Positional accuracy includes uncertainty and resolution. All observations contain elements of uncertainty. Resolution is dependent on the precision of the measuring instrument. Digital geometric data can be processed through surveying, photogrammetric mapmaking, map drawing, digitizing, and so on. Each stage in the process introduces inaccuracies. These errors can be cumulative (see equation 11.7), and can have an adverse affect on the final data.

A standardized set of evaluation criteria and procedures should be used to guarantee that the relative quality of one data set versus another can be determined. Measurement of positional accuracy may be obtained by one of the following methods, based on the U.S. Geological Survey Spatial Data Transfer Standard (1998):

1. *Deductive estimate.* Any deductive statement based on knowledge of errors in each production step should include reference to complete calibration tests and also describe assumptions concerning error propagation. Results of deductive estimates should be distinguished from results of other tests.
2. *Internal evidence.* National procedures should be used for tests

based on repeated measurement and redundancy such as closure of traverse or residuals from an adjustment.

3. *Comparison of source.* When using graphic inspection of results *(check plots),* the geometric tolerance applied should be reported and the method of registration should be described. Use of check plots should be included in the lineage portion.

4. *Independent source of higher accuracy.* The preferred test for positional accuracy is a comparison with an independent source of higher accuracy. When the dates of testing and source differ, one must ensure that the results relate to positional errors and not to temporal effects.

Requirements of accuracy are often stated with the standard deviation or RMS. As previously noted, serious errors should be rejected, and the limit is often set as three times the standard deviation or RMS for the data in question. It is also possible to specify systematic error components. Statistical methods are well developed for single points or single observations, but it is considerably more difficult to specify the geometric accuracy of lines and areas that consist of many points—not to mention volumes. Attempts have been made to show line inaccuracy as a band or a zone of uncertainty, but there is no consensus about the shape of such a band (Veregin 1999).

All geometrical data should carry accuracy designations so that GIS users may assess the consequences of the various sources of error involved. Ideally, accuracy codes should be assigned either to various objects to designate RMS or through a quality overlay. When assessing positional accuracy, actual accuracy has to be evaluated in relation to specified accuracy, especially in relation to the relevant application.

11.6 Attribute Data Accuracy

Considerations of accuracy usually focus on the accuracy of geometrical data, although in practice the accuracy of attribute data is equally important. ISO (2001) has proposed that the following elements be used for defining the accuracy of attribute data:

1. Classification correctness—comparison of the classes assigned to features or their attributes to a universe of discourse (e.g., ground truth reference data set)
2. Nonquantitative attribute correctness
3. Quantitative attribute accuracy

Attribute data often form the basis for classification. Classification correctness is determined by comparing the actual classification with the description of each class. Quality will thus be dependent on a good description of the classes and well-trained staff performing out the classification. In some cases, misclassification may involve simple sorting errors: An object of type A is put in class B. Thus a residential building may be classified as a commercial building if the demarcation between the two classifications is unclear. In other cases, the class structure may be faulty, as when there is no class C for objects containing elements of both A and B. The accuracy of classified data is best expressed with a percentage correctly classified; probable misclassification can be expressed with a matrix; deductive estimates are expressed by standard deviation.

Nonquantitative attributes are the same as nominal attributes and can, for example, include items such as type, color, name of owner, as well as characteristics such as good and bad. In general, purpose, usage, and lineage should be used to describe the nonquantitative quality of a data set. Nonquantitative correctness is determined by comparing the actual description with descriptions accepted as or being true.

Quantitative attributes, known as ratio, interval, and ordinal data, is specified with number values. Quality is controlled by comparing actual values with values accepted as or being true. Specifically, quality can be dependent on the accuracy of the measuring instruments used and the skill of the observers. Measurement instruments may be poorly adjusted, thus introducing systematic errors in quantitative data; or they may be subject to random inaccuracies, as in instruments measuring acidity and alkalinity in pH, or in satellite sensors viewing reflected and reemitted solar radiation.

The sampling rate used to collect attributes also influences accuracy of data. Here, examples include collections based on statistical distribution hypotheses or other assumptions that fail to agree with the real world, as when observations are excessively disperse. This is obvious in the choice of cell size in raster GIS, as shown in Figure 11.4, where the density of observations (cell size) is not great enough to register the phenomena desired.

Accuracy assessment for quantitative data must be performed using procedures similar to those used for positional accuracy. Accuracy tests for categorical attributes may be performed by one of the following methods, based on the USGS Spatial Data Transfer Standard (1998):

1. *Deductive estimate.* Any estimate, even a guess based on experience, should be permitted. Terms such as *good* and *poor* should be explained in as quantitative a manner as possible.
2. *Test based on independent samples.* For independent samples, a misclassification matrix should be reported as counts of sample units cross-tabulating the categories of the sample with the test result.
3. *Test based on polygon overlay.* A misclassification matrix must be reported. The relationship between the two maps must be explained. The two sources should be independent, and one should have higher accuracy.

A suitable expression of classified attribute data accuracy is to state a percentage of correct or erroneous classifications, such as "99% of all objects correctly classified." Quantitative attributes are handled as positional data regarding accuracy. Generally speaking, the accuracy of numerical attributes is verified by comparing data with randomly assigned true values and is expressed in terms of standard deviations and systematic errors.

Ideally, the accuracy of all attribute data should also be accuracy coded as geometrical data. This is seldom done, however, either because the accuracy remains unknown, the cost of assessing it exceeds the benefit of the knowledge it provides, or the user fails to appreciate the need for accuracy. Ideally, accuracy codes should be assigned either to various attributes to designate RMS (quantitative data) or through a quality overlay (qualitative data). As with positional data, attribute accuracy has to be assessed in relation to specified accuracy, especially in relation to the intended area of application.

11.7 Temporal Accuracy

According to ISO (2001), temporal accuracy should include the following elements:

1. Accuracy of a time measurement—correctness of the temporal references of an item (reporting of error in time measurement)
2. Temporal consistency—correctness of ordered events or sequences, if reported
3. Temporal validity—validity of data with respect to time

Temporal accuracy refers to the coincidence between registered and actual time or the time coordinate for the event or phenomena, stated, for example, as $+/-$ 1 hour or $+/-$ 1 day. In addition, temporal accuracy can be defined as the correctness of ordered events or sequences. The specification of temporal accuracy is dependent on the ability to measure time objectively in a common temporal coordinate system.

It is important to state how long the data are valid—*temporal validity;* without this, it is easy for data to be used incorrectly. Temporal validity is an important element in GIS and is particularly linked to timeliness. Timeliness is normally indicated by the date of registry, such as the dates on aerial photographs, photogrammetric data, and satellite data, and the dates of surveys, accidents, water leakage, and the like.

Two situations may affect the timeliness of data:

1. When the geometry and attributes of existing objects have been altered
2. When new objects, with geometry and attributes, have appeared

The degree of timeliness required depends on object type and data application. Public works agencies and utilities often require the most up-to-date information on property boundaries, buildings, roads, manholes, and other works. It is therefore advisable to update the data for such objects every two to three weeks. Zoning boundaries are typical of data that change more slowly and thus require less frequent updating. However, attributes may change more rapidly than geometry. Examples of this situation include new definitions, new registration methods, and changes of vegetation during a growing season. Another common case is that of property data, in which ownership changes more often than boundaries.

Lack of current information on maps has always troubled users. Studies (Bernhardsen 1985) have shown that the use of outdated 1:1000 and 1:5000 maps can cost each professional user an amount corresponding to one-third of the annual salary of a technical staff member. To correct such a lack, map data must be supplemented, often with in-field measurements. Given the number of professional users in the country as a whole, the aggregate cost of outdated geographic information could clearly be enormous.

The application of the data will determine what accurracy is required for temporal data. In any case, temporal data should carry accuracy designations.

11.8 Logical Consistency

ISO (2001) has proposed that logical consistency should comprise the following elements:

1. Conceptual consistency—adherence to rules of conceptual schema
2. Domain consistency—adherence of values to the value domains
3. Format consistency—degree to which data is stored in accordance with the physical structure of the data set
4. Topological consistency—correctness of the explicitly encoded topological characteristics of a data set

The abstract description of the real world features is called a *data model* or a *conceptual model*. A conceptual model described using a conceptual schema language (e.g. UML, EXPRESS, IDEF1x) is called a conceptual schema. *Conceptual consistency* thus includes adherence to the rules of the conceptual schema. This means that it should be stated when elements in a database are established on the basis of new rules, for example, new relations, as compared with what is specified in the data model/conceptual schema.

Domain consistency requires adherence of values to the value domain. This means that it should be stated if any of the data have values that exceed the specified value limits—for example, that an object type has a specified accuracy of $+/-$ 1m, whereas some objects are registered with $+/-$ 10m; or that permitted values are to lie within the area of 50 to 100, whereas some objects are registered with the values of 40 and 110.

Format consistency requires the data sets to be stored in accordance with the physical structure or the format specified. Should the data sets not meet format specifications, then they have reduced quality, which again, can cause problems for users. Any data set should therefore include a statement as to which storage format is used and any concrete deviation from this format. A number of countries are now in the process of developing standard transfer formats. Such formats are usually well specified.

Topological consistency requires the correctness of the explicitly encoded topological characteristics of a data set.

Various tests may be applied to address these questions, and the tests should be described.

A database that contains lines may have topological faults, such as:

- Overshoot and undershoot
- Polygons too small
- Polygons not closed
- Double lines
- Incorrect intersection of lines
- Lines too close

These faults affect topological consistency. Specific topological tests can be carried out to verify the following conditions:

- All links intersect at nodes.
- The cycle of chain and nodes is consistent around the polygons or nodes.
- Inner rings embed consistently in a closing polygon.

The consistency of logical relationships can be difficult to measure. In any case, quality depends on the use of a checking program as data are entered or processed through topology building; data of good quality are used to correct data entered to form topologies.

Logical connections in geometric data are essential for GIS users, although they are less important for others. Users of map-editing systems, for example, may manage perfectly well with spaghetti data, but poor-quality topology complicates or introduces error in many GIS functions, especially in overlaying and network analysis.

Consequently, data with unverified topology may cost a user dearly in terms of verification and correction work, or in terms of degraded quality of the final product. Topology consistency is best expressed as a percentage of junctions correctly formed, percentage of polygons correctly formed, and/or a tolerance used for merging adjacent areas.

Validity of values can be expressed as percentage conformance. The relations between attributes are often expressed (e.g., a building is located on a particular property, a road borders particular properties). Relational consistency is an important data quality and is best expressed as a percentage of correct relationships. The logical relationships of data must satisfy the requirements imposed on relationships between objects, which, in turn, are based on the tasks to be performed.

11.9   Completeness

Data *completeness* describes how fully data on an object type have been entered in relation to a specification—for example, whether attribute data as well as digital geometric data have been entered for all properties, all roads, all buildings, and so on, in an area.

ISO (2001) has proposed that completeness should embrace the following elements:

1. Commission—excess data present in a data set
2. Omission—data absent from a data set

Evaluation of completeness presupposes an accurate description, including a clear specification of the different object types and related attributes. Without such a description, it will be impossible to specify which object types or attributes are missing or, possibly, are surplus. Relative to the specification, completeness can refer to both under- and overcomplete data, and thus completeness is specified as either commission or omission. Redundancies are also to be found in data of poor consistency, and should be avoided. From a user's standpoint, it will be very useful to know which data may be missing or which data are in excess of the specifications. This information also states the data's applications and limitations.

Completeness may be verified by comparing the number of objects for which data are entered against the number of real objects specified in the area involved. The result of the comparison is then expressed in terms of a percentage. In general, the more attributes that are entered, the better the description; yet there are no objective measures that express how completely an object should be described. So verification of attributes is relative and expressed only in terms of a specification compiled for the purpose.

Completeness is essential for many purposes. For example, local or national regulations may require that neighbors be given notice whenever a property is divided, and GIS may be used to locate those neighbors. Should just one person be overlooked because of database incompleteness, the entire notification procedure may have to be repeated. Where a map is the data source, it should be remembered that generalizations and selections made during map production can affect completeness.

Data completeness also describes the completeness of ID coding and thematic coding (both are really attributes). Linking geometry and attributes correctly depends on the same ID codes being used in both sets of data. Consider, for example, a situation in which road geometry data are stored in one database and road attribute data in another. The roads are divided into numbered segments, and the numbers serve as common ID codes. When new roads are built, new ID codes are added in the databases, which should be updated simultaneously. If they are not, link inconsistencies may arise, diminishing the quality of the data stored jointly in the two databases. Such problems arise most readily whenever several agencies or users enter data into common databases at different times.

Consequently, data completeness should also be recorded for data entered and stored, to inform users of the limitations entailed. It is also advisable, whenever geometry and attribute data are to be linked, to compile tables of any missing links as data are entered. The consistency of linking is customarily stated in terms of a percentage (e.g., "99% successful links"). The quality report must include information about selection criteria, definitions used, and other relevant mapping rules.

The qualities of completeness and resolution are often interrelated. For example, the specification may state that all properties with an area of a half hectare or more must be registered. Some areas may be completely covered, yet the coverage of all properties within the boundaries of the area may still be incomplete. In this case, the observation density is not sufficient to depict each property.

11.10 Data Quality Overview Elements

The following data quality overview elements are particularly relevant to use in the description of nonquantitative quality data sets:

Purpose
According to ISO (2001), purpose describes the rationale for creating a data set and contain information about its intended use. For example, a land-use database can be established to produce area statistics and to *support GIS applications*. A data set's intended use is not necessarily the same as its actual use. Actual use is described using the data quality overview element. Information on the purpose of data establishment will give new users an understanding of what the data can be used for.

Usage
Usage describes the application(s) for which a data set has been used. It also describes the intended usage of the data set by the data producer or by other distinct data users. In addition, the actual use of the

data will be an important source of information for new users. A land-use database can be used as one of several sources to produce municipal area plans and national topographical maps in scale 1:50 000.

Lineage

Lineage describes the history of the data set, and origin comprises a description of the source material, including date, original application, data capture method, data producer, and treatment method (e.g., transformation algorithms). Digital map data can thus be established on the basis of photogrammetric measurements or by digitizing existing maps. Attribute data can be taken from public registers or from direct in-field registrations.

It is not sufficient to know the original data source; we also need to know something about how the data are collected—for example, what type of instruments are used (e.g., type of surveying instrument, type of aerial camera). Data are often treated before they are stored in the databases. The treatment method is important not only for the accuracy of the end result, but also because knowledge of the method makes it possible to reconstruct original data and, possibly, to carry out new (and better) calculations. For example, documentation of a transformation algorithm should be available on request by a user obtaining digital data.

Digital map data generated directly from aerial photos or entered during surveying entail fewer steps and are consequently less subject to error than data from digitized maps. From the standpoint of accuracy, original sources are *always* preferable to maps.

11.11   Accessibility

From the user's point of view, information on accessibility can be just as important as the data overview elements and other quality parameters. *Accessibility* describes where data for a particular area can be found, who has the original rights, and what other conditions and limitations (price, form of delivery, format, etc.) apply to the acquisition and use of the data. These factors are closely linked to standard formats, meta-databases, and use of the Internet. We should also reemphasize how important the quality marking of data is, with respect to positional accuracy, attribute accuracy, temporal accuracy, logical consistency, completeness, and accessibility.

11.12   Probable Sources of Error

Error is inevitable in all measurements, either quantitative or qualitative, and may range from the serious to the negligible. Yet neither error occurrence nor magnitude can be known in advance. Errors and omissions degrade the accuracy of georeferencing, attribute data, logical consistency, linking, coverage and resolution, and timeliness.

Error may be introduced in any phase of data processing. The following conditions (adapted from Burrough 1986) may affect data quality.

A. Original error sources independent of GIS processing
  1. All surveys and field registrations. Errors may be ascribed to:
     - Instrument inaccuracies
     - Satellite sensor systems
     - Aerial cameras
     - GPS
     - Surveying instruments
     - Various instruments for measuring attribute values
  2. Data processing, as when producing maps used as GIS sources. Errors may be ascribed to:
     - Mapmaking
     - Errors in computations and geodetic networks
     - Mapping instrument inaccuracy
     - Map drawing inaccuracies
     - Data editing
     - Computations
     - Enlarging/reducing and redrawing
  3. Errors due to changes in the field:
     - Registered objects change character
     - New phenomena arise
  4. Errors introduced due to lack of coverage or resolution
B. Errors introduced in GIS data processing
  5. Errors in data entry:
     - Inaccuracy in digitization
     - Equipment error
     - Operator error
     - Inaccuracies in entering attribute data
     - Human error (lack of verification routines)
  6. Errors in storing data:
     - Insufficient computer numerical precision
     - Storage medium errors
  7. Errors in data manipulation:
     - Raster to vector
     - Vector to raster
     - Generalization and thinning
     - Combining classes
     - Overlayering
     - Interpolation (contours, elevations, etc.)
     - Analyses of satellite and other data

8. Errors introduced in data presentation:
   - Plotter inaccuracy
   - Paper and other presentation media inaccuracies
C. Errors in methods
9. Errors in methods used to collect data:
   - Insufficient observation density
   - Ill-defined objects and classes
   - Insufficient expertise of data compilers
   - Uncertain borders between areas

11.13 Quality Control

Normally, it would be better to control the data during rather than after production. Data can thus be supplied with a "quality guaranty," and users will be able to assess the data's areas of applications. Control during production presupposes that the producers have established a quality control system. Such a system will contribute to the efficient production of data according to specific requirements and ensure that it is done right the first time. This assumes a product specification or an approved standard for the data sets available. A quality control system includes routines and guidelines that describe working operations, thus preventing errors, attaining the right quality, and ensuring efficient control.

It is well known that geographical data are processed (transformed, etc.) and analyzed several times and by different institutions. Ideally, therefore, each process should be subject to quality control.

A few simple and general rules may help to reduce the problems of inaccuracies:

- Employ verification routines to ensure quality.
- Verify data as early as possible.
- Verify data at several stages of their manipulation.
- Know the nature of the data, be it geometry data or attribute data.
- Be critical in all data uses.
- Apply processing results carefully.
- State inaccuracies associated with results and analyses.

Without control measures, there is no guarantee that specified quality requirements will be met. Checks should certainly be carried out during production, but they can also be necessary when the data are first being put into use by the end user. Even though the data meet specific accuracy requirements, they can still be used incorrectly because they are not understood properly or because users lack the requisite expertise. Even hypothetically error-free data may be used incorrectly. Good data descriptions and specifications of accuracy, as well as user expertise in the relevant applications, are the best assurances of correct data use.

CHAPTER 12

Database Implementation and Spatial Indexing

Within the framework of GIS, data are divided logically into two categories: geometrical data and attribute data. This division may extend to a similar division of physical storage, although the relationships between the two categories of data must be preserved regardless of whether the division is logical or physical. In current GIS, this condition is satisfied through one of five approaches to the overall database storage involved:

1. Geometrical data and attributes stored in the same file/database
2. Two separate database systems, one for geometrical data and one for attribute data in a relational database
3. A single relational database that stores both geometrical data and attribute data
4. One database for geometrical data connected to several different databases for attribute data
5. Several databases for geometrical data and attributes joined into one system—a distributed database solution

GISs that have geometry and attributes stored in the same file also have a tailor-made file-based management system (Figure 12.1). The advantage of this solution is that rapid search is usually included in the data structure. A series of operations and analyses can be performed on both geometry and attributes. The link between geometry and attributes is very close and special for each system. To increase the search and record speed, data are split into different layers and area units. Map sheets are often used as a storage unit, but it will be very difficult to change the size of the storage unit later. Because of the primitive management system, the same object is, as a rule, maintained on several files. This system can be very work-intensive and vulnerable to errors of omission. Therefore, this kind of tailor-made solution is best suited to project-related activities and, possibly, to small-scale activities requiring little updating.

Systems that store geometrical data and attribute data in two separate databases are known as *hybrid systems* (Figure 12.2). The databases are linked by unique identifiers. Systems with separate databases normally

Figure 12.1
Tailor-made file-management system for geometric and attribute data both within the same database.

employ relational DBMSs that are commercially available for attribute data and tailor-made file-based systems for geometrical data. Generally, it is possible to achieve a more rapid response time by using file-based systems than by using relational DBMSs. This is particularly important for geometrical data where large volumes of data are involved and users are dependent on rapid search and recording on the screen. The hybrid systems include both vector topological data and raster data. As in the single database solution, the geometrical data are often split into different layers and area units (i.e., map sheets, etc.) for storage. This division will normally be invisible to the user; he or she will experience the data-

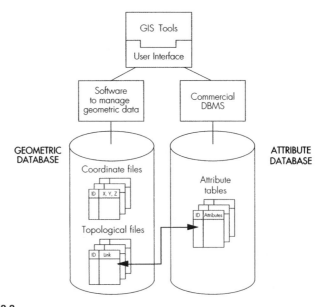

Figure 12.2
Hybrid database model, where geometric and attribute data are stored in separate databases.

base as seamless. Quad-tree-based techniques are also used in some hybrid systems for storage of the geometrical data. Moreover, with the hybrid solution the same object is maintained on several files, and IDs have to be frequently maintained in external registers.

Storage of attribute and geometrical data in the same base is known as an *integrated system* (Figure 12.3) and is realized in a relational database management system (RDBMS). Coordinates and topological information are stored in relational tables, and attribute data are stored in separate tables. Systems that combine geometrical data and attribute data in a single database employ a relational database with a geometry of nodes and links that is structured topologically. It is relatively easy to exchange data between different systems using integrated solutions because these use standard relational databases. It also makes it possible to use the flexible question language SQL, which now, through SQL3, is being extended to be able to handle spatial questions—for example, to find the object that is located closest to a given point. The data will be seamless—that is, not limited by map sheets or other units, but still with the same object maintained on several files. The DBMS tools for databases storing both geometrical and attribute data in different databases tend to be specialized in the sense that special solutions have been developed to handle geometry efficiently. There are now several products on the market that offer RDMBS functionality for geographical data types through an intermediary program in the form of a geographical database, Application Program Interface (API). The handling of topological relations is especially demanding for normal relational databases.

There are also certain systems based on object structuring with storage in relational databases. Such object-related systems have a series

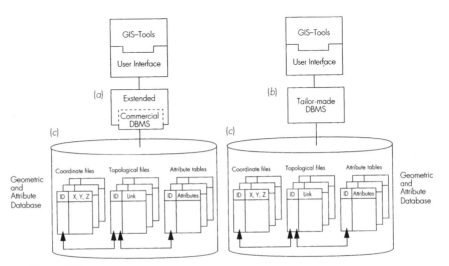

Figure 12.3
Integrated GIS model, with alternative solutions for a database management system.

of positive characteristics, such as an efficient multiuser environment, seamless administration, ease in connecting more registers, and a high degree of integrated security in respect to maintenance. However, these systems are often large and cumbersome and require particularly good working routines.

Relational databases are employed whenever attribute data alone are stored. The current trend in GIS follows the general information systems' propensity toward generalized, flexible programs and tools. This means, in practice, the increased use of relational databases. Some systems feature special functions dedicated to geometrical data, which enhance yields in manipulation and presentation. Such special functions, however, impose constraints that may complicate their use with general programs. Systems that employ a single database or commercially available DBMS tools are often more easily adapted to work with newer database systems.

Today the trend is to use standard database management systems such as Oracle, Informix, Sybase, Object Store, and DB2 for storage of common data, both attribute data and geometrical data. Often, the purpose is to store data in the simplest possible structured form. By using simple storage structures in standard DBMS, the basic data model and applications become less dependent on each other.

Normally, basic database design may be divided into four categories: stand-alone, centralized, joint, and data "viewing" systems (Figure 12.4). The choice between the four depends on design goals. The joint database and viewing designs (Figure 12.4 c and d) are the most complex, but normally provide the greatest benefit. The solution in Figure 12.4d is based on the viewing system's being able to read directly a series of different data formats. This gives many external users

DATABASE DESIGN

(a) stand-alone   (b) centralized   (c) joint   (d) direct view

Figure 12.4
Basic database design may be divided into four categories: stand-alone, centralized, joint, and direct view. The choice among the four depends on design parameters. The joint database and direct view design provide the greatest benefit but is the most complex.

simultaneous direct access to the data in the original databases. It also avoids the problem of updating both the original databases and the common database (as in solution 12.4c) and allows the user to choose the subject system that is optimal in relation to the different tasks. The solution is often based on a simple viewing tool with limited functionality. Updating and analyses of data have to be carried out in the individual theme systems.

12.2 Distributed Databases

The modern data network has made possible decentralized database solutions. Distributed database systems constitute a special decentralized solution (Figure 12.5). A system with a distributed database comprises several databases on different computers closely integrated with the assistance of a network and treated as one unit. Users experience this as if they are working with one database. The viewing solution (Figure 12.4d) is a "light" version of distributed databases.

Subdatabases may have different database management systems; copies of the same data can be found in several of these subdatabases and the data can be stored in different formats. This level of complexity makes particularly high demands on control routines in the joint database management system. Distributed solutions have special requirements with respect to common data models, routines for maintaining and checking data quality, routines for access control, and so on. All this can result in the system being relatively expensive to run. However, there are a number of advantages, such as increased total capacity, increased reliability and access to data, increased efficiency and flexibility, and the advantages of pooling (sharing) resources.

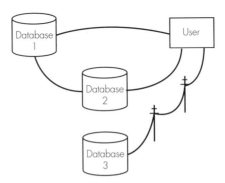

Figure 12.5
Storage can be divided among several producers or users. The manipulation (search, selection, transport, and organization) of common data is carried out by a DBMS in a data network. Distributed databases are managed as a logical database, which is stored at several locations.

12.3 Databases for Map Data, Indexing

Databases for digital map data should be able to manipulate records of varying length efficiently. For example, line length may vary considerably, resulting in a corresponding variation in the number of coordinates entered. Standard databases cope poorly with such variations, which is why many GIS suppliers have developed their own databases for digital map data. These database systems reflect geographical reality by such means as requiring that data on objects of the same type, such as the lines forming a property boundary, be stored in close proximity in the database, to speed up the response.

Tailor-made database systems using relatively simple file structures are often used for geometrical data. These databases are usually based, to a large extent, on network or relational databases or a combination thereof. To carry out analysis of the data, such as overlay, network analysis, and so forth, the data are stored in a topological structure with node tables, link tables, and polygon tables, as well as coordinate tables, as described in Section 4.2.3. As far as the geometry is concerned, the spatial search criteria also often have certain demands for special structuring (indexing) of the data. In the case of attribute data, the database system's own indexing technique is often used and can be quite efficient.

General points about indexing

A prerequisite for the efficient use of GIS is that information search is rapid. Users normally expect the interactive operation on screen to be immediate—that is, that the result must be shown within a maximum of 2 to 3 seconds. A common problem with GIS is to find a particular object within a limited area or an object that is located closest to a given point. There are often such large volumes of data involved in GIS that the use of traditional search functions take an unacceptably long time. If spatial comparisons must be carried out in addition (for example, based on criteria such as "within an area," "closest to a point," "overlapping with," etc.), then this will take a substantial amount of time if the data are not organized in an appropriate way. Arranging data in an additional structure is known as *indexing*, and for geometrical data it is known as *spatial indexing*.

The objective of indexing is to reduce the number of candidates for testing when searching through large quantities of data. Index structures can be added to the data whenever they are wanted. Familiar examples of indexing are the index of a book, the organization of a data catalog according to different keys, and the indexing of addresses on a town map. This latter is quite similar to indexing a database with geometrical data.

Geometry can be indexed by sorting coordinates in rising order, then carrying out a binary search (see Chapter 4) or by indexing with the help of geometrical areas. Separate indexing of x and y coordinates is not particularly efficient, because it produces many candidates for testing. A relatively long search time results and the method is therefore little used. A relatively simple and often used technique in GIS is indexing with the help of a regular grid with a suitable number of points in each grid area. Indexing via "quad-trees" is also an efficient and frequently used technique; however, it requires some extra resources.

12.4   Partitioning and Indexing

We have ascertained that spaghetti data require a long search time inasmuch as the data are stored in a relatively casual and unconnected sequence in the file. The time used to search for and retrieve topological data is also governed by the way in which the data are structured for storage. A rational data structure will reduce the storage volume. Special techniques have therefore been developed for dividing and structuring data.

Generally, geometric data are stored in map sheets or other geographical units, but storing map sheet data in single sequential files lengthens the response time. This has resulted in some GISs employing indexing to speed up the searching process, and enabling current map sheets to appear on screen almost immediately. Indexing specifies locations, so data are divided into sections that are distributed in such a manner as to accelerate the search. For example, *zooming* focuses on data in those sections relevant to a selected area and ignores the rest of the data volume.

Hashing

The underlying concept for hashing is that a key is used to look up, in a table, all objects that share the key value, and thereby to be able to locate the data physically on a disk. Each cell section is assigned a number, which is listed in a table of pointers to storage locations (Figure 12.6). The cell section numbers, and hence the entries in the index table, may be calculated from Cartesian coordinates.

The following is an example of the use of index tables to search in a point cloud. A single function, known as the *hash function*, can be set up based on dividing the map area into equally large rectangles. The entry value for a point in the index table can be calculated as follows:

$$Hx(x) = \left(\frac{X - X_{min}}{X_{max} - X_{min}} \right) \cdot N \qquad (12.1a)$$

$$Hy(y) = \left(\frac{Y - Y_{min}}{Y_{max} - Y_{min}} \right) \cdot M \qquad (12.1b)$$

where N equals the number of grids in the x orientation and M equals the number of grids in the y orientation. The totals should be rounded

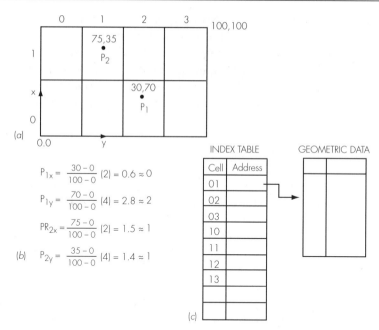

Figure 12.6
The manipulation of digital map data is made more efficient by storage in small squares according to a special addressing principle. The index table enables the system to access the data quickly once it has established in which square the data are located. To avoid conversion of the origin, the rows are numbered from the lower left corner.

down to the nearest whole number, and the origin converted if different for the Cartesian coordinates and the grid. To be able to divide the area into grids, the largest and smallest of the coordinate values must be specified. As a rule, the map area is divided into square grid cells, but rectangular grid cells allow for a more uniform division of the keys in the index table.

The time required to find all the points within a given window and bring data on screen increases as the amount of information in the geometric sections grows. In other words, the smaller the sections, the more rapid the search. But smaller sections mean more pointers, which also slow retrieval, so compromises are often necessary.

The method of dividing a map area into regular units takes its strength from its simplicity. The weakness of this method is that information is not available on the data content in each grid and that all grids must be the same size, regardless of whether they contain a small or a large quantity of data.

Trees and Morton Index
If the principle of constant grid size is abandoned, the grid's addresses cannot be calculated by the simple method described in the preceding

paragraphs. The alternative will then be to choose a tree structure. In this section we deal only with the indexing method. (Quad-tree is discussed in Chapter 4.) Each square may be assigned a number according to the *Morton index*, which consists of a successive number of squares as they are quartered (Figure 12.7). Numbers of successive quarterings are combined, so that the index number indicates the size of the square (a low number indicates a large square), as well as the relative geographical location and hierarchical location in the pointer table. Larger areas will have pointers located higher in the table so their data may be retrieved rapidly from the table. The attributes of neighboring squares on the ground are stored as neighbors in the table. In general, the search efficiency for trees is inversely proportional to the number of levels in the tree, because that is the number of disk accesses required to locate a given key.

There are a number of variations of tree indexing that are used for structuring and indexing of spatial data, such as the so-called B-tree and R-tree. These are not described further here, but one should take note that indexing with the tree technique is used for both vector data and raster data. GIS suppliers and the special database suppliers are increasingly making use of this technique.

Conclusions

The use of traditional hashing techniques and trees makes it very difficult to handle divided areas that overlap. However, routines have been developed that can handle overlapping data relatively efficiently. These have also been implemented for object-oriented solutions. In recent years, more powerful and rapid hardware has made it easier to use simple data structures for storage, but many GISs still use different "smart" solutions to obtain rapid access to data stored on the disk.

No current database system or structure completely meets the needs of database applications. There are grounds for suspecting that the excessively complex and voluminous data collections of many GISs may

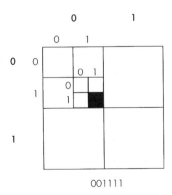

001111

Figure 12.7
Morton index with numbered cells. Black = 001111.

be ascribed to the databases employed. Clearly, further database development is in order. One goal may be to develop better object-oriented database systems.

12.5 Database Design

To solve typical tasks efficiently, GIS databases have to be designed in an appropriate way. Normally, database design comprises three main elements:

1. *Conceptual design* is hardware and software independent and specifies the end user's needs. Usually, conceptual design includes the same elements as those included in data structuring/data modeling.
2. *Logical design* is software dependent but hardware independent and sets out the logical structure of the database elements determined by the different GIS software.
3. *Physical design* is hardware dependent and includes file structure, memory and disk space, access, and speed, among other attributes.

Conceptual and logical design are described in the following section, and physical design is discussed further in Chapter 18.

12.5.1 Conceptual database design

Even though good DBMSs have been developed for GIS, the usefulness of any system will be no better than the data entered into the database. To be useful, data must be organized. Disorganized data in DBMS are unproductive; in other words, garbage in, garbage out. For example, if the intention is to show all water pipes in a network that are 10 cm in diameter, the result will be of doubtful value if the pipe data are entered sometimes as inside diameters and at other times as outside diameters. A transport analysis can be performed only if elements of the road network—its bridges, tunnels, and ferry routes—are entered as secondary parts of the primary network, not as independent elements. If all the properties on a specified part of a street are to be identified, the data must contain stable, common identifiers linking streets and properties. Similarly, if all the private roads of a township are to be identified, the definition of what constitutes a private road must be stated clearly.

Data collected and entered in a database must be processed and arranged to provide information that is meaningful for current use as well as for tasks still to be defined. This information may include the answers to questions relating to:

- The location of a particular object
- Attributes of a particular object

- Attributes of all objects in a region
- The identification of regions containing objects with attributes *A* and *B*
- The quickest route between points *A* and *B*
- The identification of all objects of type *A* contiguous with objects of type *B*

Clearly, then, data planning and structuring are essential to successful use in GIS, and almost all GIS projects should begin with careful analyses of how data are to be entered. The data model and its related structuring must be independent of the software and hardware chosen. In many ways, the problem is the converse of selecting software and hardware to suit the data model.

Data may be structured using the following procedure:

1. *Defining areas of application.* First, the real-world situation must be limited to the area of application (e.g., the operation and maintenance of water supply and sewage networks, property management, transportation planning). Second, if the database is to contain several areas of application, each must initially be treated separately.

2. *Constraints.* The tasks involved and the products to be produced are then identified for each area of application. Examples of constraints include:

   a. *Tasks*
      - Administrative and operational tasks
      - Statutory public tasks
      - Planning
      - Design tasks
      - Dissemination of information

   b. *Requirements*
      - Requirements and needs for standardized map products
      - Requirements and needs for standardized reports

3. *Criteria for assigning priorities.* As each constraint identifies a number of object types with their attendant attributes, only relevant data should be entered in the database and the numbers of object types must be constrained. Thus, the data on a road surface are important, but not as important as the width of the road, which, in turn, is not as important as its geometry and location. Objective criteria must then be chosen to determine the objects included. Here, cost-benefit analyses are often a good starting point. In practice, however, the problem is often the converse: How can the necessary data be acquired?

4. *Descriptions of object types and attributes.* The object types chosen and their attributes must be described. Each type of object comprises a

collection of individual entities to be treated equally, given names, and defined. For example:

- Object type: road
- Definition: All public and private roads passable for cars, excluding detached roads and roads shorter than 50 m

Attributes are unique descriptions of objects. Each attribute has a definition and a permitted range of values. For example, the attribute description for the object road may be:

- Attribute: load capacity
- Permitted range: minimum 0 kg, maximum 20 tons

5. *Coordinating terms and definitions.* The various object terms and definitions must be coordinated, particularly for common data. Disparities, of course, are inevitable. For example, although parts of roads may be regarded as property boundaries in property administration applications, they would certainly be defined as roads in road and transport applications.

Attribute overviews must be uniform. All attributes for a single object type, such as for roads, should be assembled and globally assigned permitted values. Uniform terminology may be difficult to achieve. For example, does a pedestrian area that is accessible to emergency vehicles classify as a road?

6. *Geometric representation.* Rules must be established to determine how objects are to be represented geometrically and which basic geometric elements are to be used. In principle, the choice is between:

- Vector representation (points, lines, areas)
- Raster representation

Entities comprising several objects must be allowed.

In principle, the areas of application determine how objects are to be represented, but other elements, such as costs and ease of updating, must also be considered.

7. *Relations.* Relations between objects must be defined before tasks can be addressed. Relations may include:

- Composition of objects (e.g., a country comprises counties, a county comprises townships, and a township comprises statistical units)
- Locations of objects (e.g., a particular building is located on a particular property, or a planned development is in a recreation area)
- Affiliations of objects (e.g., which addresses are particular to a street, which stop valves are parts of which water pipes)
- Neighbors of objects (e.g., which properties border on a particular property, which type of land use borders on a preservation area)

Some relations may be computed from digital map data, as when forming topology, whereas others must be entered as attributes.

Sometimes it may also be necessary to distinguish between actual relations and potential relations. Concrete user needs should dictate the selection of the relations described. Relations that are not used need not be described.

8. *Quality requirements*. The data must be subjected to quality requirements, including:
   - Positional accuracy
   - Attribute accuracy
   - Temporal accuracy
   - Logical consistency
   - Data completeness

9. *Coding*
   - Code lists of the geometric object type designations must be compiled. The code system should be based on the data structure defined in steps 4, 5, and 8.
   - Code lists may also be compiled for attributes and for descriptions of relations.
   - The identifiers linking geometry and attributes must be designated, and if possible, their coding delineated.

Completion of steps 1 to 9 results in an object catalog that contains descriptions and codes for all objects to be entered in the database. Parts of the data structuring (steps 4 to 8) can be perceived as data modeling and are shown by the use of graphic diagrams. One of the most widely used approaches to forming a conceptual model is known as the *entity-relationship* (ER) *model*.

The ER method involves defining the entities that are to be included, together with the relationship between them. In the ER model a standardized diagram is used whereby entities are expressed as rectangles and attributes as circles. The relation type (1:N, N:1, etc.) is shown by arrows on the connecting lines between the figures. In the ER model, a construction of the data model is made, which consists of many sketches (ER diagram). To illustrate the ER model, one can use the example of the object type "built-up area" (see Figure 12.8).

Each element of an ER diagram has to be described. The following descriptions apply to Figure 12.8:

1. *Definitions*
   a. *Built-up area:* a chosen set of buildings spatially described by a generalized area outline (a subset of the "land" class).
   b. *Building:* any manufactured construction except those related to transport and utilities, which are described under the object type "transport and utilities."
   c. *Building geometry:* a set of geometry primitives (polygon) describing the outline of the building.
   d. *Building center:* a point representing the spatial position of the building.

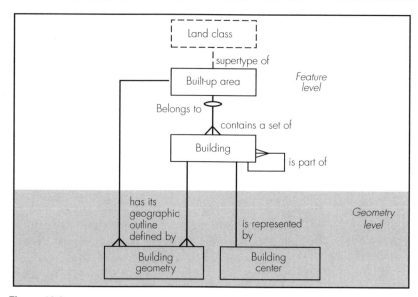

Figure 12.8
Data model for buildings, based on the entity-relationship (ER) method. (From National Standard Zimbabwe, Digital Geographical Information.)

2. *Description of the model.* A building is normally described by its roof outline. A small building can be represented by a center point that will represent the spatial position of the building. Built-up areas are used for representing groups of buildings where the density at the actual scale does not make it possible to represent each building.

Once the data have been structured in a theoretical model, it remains to:

- Realize the logical structure in a GIS with a concrete database application system—*logical design.*
- Choose the physical files to be structured on disks for optimum utility—*physical design.*

Experience indicates that structuring entails problems, the most common of which are that:

- Existing maps are not always good data models.
- Terminology is imprecise and not based on common concepts.

The object-oriented model is an extension of the ER model.

12.5.2 Logical database design

Logical database design is software dependent. In most GIS database systems, data are organized in geographical units as map sheets, administrative areas, or other delineated areas. The sizes of the areas involved may vary considerably within a single database. However, from the user's viewpoint the data involved appear seamless; that is, the user does not see the borders between adjoining areas or sheets.

The size and shape of storage units affect data storage efficiency and accessibility and the ease with which data may be updated. Fixed unit size is most efficient whenever data are spread evenly over the entire region covered by a database. Small units are an advantage whenever data density is large. The scale of presentation and the sizes of areas that may be processed simultaneously also influence the choice of unit size. In practice, units are often chosen to correspond to the map sheets of the region covered by a database. The geographical storage structure should be chosen with care, as established database structures are difficult to modify.

In most GISs, data are also organized in thematic layers, which correspond to the overlays of traditional map production. The choice of topics for thematic layers determines how efficiently their data may be used. Each thematic layer comprises types of objects that are to be processed together, so data use determines the allocation of objects to layers. For example, one application may benefit most from having all linear objects (e.g., roads, rivers, coastlines) in a single thematic layer, whereas another application may separate rivers and coastlines into separate thematic layers.

The greater the variety of uses of the data in thematic layers, the greater the number of thematic layers. However, data may be collected in a common thematic layer if they are:

- Always used together
- Logically related
- Updated simultaneously
- Valid for several functions (e.g., lines indicating roads, property boundaries, land-use boundaries)
- Used together in presentation (e.g., as all 0.3-mm-wide lines, all areas colored red, all place names)

Logical database design also includes specification of the coordinate system and various tolerance definitions, such as those that coordinate movement or the presence of weeds.

The uses of a database may change with time, so databases should be designed to be modified to incorporate new object types and attribute types, as well as to allow for enlargement.

12.5.3 Physical database design

Physical database design is hardware dependent. Database design depends on computer hardware with respect to:

- Data volume
- Storage medium

- Storage capacity
- Accessibility and speed
- Load of data (e.g., large and dense maps)
- Single or multiple users
- Number of workstations
- On-line use
- Off-line use
- Storage medium
- Network

Different systems employ different approaches to data storage, and storage methods often may be modified or expanded. For example, at present, the storage capacity of stand-alone PCs is limited by the storage capacities of hard disks. Memory capacity in the future may be considerably greater as CD-ROM, optical disks, and other storage media are employed in PCs. Even with current technology, effective PC storage capacity may be increased considerably by connecting a PC to a local-area network that has a larger server with greater storage capacity. In many cases, the constraints imposed by data storage facilities are more administrative than technical, as organizations implementing new systems frequently wish to employ existing data processing infrastructure. Physical design is discussed further in Chapter 18.

CHAPTER 13

Housekeeping Tools

From the user's viewpoint, the ideal GIS should include enough functions to perform all conceivable manipulations of geographical data. In practice, user needs comprise various tasks. So, as listed in Figure 13.1, an overview of user tasks defines the overall GIS requirement (Larsen 1991). The first two tasks in the table are customarily associated specifically with GIS and hence are often incorporated into infrastructures under the jurisdiction of national map agencies, military agencies, or public utilities. The remaining tasks are supported to a greater or lesser degree by the various GISs now on the market.

As a software system, a GIS usually comprises a set of software tools, including:

- Database and database management system (DBMS)
- User interface
- Query language (QL)
- Application functions and programs

Databases, DBMS, user interfaces, and QL have been discussed previously. The basic application functions employed in GISs are discussed in this chapter; the more specialized analysis functions are the subjects of Chapters 14 and 15.

The basic application functions often employ standard functions supported by the DBMS, the QL, and the user interface. The wide range of needs encountered requires numerous special GIS functions. Only a few GISs have all possible functions. In all cases, each function may be executed by one or more commands. The basic GIS functions may be divided into four main categories:

1. Functions for data registering, entering and storing
   - Registration and verification
   - Entry (including format transformation and similar operations)
   - Organization of storage operations
2. Functions for correcting and adapting data for further use
   - Geometrical manipulation
   - Editing
   - Edge matching
   - Map projection transformations

User need (task)	Data manipulation required
Access catalog	Locate relevant data (e.g., metadatabase)
Requisition	Send data requested
Enter	Enter and/or register in the user system data selected
View	Study and evaluate data received or registered
Select	Choose data required
Correct	Rectify errors and omissions
Compile	Compile dissimilar data to a model or image of the real world
Overview	Use model or image to provide a general overview of an area
Monitor	Use model or image to monitor occurrences in an area
Navigate	Use model to locate position of vehicle or ship and to determine quickest route between designated points
Analyze	Evaluate connections, possibilities, conflicts, and consequences
Make decision	Ensure that data and data processing support choices of alternative actions
Present	Present overviews, analyses, and decisions in graphic form (maps and/or reports)

Figure 13.1
User tasks required by GIS.

- Coordinate system transformations
- Attribute editing

3. Functions for processing and analyzing data
   - Query and report extraction
   - Processing geometric data
   - Processing attribute values
   - Integrated processing of geometry and attributes

   Processing and analyses comprise, among other factors:
   - Measurement of length, areas, peripheries, etc.
   - Altering attribute values
   - Deleting and compiling objects
   - Line thinning and smoothing
   - Statistical computation on attribute values
   - Topological overlay
   - Network analyses
   - Connectivity operations
   - Terrain surface operations

4. Functions for data presentation
   - Use of symbols (cartographic variables)
   - Text insertion
   - Perspectives and other drawings

The functions for data presentation are dealt with in Chapter 16.

The four main categories in the preceding list are based primarily on the various phases of most GIS projects, but in practice the categories may be mixed. For example, functions for editing geometry may be employed both in editing and in presenting data.

13.2 Data-Entry Functions

A GIS may be implemented with software supporting various forms of data entry, including entry from:

- Digitizing table and on-screen digitizing, including line following
- Surveying stations and GPS, including computation of coordinate geometry
- Scanning, vectorization, and pattern recognition
- Miscellaneous entry program

The relevant programs include facilities for expressing original data in ground coordinates, for entering codes, and for performing quality checks. The software systems also have flexible modules for entering assorted types of attribute data, including the presentation of various screen forms.

13.3 Importing Existing Digital Data

Most GISs have characteristic storage formats, specially for the geometry. Thus, all GISs contain software that permits the import of existing digital data. The relevant processes entail transforming various exported formats to the GIS internal file format. Many GIS programs can have international standard formats such as ASCII, DIGEST, SIF, DXF, SHAPE, SIF, and HPGL for vector data, and TIFF, JPEG, and CGM for raster data. In addition. national GIS suppliers/agents have often developed programs to read special national formats. Most GISs can import attribute data from standard relational databases and from spreadsheets and word processing systems.

The ability to read different formats can often significantly enhance a program's user-friendliness, as most GIS users will find that they have to collect data from many different sources. Suppliers of GIS software have started to take this into account, and today it is possible to find solutions that are able to handle simultaneously many types of data warehouses with different formats. International GIS suppliers have now formed the Open GIS Consortium, with the aim of making systems less dependent on system-specific storage formats. In addition, in many countries standards are being developed for national exchange formats.

13.4 Organization of Data Storage Operations

Software systems often organize data to ensure effective use. Such organization may involve various logical paradigms concerning the grouping of object types and the divisions of geographical areas. The physical limitations of system file capacities may also be a practical reason for thematic and geographic divisions. A list of all maps in a system, organized by location and theme, forms a map library from which the user can select the map he or she needs and store it in the workspace of the computer.

Thematic layers

Data in most GISs are organized in layers (levels), much like the overlays of conventional mapmaking (Figure 13.2). Similarly, individual data layers are stored in individual data files. This is done to avoid the problem with overlapping polygons, to reduce the actual data volume with which the computer has to work simultaneously, and to structure the data. These layers may, for example, contain object types intended to be processed together, such as points in one layer, lines in another, and polygons in a third. Alternatively, the individual data layers may be organized by theme—perhaps one layer for topography, another for property boundaries, others for roads or types of land use, and so on. Furthermore, each layer may contain subsidiary layers in a hierarchy. Thus, a layer for roads may encompass subsidiary layers for national, county, urban, and private roads. Layers of this type are often called coverages.

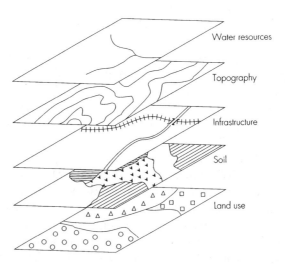

Figure 13.2
How data are organized. Many systems organize data into different layers (files). Their efficient use often depends on the well-planned organization of data into a layer system.

Collecting logically similar objects can reduce the amount of data required to describe an individual object. Objects that represent several themes, such as lines that are simultaneously roads and property or land-use area boundaries, may be collected in one layer. The line geometry of that layer may then be transferred to other layers as needed. Objects that are updated frequently or from the same source of information may also be collected in a single layer to facilitate updating work. The cartographic effects of plotting are frequently dependent on the sequential plotting of layers containing like objects.

The separation of data into layers may seem analogous to the traditional separation of map information into overlays, and therefore not always a realistic data model of reality. One of the reasons for this layered storage is, as mentioned, that many earlier systems have not been able to store overlapping polygons in the same layer. Today, however, there are systems that can handle this problem. These GISs circumvent "map overlay thinking" by being more object oriented; that is, each object is manipulated as an independent entity with regard to both its geometry and its attributes.

Partitioning the area

In GIS it often necessary to work with very large quantities of data. Many GISs have facilities that will divide surfaces (geometry) into smaller units to reduce the actual data volume with which the computer has to work. Individual surface segments are then stored in individual files, division by map sheets being the most common. Some GISs support divisions of data structures into projects, each of which may then be further divided into subprojects. This kind of division will also facilitate updating of the data.

The manipulation of data for a larger area often involves combining data for their constituent segments. This is done either manually by the user or automatically by the system. In spite of this division, users often, in practice, experience that data as seamless, even though the stored data may be divided into map sheets (Figure 13.3), which in

Figure 13.3
How data are organized into separate map sheets. Many systems organize data into surface units such as map sheets. Careful consideration must be given as to which storage element is most suitable, because restructuring to other elements is difficult once the data have been organized in a specific way.

turn may be divided into indexed cells in grids (see Chapter 12). During the drawing of larger areas on the screen, however, the map sheet structure will normally come into view.

Users must select the most suitable elements for storing data, such as the map sheet sizes and area divisions. Such choices are vital for two reasons. First, the organization of data storage elements can have a considerable influence on the efficiency with which the data are used. Second, once the storage elements are chosen and the data have been stored, restructuring to other storage elements is extremely complicated.

Hybrid systems, with the geometry in its own database, organize the data in map sheets, projects, and so forth. Integrated systems, however, with both attributes and geometry stored in the same relational database, have seamless storage. Both solutions can have, in addition, different types of geometry indexing to increase efficiency, but this does not affect the users.

13.5 Functions for Correcting and Adapting Geometric Data for Further Use

Data are ready for use only after they have been verified and, if necessary, corrected. All GISs contain programs for adapting data. These include:

- General utility functions
- Editing and correction of errors
- The ability to create topology
- Transformations to a common datum and a common coordinate system
- Adjustments at map edges or between other storage elements
- Coordinate thinning and line smoothing

13.5.1 General utility functions

Like all computer systems, GISs have collections of programs that provide such useful functions as file manipulation, editing, and program cross-referencing. GISs incorporate other utility functions specific to GIS tasks. Zoom and panning are examples. Zoom may be activated either in the main GIS software or in the screen software. *Screen zoom*, which allows either direct enlargement or reduction, is the more rapid of the two. Although slower, GIS *program zoom* can include various intelligent functions that alter texts, line widths, and other features in proportions differing from those of the overall scale magnification or reduction. In most systems there is a limit to how much an image may be enlarged. Another GIS utility can highlight selected themes or objects, thus easing on-screen work with complex maps comprising many elements.

13.5.2 Editing and correcting errors and omissions

Most GISs support geometric map editing functions, including:

- Supplementing
- Copying
- Deleting
- Moving
- Rotating
- Dividing lines
- Joining lines
- Altering form

Raw digitizing data always contain some errors and omissions, such as lines crossing erroneously, unintentional wiggles in lines, missing points and lines, and so on (Figure 13.4). Omissions are corrected most easily by entering data directly from a digitizer. Errors are corrected most efficiently using a keyboard or a mouse.

Lines may be divided or line segments deleted to allow text or symbols to be placed on maps. Lines and composite map objects may be moved whenever objects are digitized incorrectly. Line segments to be moved or deleted are marked, as well as any new positions (Figure 13.5). In some cases, objects are rotated through prescribed angles. There are also functions to change the shape of graphic objects, such as automatic correction of buildings so that all angles are right (90 degrees).

As previously mentioned, if old data are changed during updating, historical data will normally be lost. The correction of both geometry and attributes to update the data is in many ways a difficult and risky operation, which most GISs do not tackle well enough.

Other utilities support text editing for altering or amending place names, road names, and other names (Figure 13.6). Many systems support the direct printout, for example, of such attribute values as addresses and property registry numbers on maps. Some systems treat all text as attributes. Raster-based systems have programs with editors dedicated to editing raster data. Hence, systems that can manipulate

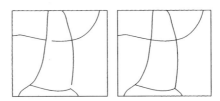

Figure 13.4
Polygons that are not closed can be corrected by joining lines automatically where the gap is below a given parameter. It should be noted, however, that automatic closing can have undesirable consequences should the closing parameter be set too high.

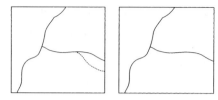

Figure 13.5
Digital map data can be edited by, for example, the addition of new data and removal of old data, so that objects retain their correct shape.

both raster data and vector data have separate raster and vector editors. Coordinate values may be modified using an ordinary text editor.

13.5.3 Creating topology

Corrected data may result in an unacceptable map image, and even when such data are verified, the image may contain errors that are not evident until logical connections are established. Consequently, many GISs have functions that compute nodes and links automatically and compile topology tables. By creating topology it is possible to identify such errors as:

- Polygons that are not closed
- Disconnected lines (dangling ends)
- Missing or repeated ID codes
- Polygons lacking, or having too many, ID points (in some systems, polygons are assigned ID points with codes and coordinates)
- One or more missing polygons

Errors such as these may be reported by a dedicated error reporter or marked with specific symbols on screen plots; such errors may be corrected manually or automatically. Errors in ID codes and points are corrected manually via a keyboard.

Small dangling ends can be deleted automatically by setting a parameter that stipulates the sizes of the ends to be deleted. Gaps may also

Figure 13.6
Normally, text can be freely placed with regard to beginning and end, rotation, following lines, and so on. Font types are generally system defined.

be closed automatically. However, if the gap-closing parameter is set too high, automatic gap closing may introduce errors by connecting lines that ought not to be connected. It is always advantageous, therefore, to digitize line intersections in order to reaffirm line connections. If the gap-closing parameters are set too low, it will not be possible to close all the faulty gaps (Figure 13.7).

Small, meaningless polygons (slivers) can also be removed automatically by using the area size. Here it is also possible to introduce errors. If the test value is set too high, polygons that should have been included will be removed. Generally speaking, the automatic correction functions should be used with great care.

Many GISs are based on the principle that inside each polygon there must be a single ID point or label point to which attributes can be attached. Through topology creation, polygons are identified which are lacking ID, or have too many ID points (Figure 13.8)

If there are two label points within an area, the reason may be that a dividing line is missing or that there is a gap in a line that should divide two polygons. If there is no label point inside a polygon, the reason may be that the point is wrongly placed or that the polygon is wrongly divided in two by a line.

Quality control of logical consistence with the help of topological test procedures is available in most GISs.

13.5.4 Transformations to a common datum, map projection, and coordinate system

Because GIS often requires data to be collected from a series of different sources, users will have to transform geometrical data into a common datum, common map projection, and common coordinate system. Many GISs have therefore integrated these transformation options.

Transformations to a common datum
As previously mentioned, there is no unique mathematical connection between different datums. The parameters for transformation from

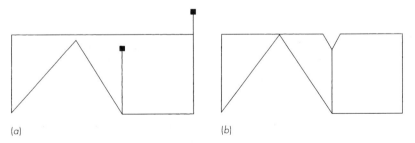

(a) (b)

Figure 13.7
(a) Various types of data error can often be identified during topology creation.
(b) Automatic error correction can be risky, because it can introduce completely new, highly undesirable errors.

Figure 13.8
Topology creation can reveal which polygons are missing or have more than one ID point /label point.

one datum to another are normally determined on the basis of measuring in fixed points and empirical formulas. Transformation programs with formulas and parameter values are usually available from the national mapping authorities in each country. These programs should also be available via GIS.

Transformations to a common map projection
Data from a variety of sources are useful only if referenced to a common map projection. However, data from differing sources are often referenced to differing map projections. One common disparity is that survey points are entered in coordinates of latitude and longitude in an azimuth projection, whereas other map data are entered with reference to a national rectangular coordinate system based on a cylindrical projection. Consequently, most GISs support various transformations for converting data from one projection to another. Some systems use latitude and longitude coordinates in degrees, minutes, and seconds, and others require that they be expressed in decimal degrees (fractions of degrees). The formula for converting from degrees, minutes, and seconds to decimal degrees is:

Decimal degrees = degrees + minutes/60 + seconds/3600

Mapping hemispheres and other major land areas usually entails selection of the map projection that most closely suits the data presented. In some cases, this can mean that entire sets of data must be transformed.

These transformations are based on the fixed mathematical relationships that describe the various map projections. Thus, a transformation converts the coordinates of one system to the coordinates of another. As the transformations themselves are digital, ancillary operations are necessary whenever positional information is not decimal, as when the coordinates are in degrees, minutes, and seconds.

Transformations to a common coordinate system
Coordinate transformations involve mathematical functions that relate coordinate geometries to each other. A conversion function also con-

tains computation of parameters based on a knowledge of the respective coordinates of a number of reference points that are common to the initial and final systems of the transformation involved. When working in areas that border two different zones, for example in UTM (Universal Transverse Mercator), it is often necessary to transform all the data to the same zone. Many GISs also have this kind of transformation function.

Digital data from various Cartesian coordinate systems, as from local ground system, map digitizing, and field surveying, may be used together only if referenced to a common coordinate system (Figure 13.9). Sometimes systematic errors in such themes as displacements, rotations, and scale errors can easily be compensated for if the data involved are transformed to an error-free base. In other cases, the data of a particular thematic layer must be transformed to be compatible with the data of other thematic layers. Consequently, most GISs support a variety of coordinate system transformations, such as conform and affine transformation. In GIS this kind of transformation process is often called rubber sheeting.

13.5.5 Adjustments at map edges and between neighboring areas

As anyone who has glued neighboring map sheets together to make a larger map knows, lines that *should* meet at the joined map edges often do not. Digitizing errors, scanning errors, and nonuniform paper shrinkage are often the causes of this common and vexing problem. The problem is compounded whenever map sheets are digitized,

Figure 13.9
Data registered in a coordinate system *(a)* can be converted to another system. *(b)* to a common and complete map picture *(c)*.

261

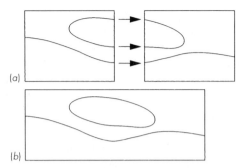

Figure 13.10
Map sheets can be aligned automatically and combined into a new, seamless unit.

because digitization may also introduce error. Most GISs therefore support functions for automatic adjustment of disparities along neighboring map sheet edges (Figure 13.10). This is a time-consuming process, because for each disparity along the meeting edges, a decision must be made to move either one or both lines. When the process has been completed for two adjoining edges, the polygon representing the edge may be deleted.

13.5.6 Thinning coordinates and smoothing lines

Thinning coordinates

Digitization often produces more data than are needed for the immediate task, particularly with linear objects such as contour lines, borders for types of soil and vegetation, and so on. Consequently, most GISs have routines for *thinning*, deleting superfluous data. Thinning increases data efficiency, reduces the storage capacity required, and increases the plotting speed. In some cases, data volume may be reduced by 60 to 95% without reducing the data quality.

Various methods are available for thinning points on lines. The simplest approach is to delete every nth point—every other point, every third point, and so on—up to the number n chosen (Figure 13.11). A more advanced approach is to remove points that exceed a given value

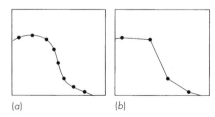

(a)   (b)

Figure 13.11
Lines *(a)* can be thinned by removing points *(b)*, based on certain functions.

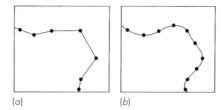

Figure 13.12
Height contours and other linear elements *(a)* can be smoothed by the automatic addition of new points *(b)*, based on certain functions.

for the distance between points on a line and straight lines that are stretched in a certain system between different end points on the line (Douglas Peucker algorithm). The number of points needed to describe a line may also be reduced by moving forward a corridor of a given width until it touches the digitized line (see Chapter 4). All the points on the line that lie within the corridor, apart from the first and last, are thereby deleted. This process is repeated until the entire line has been trimmed. Lines may also be replaced by polynomial spline functions. Good thinning routines remove points without having any noticeable effect on the geometric accuracy.

Line smoothing
Lines delineated by points are never completely smooth, so line smoothing is often used to improve the graphic quality of lines. Smoothing functions are usually associated only with data presentation. Line smoothing means increasing the total number of points delineating a line. Several smoothing methods are available, but those used most frequently are based on curve fitting, using third-degree polynomials, as illustrated in Figure 13.12.

Smoothing affects appearance only. Although a smoothed line may seem more accurate, it is no more accurate than the original data for the line. Because smoothing increases the number of points, this function is often linked to the plotting itself and thus does not affect the storage volume.

13.6   Editing and Correcting Attribute Data

Like digital map data, attribute data must be edited and corrected. These operations include error correction, updating, and amending. The editing tasks may be carried out by using standard editing tools, such as those available in word processing, or specific GIS commands. Some GISs use SQL (Structured Query Language) to manipulate attribute data in relational databases. Specific GIS commands include commands for changing object thematic codes and switching codes between objects, as well as for editing thematic codes containing texts.

The guidelines for entering data may change with time, mandating changes in the codes of older data. Usually, common mathematical signs such as $+$, $-$, $*$, and $/$ are used for this purpose. The currencies in which prices in attributes are expressed may be changed [e.g., from U.S. dollars (\$) to pounds sterling (£)] by entering an exchange rate. Relational and other databases used to store attribute data usually incorporate effective editing tools. These permit a variety of operations, including searching for members of a prescribed class and then editing one by one, or assigning new values to an entire class using a single command.

Incorrect attribute values are often very difficult to detect, both in raster and in vector data. In raster data attributes are an integrated part of the geometry, and erroneous single values can be relatively easy to spot visually. In many raster GISs one can also use a run-length encoding strategy to edit incorrect values. In vector data incorrect attribute values are more difficult to identify, because they often do not appear visually on the map and because they can behave in many different ways. Logical errors, such a value too low or too high, can be automatically identified by the software. It is more difficult to detect wrong names and similar errors. Peter Anderson can be registered as the owner of a property, when the correct owner is John Anderson. In this situation, manual control of the original data is the only way to identify and correct the error. Moreover, active use of the data increases the likelihood that this type of error will be discovered, which, again, can lead to improvement in the quality of the data.

CHAPTER 14

Basic Spatial Analysis

14.1   Analysis of Spatial Information

Even though most of the real world can be observed with the naked eye, it is often difficult to interpret and systematize what is observed. It becomes even more difficult when the image of that reality is stored in digital form as geometric data and attribute tables. To bring out the patterns, connections, and possibly, the causes of variations in the data, various computer-based techniques are applied to analyze the data. Spatial analysis techniques are an attempt to imitate what concerns the human brain (i.e., to create an understandable image of reality). These are the techniques discussed in this and the following chapter.

Technology can help only to a limited extent, however. Stating problems and delineating the approaches to solutions together comprise one of the most difficult steps in GIS analyzing. This has to be solved by the individual user or operator before the technology can be put to use, based on his or her professional knowledge in the field of agriculture, environmental protection, planning, or other area, supplemented with knowledge of GIS.

Analyzing data normally comprises three principal phases:

1. Stating the problem
2. Choice of data
3. Analyses of the data chosen

Stating the problem is generally the most difficult step in using GIS for analytic analysis. Considerable specialized knowledge is often required—for example, in the fields of biology, physical planning, and geology—for identification and formulation of the problem. Once the problem has been formulated, however, it is necessary to choose the right GIS tool, and for this consideration knowledge of GIS is required.

All GIS provide functions for analyses of data chosen and for storing results of such analyses. Data may be selected according to:

- Geographical location and/or
- Thematic content

Most GISs permit defining the criteria for selection. These are often based on SQL or are in menus with provisions for generating SQL queries. Some GISs provide predefined selection criteria; other systems use a macro language to set up the selection criteria. Specific systems have predefined menus dedicated to the relevant applications. In most systems, selection criteria may be stored for subsequent use. All GISs provide functions for analyses of data chosen and for storing the results of such analyses. Data may be analyzed at various levels:

1. Data in attribute tables are sorted for presentation in reports or for use in other computer systems.
2. Operations are performed on geometric data, either in search mode or for computational purposes.
3. Arithmetic, Boolean, and statistical operations are performed in attribute tables.
4. Geometry and attribute tables are used jointly to:
   a. Compile new sets of data based on original and derived attributes.
   b. Compile new sets of data based on geographical relationships.

Within each of these levels, the operations used may be logical, arithmetic, geometric, statistical, or a combination of two or more of these four types. Operations may be performed on individual points or on areas, and may involve considerations of proximity or of changes over time. Numerous operations may be performed on line networks. The more important commands are discussed in the following sections. The functions implemented vary from one GIS to another, and some GISs contain functions not discussed here.

14.2   Logic Operations

Logical searches in databases normally employ set algebra or Boolean algebra. Set algebra uses the three operators equal to, greater than, and less than, and combinations thereof:

$$=, >, <, \geq, \leq, <>$$

These operators are included under SQL.
   Practical applications include:

- Identifying extrema, such as finding attribute minima or maxima within various polygons and, as a result, delineating a new thematic layer (new row in the attribute table)
- Selection or isolation, where particular values are selected for subsequent ranking in a new thematic layer

Boolean algebra uses the AND, OR, NOR, and NOT operators to test whether a statement is true or false. AND, OR, and NOT are used in SQL. For two items, A and B, we might have any of the following statements:

A AND B, A OR B, A NOR B, A NOT B

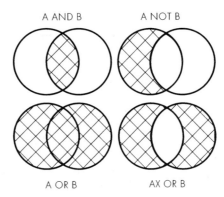

A AND B   A NOT B

A OR B   AX OR B

Figure 14.1
Logical operations shown in a Venn diagram. Similar operations can be carried out on geographical data. The shaded areas represent true statements.

Such statements may be illustrated in a Venn diagram, which is a schematic representation of a set in which magnitudes illustrated by surfaces are superimposed, as shown in Figure 14.1. The shaded areas represent true statements. This technique is well suited to analyzing geographical data. For example, potential conflicts between forestry and cattle farming can be illuminated by assigning A to potentially productive forest tracts and B to known grazing areas. The test A AND B on the two operands will identify conflict areas that can be assigned special symbols and drawn on maps.

Logic operations are particularly powerful when the relationships are complex. In GIS, logic operations may be performed simultaneously on more than two themes and may involve several operators.

14.3 General Arithmetic Operations

Arithmetic operations are performed on both attribute and geometric data. All GISs support the customary arithmetic operations of addition, subtraction, multiplication, division, exponential, square root, and the trigonometric functions:

$$+, -, \times, /, {}^{n}, \sqrt{}, \sin, \cos, \tan$$

These operators may be used for many purposes, including assigning new thematic codes. Typical examples include:

- Reclassification of soil types, in which areas are to be converted from decares to hectares by dividing all area figures by 10.
- Conversion of distances along roads to driving times, by dividing all distances by a specified average vehicle speed. The result is a new set of attributes that are useful in transportation planning.

Arithmetic functions are used in all geometric computations involving coordinates, as in calculating distances, areas, volumes, and directions.

14.4   General Statistical Operations

Statistical operations are performed primarily on attribute data, but may also be effected on some types of geometric data. Most GISs support a range of statistical operations, including sum, maxima, minima, average, weighted average, frequency distribution, bidirectional comparison, standard deviation, multivariate, and others. The computation of averages requires averaging two or more attribute values and stating the result as a new attribute. Frequency distributions are used to compile *histograms,* charts comprising rectangles whose areas are proportional to relative frequencies and whose widths are proportional to class intervals. The data used to draw a histogram may also be employed to plot a curve. Other statistical operations in common use include least-squares computations of transformation parameters from regression models, with the standard deviation as an expression of accuracy. Bidirectional comparison involves point-by-point correlation of two themes to produce a new statistical thematic layer and hence a new attribute.

In geometry, for points one can use statistical methods to look at the pattern in their spatial distribution (evenly distributed, randomly distributed, distributed in groups). To do this, nearest neighbor analysis is used. This is based on the middle value of all distances between each point and its closest neighbor. Curve shape can be expressed as the relationship between the straight distance and the real distance. The shapes of surfaces can be calculated by use of the so-called Euler number, which gives an index based on the number of holes/islands in the area and the number of fragmented areas. The relationship between the size of the area and the length of the periphery is also used as an index for the shape.

Satellite data are usually analyzed statistically in dedicated image processing systems, which are often connected to GIS facilities. Some vector GISs support image processing. Multivariate operations, such as cluster analyses, are vital in image analyses. These operations assign new classes to entities on the basis of statistical selection criteria. Pattern recognition based on statistical models is incorporated in some GISs.

14.5   Geometric Operations

Operations on geometric data involve the customary arithmetic operations in computations of centroids, intersection of lines, distances, areas, volumes, and directions. Most spatial analyses in GIS are based on

Figure 14.2
Test to see whether a point lies in an area: *(a)* Odd (one) intersection—the point
lies inside the area. *(b)* Even (two) intersections—the point lies outside the area.

these relatively simple geometric basis operations, where the most important elements are the calculation of distance and calculation of whether a point is located within a given polygon.

14.5.1. Measurement of centroides and intersections

The gravity of a surface is calculated with the help of the formula:

$$X_c = \frac{\sum_{i=1}^{n} X_i}{n} \tag{14.1a}$$

$$Y_c = \frac{\sum_{i=1}^{n} y_i}{n} \tag{14.1b}$$

Where x_1, y_1 to x_n, y_n are the clockwise coded coordinates for the break points of the polygon round the area, X_c, Y_c are the coordinates of the centroid, and n is the number of points. The polygons can have widely varying shapes, and this kind of calculation is no guarantee that the point is located within the area. Therefore, each case must be tested as to whether the point is located within or without the area. Such a test is carried out by pulling a right line in any direction (preferably vertical) from the point and checking on how many times this line cuts through the area's boundary line. If the number of crossing points is odd, then the point is located within the area; but if it the number is even, then the point is located outside (Figure 14.2).

Centroids are used, among other things, to establish ID/label points. The technique for testing whether a point lies within an area is also used in the overlay operation "point in polygon," which is discussed later in this chapter.

The intersection between two straight lines is calculated by loosening the equation system for both lines, whereby they are represented with the formulas (Figure 14.3):

$$y = c_1 + a_1 x \tag{14.2a}$$

$$y = c_2 + a_2 x \tag{14.2b}$$

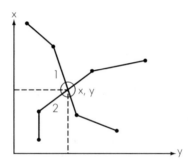

Figure 14.3
Intersection between two straight line segments (1 and 2).

Where x and y are the coordinates for the common intersection, c_1, c_2 are the single lines intersecting with the x axis and a_1, a_2 are the lines' ascending coefficients. Lines normally consist of many straight line segments; thus, all the line segments in each line have to be tested in pairs against each other to find out which segments intersect each other.

These calculations have to be carried out to test whether a point lies inside or outside a polygon, as well as in all topology creation to establish nodes. There are a number of smart methods (minimum bounding rectangle [MBR], monotonic section, etc.) that are used to reduce calculation time.

14.5.2 Measuring distance

In principle, distances should be measured using a uniform unit (m, cm, etc.), a starting point, and a fixed procedure, as given:

Distance = measured unit $\times$ number of units

Distance in vector
The distance between two points can be calculated in two ways, both based on a right-angled triangle:

1. *Euclidian distance*, with the help of Pythagoras (see Figure 14.4a).

$$D = \sqrt{(x_1 - x_2)^2 + (y_1 - y_2)^2} \qquad (14.3)$$

2. *Manhattan distance*, calculated by summarizing the two shortest sides in the right-angle triangle. The sides have to be parallel with the axis in the coordinate system (see Figure 14.4b).

$$D = a + b \qquad (14.4)$$

The length of a line is calculated by summarizing the distance of all the line segments in the line. The periphery of a polygon is calculated

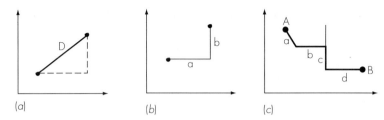

Figure 14.4
Calculation of distances: *(a)* Euclidian distance, *(b)* Manhattan distance, *(c)* distance in network.

in the same way. Where there are several right-angled triangles, the cathetus (shortest sides of a right-angle triangle) are summarized to find the total Manhattan distance. Euclidian and Manhattan distance calculations can both be employed on vector data and raster data. The Euclidian is the most commonly used method for calculating distance in GIS.

The length of a line section is calculated by summarizing the distances between single points on the line. The length between the start and the end point in a line network is calculated by summarizing the length of the line sections that form the continuous connecting line between the start and the end point (Figure 14.4c). When calculating topographical distance, one also has to allow for height variations, and that presupposes that one has the x, y, and z coordinates for points on the line. Calculations of distance on the curved surface of Earth are based on spherical formulas, taking into account the Earth's varying curve radii.

In GIS analyses it is often interesting to calculate the distance between points, lines, and areas. Distance here generally means the shortest distance between objects (Figure 14.5). For calculations of this kind, it is occasionally possibly to use the same set of formulas as used for calculating the intersection point, but the geometry can be so complex that more advanced calculations must be performed.

In transportation analyses, results may best be stated in travel time or travel costs, rather than in meters or other units of length. Because the starting point will always be coordinates, which, again, give only

Figure 14.5
Proximity of geometrical objects: *(a)* shortest distance between point and line, *(b)* straight distance between point and line, *(c)* shortest distance between areas.

the basis for calculation of length, a mathematical link must be made between the unit of length and unit of time (for example, speed expressed in km/hr) and possibly between the unit of time and unit of cost.

Distance in raster
The calculation of distance in raster data is, in principle, carried out in the same way as for vector data. In a raster GIS, the measuring unit may be cell width or diagonal, and it is possible on this basis to calculate Euclidean distance or Manhattan distance (Figure 14.6)

To calculate the length of lines in raster data—for example, the periphery of an area—the line structure must first be identified. This is no simple task, and it requires that all neighboring cells be examined step by step through the structure. Secondly, the total distance is represented by the sum of all the diagonals in the line structure, or by all the cell widths.

14.5.3 Measuring areas

Vector
The surfaces of areas in vector data can generally be calculated with the formula:

$$A = 1/2 \sum_{i=1}^{n} y_i(x_{i+1} - x_{i-1}) \tag{14.5}$$

The coordinates x_i, y_i represent the breaks in the polygon that surrounds the area. This assumes that the points are coded clockwise. Areas can take very complex forms. In order to simplify calculations, areas can be split up into simpler shapes and the area is calculated individually for each part. There will often be islands within the polygon, which should not be included in the calculation. These have to be identified and their areas calculated and deducted from the total area.

2.8	2.4	2.0	2.4	2.8
2.4	1.4	1.0	1.4	2.4
2.0	1.0		1.0	2.0
2.4	1.4	1.0	1.4	2.4
2.8	2.4	2.0	2.4	2.8

Figure 14.6
Distance in raster data. Distances are calculated from the center in the outer cell, and in relation to the center of the neighboring cells. The cell unit (1) is used as unit for measurement.

Figure 14.7
The volume in a digital terrain model can be calculated to total the volumes of all affected cell elements. The volume of each single cell element is calculated by multiplying the size of each cell by the distance to the relevant level surface.

Raster

The area of surfaces in raster data can be calculated by summing the number of cells in the surface and multiplying by the cell size. Just as with line calculation in raster data, the peripheral line must first be identified in order to find the boundary of the surface.

In many GISs, the periphery and area are computed automatically for each polygon and connected to objects as attributes.

14.5.4 Measuring volumes

Volume calculations can be carried out in digital terrain models. If the model is based on the raster principle, then the volume of each affected cell can be calculated by multiplying the size of the cell as with distance (down) to the level to which the volume is to be calculated, see Figure 14.7. The total volume is calculated by summarizing the volumes linked to the individual cells.

As mentioned previously, height is generalized as being constant in each cell. This also applies to new level surfaces in relation to which volume is to be calculated—for example, a planned road with a certain gradient.

Volumes and directions are calculated in models manipulating terrain, as in determining excavation volumes, terrain slopes, exposures, and the like. Flow analyses perform geometric computations to find the shortest routes between designated points. This is discussed in the next chapter.

14.6 Search and Report Generation from Attribute Data

Inquiry in attribute data may be the primary GIS task in daily practical applications, particularly in the operational, maintenance, management, and service sectors. Inquiry calls for a human–machine dialogue in which the user employs various input/output devices and their supporting software. Tailor-made software is employed in some

GISs. In others, standard routines such as SQL are employed, particularly in those that access relational databases.

Inquiry of and search for data may be based on logical and arithmetic operations and specific relational database functions, as with projection, selection, and joining. The query criteria may be complex and cover several attribute tables. In practice, responses to inquiries should be displayed in a comprehensible manner, not simply as brief on-screen information. This is particularly true when attribute data are used as a reference source. Consequently, many GISs support a standard report format with fixed headings, column locations, and other layout features. In addition, users may define their own report formats.

Some GISs support report formats that are tailor-made to each application, whereas others use the report function and report form storage facilities of the report generator supported by the DBMS in use. In many cases, reports may be simple. For example, an inquiry may be for the name and address of the owner of a property at a particular address, or for a list of all the properties registered on a specific street. Reports in various formats may also be stored in separate files for use with other system files or for export to other systems for processing, as in statistical analyses.

14.7 Geometric Data Search and Retrieval

Geometric data may be searched using specified criteria, employing a dedicated query language or SQL for spatial data. Searches start from coordinate values, and in principle, any part of a database may be searched. Seamless databases permit searches anywhere within a geographical area. Databases based on map sheets, however, limit searches to one map sheet at a time, although contiguous map sheets may be combined for search purposes. In this case, the combined map sheets function as a partial copy of the database as a whole. For example, a search may be conducted within an area specified by the coordinates of its lower left and upper right corners, without involving the rest of the database.

Geometric data usually carry thematic codes for various types of objects, such as roads, administrative borders, water supply systems, buildings, properties, and so on. In searching these codes, the various map themes may be turned off and on, or shown in different colors, for the different types of objects.

A typical task may require the display of all map details within the limits of x minimum = 10,000 to x maximum = 12,000 and y minimum = 5000 to y maximum = 7000. The query program then reads and tests all coordinate values in the database against these constraints and retains those that lie within the limits stated. Proximity operations often necessitate searches in map data.

In such searches it is often presupposed that certain spatial relations have to be met. For example, a point is specified on the screen and a search is made for the object lying nearest the point; or all objects that lie within a given surface or distance are to be identified. In such situations the geometric operations described earlier are used, whereby the distance to all possible objects in the database is tested.

14.8 Complex Operations of Attribute Data

Mathematical, logical, and statistical operations may be performed on attribute data (Figures 14.8 to 14.11). The mathematical operations include addition, subtraction, multiplication, division, exponential, square root, and the trigonometric functions:

$$+, -, \times, /, {}^{n}, \sqrt{}, \sin, \cos, \tan$$

The logical operations require the operators of set algebra: equal to, greater than, and less than, and combinations thereof:

$$=, >, <, \geq, \leq, <>$$

as well as those of Boolean algebra:

AND, OR, NOR, and NOT

The statistical operations include sum, maxima, minima, average, weighted average, frequency distribution, bidirectional comparison, standard deviation, multivariate, and others.

Attribute data may be either quantitative and expressed numerically, or qualitative without numerical magnitudes. Numerical operations may be performed only on numerical data and hence mainly on quantitative attribute data. The numerical treatment of qualitative attribute data is limited to counting operations, as in determining the

ATTRIBUTE TABLE

ID	Area value $
1	7.24
2	10.13
3	19.05
4	1.58
5	22.01

(a)

ATTRIBUTE TABLE

ID	Area value £
1	10.86
2	15.20
3	28.58
4	2.37
5	33.02

(b)

Figure 14.8
Arithmetic operation on attribute values. Here U.S. dollars (a) are converted to British pounds (b) by multiplying by the exchange rate (1.50).

ATTRIBUTE TABLE

ID	Area value $
1	7.24
2	10.13
3	19.05
4	1.58
5	22.01

(a)

ATTRIBUTE TABLE

ID	Area value $
1	7.24
4	1.58

(b)

Figure 14.9
Logical operation on attribute values *(a)*. All values below 10 are selected *(b)*.

ID	Attributes			
	A	B	C	D
1	10	10.1	0.3	55
2	11		0.7	55
3		98	0.8	51
4	6	116		50

(a)

ID	Attributes			
	A	B	C	D
1	10	10.1	0.3	55
4	6	116		50

(b)

Figure 14.10
Boolean algebra on attribute values *(a)* identifies all objects that contain both attributes A and B *(b)*.

ID	Attributes
1	7.24
2	10.13
3	19.05
4	1.58
5	22.01

(a)

Sum $	59.91
Average $	11.98
Max. $	22.01
Min. $	1.58

(b)

Figure 14.11
Statistical operations on attribute *(a)* values make it possible to find the sum and the average, maximum, and minimum values in the table *(b)*.

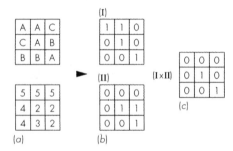

(a)

(b)

(c)

Figure 14.12
Operations on raster data: finding cells in the two raster layers *(a)* I and II which contain both A and ℒ The cells are reclassified *(b)*, and by using the logical operations AND a new layer is formed *(c)*.

number of classes into which data may be ranked or how many observations fall within each class. For example, four classes— A, B, C, and D—may be identified and found to contain differing numbers of observations: 12 in A, 23 in B, 2 in C, and 9 in D. Attribute data are frequently processed to discern new patterns and relations in the data.

Geometrical, logical, and statistical operations are carried out directly on the cell values in raster data. Thus, a new raster layer is created having a mathematical connection to the original layer. For example, it is possible to create a new layer by dividing by 10, so that values will be expressed in meters instead of decimeters.

With binary data it is quite simple to perform logical operations. The logical operation AND is carried out by multiplying two layers with each other in order to see which cells appear in both layers (Figure 14.12).

14.9 Classification and Reclassification

To obtain a better view of the attribute data, they are often divided into classes. In the classification process, attributes are grouped according to limits set by the user. For example, three classes may be set for the attribute "year": A = before 1970; B = 1971–1980; C = 1981–1990. Each object with a year attribute, such as a water supply pipe, is then assigned a new year-class attribute, A, B, or C. Plotting the classes in distinctive symbols or colors may reveal new patterns—showing, for instance, that water supply pipes in class B are those most subject to leakage.

Where the whole observation material is available, it is simple to divide the data into other classes, through reclassification. Reclassification can be desirable in order to see a new pattern and connections. In the preceding example, it can be possible to discover new patterns if one looks at a five-year period instead of ten-year period. Reclassification involves changing attribute values without altering geometries (Figure 14.13). Arithmetic and some statistical operations are used to assign new attribute values. In many ways, reclassification may be compared to changing colors on a map, in that the reclassified attributes may plot out in new colors. Reclassification may be used to isolate object types. For example, in a raster GIS, a "built-up area" characteristic may be isolated by assigning all other areas a value of zero (Figure 14.12). When the data are plotted out, only the built-up areas will then be shown. The boundaries between polygons of the same type may be deleted to combine the polygons as larger units. Polygons of dissimilar types must be reclassified prior to being combined (Figure 14.13).

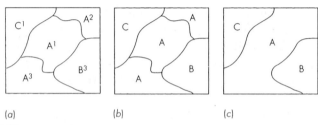

Figure 14.13
Reclassification of polygons *(a and b)* and combining polygons *(c)*.

14.10   Integrated Processing of Geometry and Attributes

One of the simpler forms of integrated processing of geometry and attributes in a hybrid GIS involves pointing to the location of a building displayed on screen and requesting retrieval of all information stored on the building. On receiving the query, the GIS searches the map database to find the building corresponding to the coordinates that have been pinpointed. Using the building ID number stored with the coordinates, the system then searches the attribute database for all

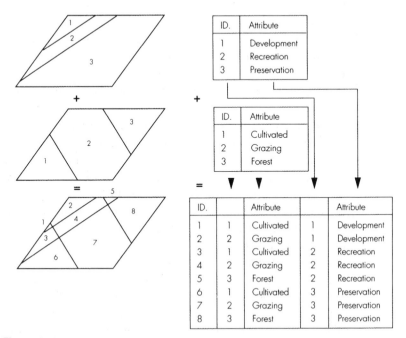

Figure 14.14
A computerized superimposition can, in many ways, be compared to a series of map overlays. However, this also leads to an expansion of the attribute table, in addition to the geometric changes. Arithmetic, logistical, and statistical operations can be carried out in the expanded attribute table.

available information, which can then be displayed or printed out. It is possible to go in the opposite direction and first search for the object in the attribute database by using SQL, then identify the object in the map database with the help of common ID code, and finally draw the object on the screen. Where particular surfaces are to be identified, topology has to be created in the map data. In integrated GIS where all data lie in a relational database, it is possible to use spatial SQL for this type of search.

More advanced integrated processing is also based on the condition that each object type (cultivated land, deciduous forest, protected area, etc.) is represented both in geometry and in an attribute table. The geometry concerned may be likened to a single thematic map. Single maps may then be superimposed to integrate with each other and thus produce a multithematic map containing information from each of the initial thematic maps. The integrated map comprises comparable units [integrated terrain units (ITUs)] and a new attribute table is compiled, as illustrated in Figure 14.14. Arithmetic, logical, and statistical operations may be performed in the new attribute table. The geometry and the attributes may then be used to compile a new thematic map. This operation is known as overlay.

14.11   Overlay

Overlay is used in data integration and is a technical process, the results of which can be used in realistic forms of spatial analysis.

14.11.1   Polygon overlay

Polygon overlay is a spatial operation in which a first thematic layer containing polygons is superimposed onto another to form a new thematic layer with new polygons. This technique may be likened to placing map overlays on top of each other on a light table (Figure 14.14). The corners of each new polygon are at the intersections of the borders of the original polygons; hence, computing the coordinates of border intersections is a vital function in polygon overlay. The computations are relatively straightforward, but they must be able to cope with all conceivable geometric situations, including vertical lines, parallel lines, and so on. Computing the intersections of a large number of polygons can be very time-consuming.

If areas are stored as links in a topological model, fewer intersections need to be computed, thus reducing the computing time. The new intersections are identified as nodes and the lines between the nodes as links. The new nodes and links then constitute a new topological structure. Consider as an example polygon C4 in Figure 14.15, which is a combination of polygons C and 4. The system cannot associate attributes with C4 unless topology is associated with either the original or

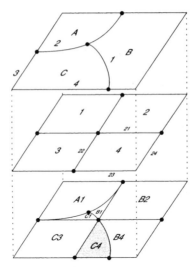

Figure 14.15
It is necessary to create topology in order to identify the attributes that the new
polygon has been given after the overlay operation. (From ESRI 1995.)

the new data. With topology for the new data, the system recognizes
that polygon C4 comprises line 22 and parts of lines 23 and 1. Furthermore, the system is aware that polygon 4 is on the right side of line
22 and on the left side of line 23, and that polygon C is on the left side
of line 1. Hence, the system now knows that polygon C4 is a composite of polygons 4 and C both geometrically and in terms of attributes.

Each new polygon is a new object that is represented by a row in the
attribute table. Each object has a new attribute, which is represented
by a column in the attribute table. Superimposing and comparing two
geometrical data sets of differing origin and accuracy often give rise to
a large number of small polygons. For example, two polygons representing land areas may have slightly differing geometric borders on a
lake, yet on a piecemeal basis the borders may coincide. Superimposing the two polygons can then produce an unduly large number of
smaller polygons, called slivers (Figure 14.16). This proliferation of
smaller polygons may be counteracted automatically by laying a small
zone—fuzzy tolerance—around each line. If two zones intersect when
the polygons are superimposed, the lines they surround combine into
one line. Smaller polygons may also be removed later, using area size,
shape, and other criteria. In practice, however, it is difficult to set limits that reduce the number of undesirable small polygons while re-
taining smaller polygons that are useful.

In addition to performing the overlay computations, the system can
present a new image of the new structure, borders between polygons
of like identity being removed to form joint polygons. This combination process may be either automatic or controlled by the user.

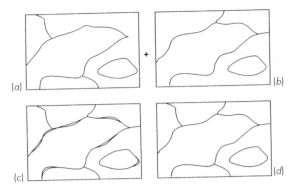

Figure 14.16
The overlaying technique *(a* and *b)* can easily lead to the creation of many small meaningless figures—slivers *(c)*. These can be removed automatically *(d)*.

The overall procedure for polygon overlay is to:

1. Compute intersection points.
2. Form nodes and links.
3. Establish topology and hence new objects/new IDs.
4. Remove excessive numbers of small polygons where necessary and join like polygons.
5. Compile new attributes and additions to the attribute table.

Polygon overlay is a sizable operation for which even on the most powerful computers may require relatively long processing time. The process may be used to clip geographical windows in a database. For example, a township border in one thematic layer may be used to clip all other thematic layers in order to produce a collection of data for that township only (Figure 14.17).

Overlay produces a thematic map comprising thematically comparable homogeneous units [integrated terrain units (ITUs)] and an expanded attribute table. Arithmetic, logical, and statistical operations may be performed on the attributes to reveal patterns, connections,

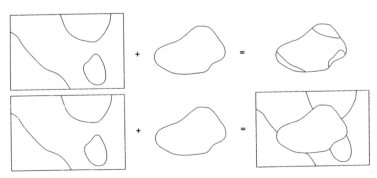

Figure 14.17
Various overlay operations can be used to clip out geometric windows.

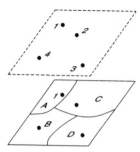

ID	Building no.	Polygon	Property
1	660	A	44/110
2	659	C	44/95
3	610	D	44/121
4	665	B	44/81

Figure 14.18
Superimposing points on polygons.

and conflicts, for example, when simulating alternatives and studying consequences. The advantages that GIS offers in these analyses are that the number of alternatives considered are not limited by system capacity, that all alternatives can be based on the same data, and that all available information may be used.

14.11.2 Points in polygons

Just as polygons may be superimposed on other polygons, so may points be superimposed on polygons (Figure 14.18). The points are then assigned the attributes of the polygons on which they are super-imposed. The relevant geometric operation means that points must be associated within polygons, and there will thus not be created new geometry in point-in-polygon operations. The approach requires computing the intersection of the polygon border with lines through points, to test whether the points lie within the polygon. Attribute tables are updated after all points are associated with polygons.

14.11.3 Lines in polygons

Lines may also be superimposed on polygons (Figure 14.19), with the result that a new set of lines contains attributes of both the original

Attributes

ID	Line	Road no.	Polygon	County
1	1	Rv. 410	B	Akershus
2	1	Rv 410	C	Oslo
3	2	Rv. 9	C	Oslo
4	3	E 18	C	Oslo

Figure 14.19
Superimposing lines on polygons.

lines and the polygons. These particular computations are similar to those used in polygon overlay: Intersections are computed, nodes and links are formed, topology is established, and attribute tables are updated. The new attributes can be analyzed after the geometrical operations are carried out.

14.12 Buffer Zones

Buffer zones are used to define spatial proximity. These comprise one or more polygons of a prescribed extent around points, lines, or areas (Figure 14.20). Using the buffer zone technique, points and lines can be given an appropriate surface representation. The new polygons have the attributes of the original objects, or they can be given new attributes. For each element, coordinates are calculated for the buffer zone limit. The limit will normally be circles around points and parallel lines on each side of the line. In addition, single lines are limited at each end with a half circle. Where there are overlapping buffer zones, these can be merged into one surface. Around periphery lines of surfaces, a single limit can be made at the inside or the outside of the periphery, known as a setback. Many GISs support the automatic compilation of buffer zones. Here, the operator interaction usually consists of keying in specific zone parameters, such as stipulating a 50 m zone width on either side of a road, in order to be able later to determine which objects lie within and, possibly, outside the buffer zone.

The creation of a buffer zone is not in itself an analysis, but the new polygons that are created can be used in analysis. Buffer zone polygons may be processed in the same way as polygons generated during operations such as overlay, arithmetic, logical, and statistical computations in which attribute values come within the respective zones.

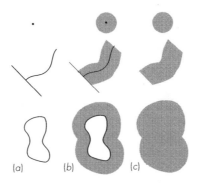

Figure 14.20
Buffer (b) zones can be established around points, lines, and polygons (a). The buffer zones are the basis for the creation of new polygons (c) that have the attributes of the original objects, to be used in further analyses.

14.13   Raster Data Overlay

Raster data may also be overlaid. Indeed, raster overlay is often more efficient than vector overlay (Figure 14.21), because the extent of the calculations is much less. The positions of the overlaid thematic layers need be tested only to see whether they contain cell values. The resultant cell-to-cell comparison presupposes that all cells in each thematic layer are queried, regardless of their values. The total number of cells therefore has an effect on processing time. The new composite cells are assigned attributes composed of those from the original cells. These new cells are registered as a new thematic layer.

Raster data consist of equally spaced cells of equal size (assuming that the various thematic layers cover the same area or have been modified to do so). Consequently, there is no formation of smaller erroneous polygons as with vector data overlays, and there is no need to distinguish between polygons, lines, and points, because all raster data comprise cells. However, an important condition is that both raster layers have identical geometry. That is, cell size must be the same and there must be no relative rotation or transfer between the grids. Deviation is corrected by transformation and resampling to the same cell size.

In raster data, attributes are not usually listed in tables as in vector data, but are represented by thematic layers. Therefore, arithmetic operations and some logical and statistical operations may be performed directly during the overlay process; two thematic layers may be combined, subtracted, multiplied, and so on. If, for example, an attribute of volume in liters is to be modified to deciliters, the thematic layer of volumes merely needs to be multiplied by 10 in each cell of an ancillary thematic layer.

The arithmetic operations on two thematic layers, A and B, produce a new thematic layer, C, through the operations

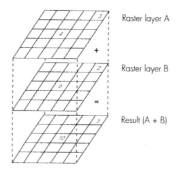

Raster layer A

Raster layer B

Result (A + B)

Figure 14.21
Raster data overlay is simpler than vector data overlay and can be carried out directly on the attribute or cell values.

$$C = A + B, C = A - B, C = A/B, C = A \times B$$

Typical logical operations might be

if $A > 100$, $C = 10$; otherwise $C = 0$

or

C = max. (or min.) of A or B

Boolean operators, such as AND (set intersection) and OR (set union) can also be used. Some GISs support logical operations in the resultant C layer but not directly in the original A and B layers.

As raster overlay is far more efficient than vector overlay, many GISs support functions for manipulating both raster and vector data. Vector data may be converted to raster data so that overlaying can be performed, and the results can then be converted back to vector form.

14.14 Procedures in Integrated Data Analyses

We have already seen that overlay and the creation of buffer zones are mechanical operations and should be looked upon as a link in an integrated data analysis. Integrated data analyses may be relatively complex, so fixed procedures should be employed to rationalize the various tasks, as follows:

1. State the problem.
2. Adapt the data for geometric operations.
3. Perform the geometric operations.
4. Adapt attributes for analysis.
5. Perform attribute analysis.
6. Evaluate the results.
7. Redefine and instigate new analyses if needed.

Stating the problem
Stating problems and delineating the approaches to solutions normally constitute the most difficult steps in using GIS analytic tools. Consider, for example, a typical assignment: the location of a recreational area that can provide a wilderness experience. The selection criteria may include:

- Remoteness, at a specified distance from manufactured facilities
- Reasonable accessibility
- Lakes and streams
- Varied topography
- A variety of vegetation

Each criterion may be analyzed using layering, buffer zones, and other GIS analytical functions, so before the analyses begin, the criteria to be used must be chosen carefully. Particular regard must be given to the potentials and limitations of the original data: data quality, object definitions, the bases for divisions into classes, and so on.

Adapting the data for geometric operations

As a rule, data available from a database must almost always be modified before they can be used for geometric operations. Map data, attribute data, or both may, for example, be modified by clipping a selected area out of a map database, modifying the attribute statements of the area from hectares to square kilometers, or removing borders to join polygons in order to reduce the number of land-use classes involved.

Performing the geometric operations

Geometric operations, employed to sort out the objects to be analyzed, include specification of buffer zones, overlaying, search and retrieval, joining polygons, and other tasks. Each operation results in new data for further processing. For example, in the location of a recreational area, broad buffer zones may be set up around roads and other manufactured facilities. The untouched areas outside these zones may then be sorted out and overlaid with vegetation, hydrologic, and other data relevant to the selection. For each task, the geometric operations must be defined on the basis of the analytic criteria involved. In the process, new polygons are formed, net topologies are established, and new attributes are added.

Adapting attributes for analysis

Most analyses require prepared sets of geometric and attribute data, and just as geometric data must be adapted for processing, so must attribute data be prepared for the arithmetic, logical, and statistical operations that have to be performed. Attribute tables must contain an adequate number of empty rows and columns for new entries. For example, in selecting a recreational area, a new attribute labeled "suitable characteristic" may be compiled to hold codes that indicate the degree of suitability of the various combinations of thematic layers.

Performing attribute analyses

Arithmetic, logical, and statistical operations are performed on attribute data associated with the objects chosen in the geometric selection phase. For example, for the recreational area discussed earlier, an analysis may be performed to identify all areas outside the buffer zones around manufactured facilities. Another analysis may aim to select areas classified as moderately hilly. Yet another may try to identify minor and major streams and rivers. In all cases, data may be sorted with respect to extrema stipulated. The final results of the operations are identifications of georeferenced characteristics that satisfy the selection criteria.

Evaluating the results

The results of analyses must always be evaluated with respect to accuracy and content. In short, do the results seem trustworthy, and do they make sense? These evaluations are made most easily using maps and written reports.

Redefinitions and new analyses

Unacceptable results must be modified, or the analyses that produced them improved and performed again. In some cases, new data may be required and the initial criteria may have to be changed. Similarly, new analyses may be required to provide results with a greater selection of alternatives. New analyses usually start at a suitable stage in the overall analytical process. Here again, GIS provides an efficient tool because GIS systems can process large quantities of data and perform complex computations rapidly to arrive at and compare alternatives.

Presenting the final results

Analytical results are best presented in easily read maps and written reports.

CHAPTER 15

Advanced Analysis

In this chapter we have collected a number of special GIS functions and complex operations in which functions are selected according to certain rules. The term *advanced* will not always be adequate, because although individual functions may be relatively simple, they can be used in a complex context.

15.1 Network and Raster Connectivity Operations

Connectivity operations, which utilize sets of functions that exploit the connectivity in data, include:

- Network operations
- Contiguity
- Spread functions
- Streaming

Network operations are performed on vector data. Spread functions, streaming, and some connection operations are performed primarily on raster data.

15.1.1 Network operations

Many GISs support analyses in networks, which are systems of connected lines represented in vector data (Figure 15.1). In the real world, networks consist of road systems, power grids, water supply systems, sewerage systems, and the like, all of which transport movable resources. Network operations are based on:

- Continuous, connected networks
- Rules for displacements in a network
- Definitions of units of measure
- Accumulations of attribute values due to displacements
- Rules for manipulating attribute values

GIS network operations usually include:

- Displacement of resources from one place to another (route optimization; Figure 15.2)
- Allocation of resources from/to a center

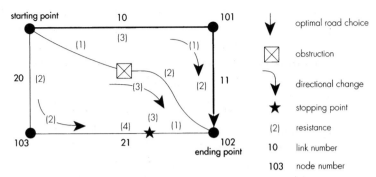

Figure 15.1
Network analyses can enable a system to find the fastest route between start and end points. Resistance (time delay) is entered as an attribute, and the route with least resistance is selected. (Adapted from ESRI 1995.)

These operations are best described through the example of transport analyses.

As is the case for other GIS analytical tasks, network analyses depend on the existence of topology in the data. The system must, for example, know which roads can accept traffic flow at intersections. Every link and every node in a network must have a unique identity, and the program must contain specifications of where roads begin and end.

An analysis may include the simulation of moving resources—cars, people, refuse, and so on—along the lines representing roads. The resources moved along these lines encounter resistance in the form of speed limits, roadwork, peak traffic, barriers, one-way streets, weight limits, traffic lights, bus stops, and sharp curves, to name only a few.

Figure 15.2
The figure simulates a given bus route. At stop 1, 10 passengers board the bus; at stop 2, 30 passengers get on board; at stop 3, 25 people dismount; stop 4 is omitted; at stop 5 the remaining passengers get off and the bus carries on to the depot (end point). (Adapted from ESRI 1995.)

289

The network system elements comprise links, barriers, stopping points, and centers. Assigning attributes to such elements permits the simulation of realistic situations. For example, attributes representing resistance in the network may be expressed in seconds or minutes—perhaps a delay of 20 seconds at each traffic light or of 1 minute at each bus stop. Resistance in the network may also be computed on the basis of the accumulated distance and stipulated speed limits. A section of a road that is 1 km long and has a posted speed limit of 60 km/hr represents a delay of 1 minute. A barrier can, for example, be described as a one-way street, left turn prohibited, or limited height under a bridge.

Once all the attributes have been allocated, the system can assess the movement of resources through the network. A unit of measure (distance, time, monetary unit, etc.) is then employed so that the system can evaluate alternatives and select the route of least resistance. In this manner, transport analyses may be conducted to optimize driving time and minimize travel distance and costs. In practice, analyses usually result in several alternatives that illustrate the consequences of the foregoing simulations. There are different algorithms for calculation of the shortest, and possibly fastest, road between two points in a network. There will be many alternatives for testing in a larger network. Therefore, it is usual to choose a strategy that is limited to the most probable alternatives. There are also solutions to finding the fastest/cheapest road where there are a series of fixed points along the road—known as the "traveling salesman."

Other common network analyses include the distribution of resources from one or more centers, or the assignment of resources from an area to one or more centers. Such resources, flowing through a network to or from centers, must be associated with network lines or points. For example, all pupils living in a school district are associated with their respective home addresses.

The computational processes necessary to move resources out from a center continue to a prescribed limit. A limit of 10 minutes from a center in a street network will define those streets as having a total resistance of 10 minutes from the center. The result may be a plotted map of the streets involved (Figure 15.3). Resources displaced from or to a center may move along nonbilateral lines, perhaps one offering a resistance of 10 seconds in one direction and 15 seconds in the other; several centers may be involved in the same computational system.

Centers can delineate areas even further. An example is an area within 5 minutes driving time of a school. Thereafter, the area delineated may be employed either in a proximity analysis of the objects it encompasses, or in an overlaying function.

15.1.2 Connectivity operations on raster data

A few GISs support connectivity operations in raster data, a process that requires discrete cell-by-cell displacements, originating from a sin-

Figure 15.3
Displacement of resources (travel time) from a center. (From ESRI 1995.)

gle starting point (Figure 15.4). The cells have to contain values—friction values—which are significant for how one can move on the surface. Thus, the whole raster layer will represent a friction surface. Each cell has eight neighbors (fewer on the outside edge), and before moving, all eight alternatives have to be tested to find the direction that gives the least resistance (Figure 15.5), and the easiest direction to move.

The process paradigm involves the choice of units of measure and of other measurement parameters. The attribute values of the cells are accumulated during the transit of each cell. By combining this with distance measurements in the raster it is possible to find the shortest or fastest way between two points on the surface. The result will be two raster layers, one with accumulated resistance and one with the corresponding choice of route.

The technique can also be used to compute precipitation runoff from a terrain surface, as discussed in Section 15.7, "Hydrologic Modeling." The accumulated cost surface can also be used in dispersion modeling and least cost analyses.

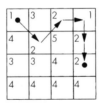

Figure 15.4
Optimized route location on raster data.

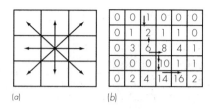

(a)   (b)

Figure 15.5
There are eight possible flow directions from a single cell (a). The number of cells that run into each other are accumulated (b). The cells with the highest numbers (values) represent the watershed network.

15.2   Spatial Interpolation and Proximity Operations

All GISs support some form of proximity operation, in which attribute values are assigned to new points on the basis of the values of existing neighboring points or observations. In ordinary circumstances, proximity manipulations (which include topographic functions associated with manipulations in digital terrain models) involve routines for area search and interpolation.

Search operations take place within a specified area, usually from a defined position to a new point, as within a circle around a point (Figure 15.6). Existing attribute values within the area are treated in accordance with the criteria for determining the values of new points. The procedure for proximity operations usually comprises four main steps:

1. Identify a base point.
2. Define or compute the search area.
3. Select or search for objects.
4. Manipulate the attribute data in accordance with the selection criteria.

The manipulations of attribute data may comprise statistical computations, including:

Figure 15.6
An area search operation can be carried out from a given point and within a circle of a given radius.

- The sum of all values
- The average of values
- The greatest or least value
- Interpolation of values from those of neighboring objects
- Identification of the most frequently occurring value
- Statistical distribution of values

Area search and the subsequent computation of elevations of grid points from elevations of points within grid cells are used to compile digital terrain grid models. The search criteria for attribute data may be stipulated at the same time as the delineation of the search window, which may simply be a specification of the coordinates of the upper left and lower right corners. For example, in a search covering properties, the criterion may be all residences on lots larger than 5 decares. Such searches often combine data from several thematic layers, thus necessitating searches among them. Another area search criterion may be a random polygon, perhaps all the cultivated fields within a given window that correspond exactly, within a given polygon. Hence, in many respects, proximity operations resemble overlaying.

Interpolation and extrapolation

There are many situations in which the available observations are fewer than needed to cover an area satisfactorily, whether it be attribute data or geometric data. With the use of both interpolation and extrapolation techniques (Figures 15.7 and 15.8), it is possible to calculate new values for points lying within or outside the applicable data area. The various inter-/extrapolation methods are based on different assumptions about the "nature" of the observations or their different modes of behavior. The methods thus describe as well as possible a reality that, to a certain extent, is unknown. The task is to calculate the most likely value of the new point based on available observations. The most usual techniques are:

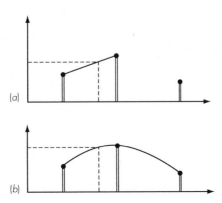

Figure 15.7
Interpolation in data: linear *(a)* and spline *(b)*.

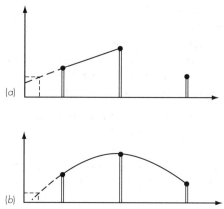

Figure 15.8
Extrapolation of data: linear *(a)* and spline *(b)*.

- *Closest-point inter-/extrapolation.* The new point is given the same attribute value as the closest known point/observation.
- *Linear inter-/extrapolation.* The new point is given a value calculated from a straight line between the two closest known points/observations.
- *Spline inter-/extrapolation.* The new point is given a value calculated from a curve between the three closest known points/observations.

Different variations of these techniques are available. For example, several known points can be used and the values of the known points can be given different weights in the calculations. It is often the case that the closest known points represent the new point better than points located at a distance; thus, close points can be given greater weight in the calculations than distant points. This method is known as *weighted average* (Figure 15.9). It is a relatively simple method, but the result is influenced by how weighting is determined, how many observations are included, and how the observation is geometrically divided.

We have previously seen that splines can be used to smooth curves and compress data. This method can be used to interpolate new values in space. The use of splines is based on polynomials being adapted to local parts of the mass of points/observations by a mathematical adjustment. Following the adjustment, the new surface may not necessarily go through the observations, but will be adapted as well as possible.

Another variation is known as *kriging*, after a South Africa mining geologist, D. G. Krig. Kriging is based on the theory that the same pattern of variation can be observed at all locations and uses a mathematical function to model the spatial variation in values within the in-

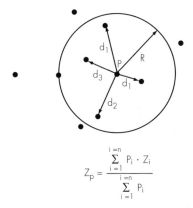

$$Z_p = \frac{\sum\limits_{i=1}^{i=n} P_i \cdot Z_i}{\sum\limits_{i=1}^{i=n} P_i}$$

Figure 15.9
Weighted average: where the value (Z_p) of the new point (P) is based on the weighted average of the value (Z_i) of all known points within the circle. The known points that lie closest (d_i) are given the highest weight.

put sample points. The variation is measured using the semivariance, which is half the average squared difference in values between pairs of input sample points; it is used to determine the optimal weight for interpolation.

Borders may also be interpolated around points to form surfaces, which themselves may be formed using *Thiessen polygons,* in which the borders around a point are delineated equally distant from all neighboring points (Figure 15.10). The polygons are formed by pulling lines successively between the points, and in the middle of the each line a normal is established. These center-normals are extended until they cross each other and form corners in the polygon around each point.

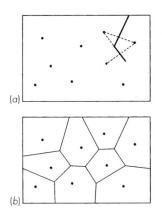

Figure 15.10
The Thiessen polygon is created around a point in such a way that the polygon's border will be located equidistant from each adjacent point. This is done by establishing normals in the middle of the lines drawn between two neighboring points.

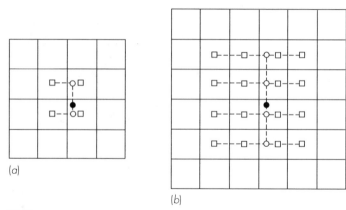

(a)

(b)

Figure 15.11
The value of a new point (●) in the raster can be interpolated by (a) linear interpolation from the centers in the 2 × 2 closest neighbor points, or with the help of (b) polynomial interpolation from the centers in the 3 × 3 or 4 × 4 closest points. Temporary auxiliary points (○) are established during the interpolation. Different strategies can be used for the choice of interpolation points: the closest center point, as in the figure, or closest corner point.

Interpolation and proximity operations on raster data
Raster data are well suited for interpolation and contiguity operations. A raster GIS can assign new cell values that are dependent on the values of the neighboring cells of the original layer. Values for the new pixels can be interpolated from the old ones with the help of various methods. The two most common interpolation methods are *bilinear interpolation* and *cubic convolution*. With binary interpolation the value is determined by linear interpolation from the 2 × 2 closest pixels, and by cubic convolution the value is determined by polynomial interpolation from the 3 × 3 or 4 × 4 or closest pixels (Figure 15.11). One can base interpolation on either the center point in a raster cell, as in Figure 15.11, or the corner points of the cell.

This may be accomplished, for example, by step-by-step scanning using, for example, a 3 × 3 cell window or a filter (Figure 15.12). The value of the cell in the middle of the window may be computed from

3 x 3

			2	2	2	2	2	2
	2		2	2	2	2	2	2
			3	3	3	2	2	3
1	1	1	3	3	3	2	3	3
4	4	4	4	4	4	3	3	1
4	4	4	4	4	3	3	1	1
5	4	4	4	3	3	1	1	1
5	4	4	3	3	1	1	1	1

Figure 15.12
A contiguity operation using step-by-step window scanning of raster data. The value of the center cell within the window can be manipulated in various ways.

the values of all nine cells viewed. Different windows' filter sizes can be selected. Once the filter size is stated, the operations to be carried out on the cell values covered by the filter can be specified. The most usual operations are calculation of the mean value, the most frequent value, standard deviation, or interpolation with the help of a linear function or with a polynomial.

15.3   Fuzzy Analysis

In the traditional discrete model, objects are defined as being either outside or within specific classes. It areas with a population density of more than 1000 persons per square kilometer are defined as high density, areas with a density of 999 persons per square kilometer would be classified as medium density, even though the real difference is very slight. These problems can be solved by using *fuzzy set theory*, which allows an object to belong only partially to a class. Each object can be given one or more additional values between 0 (nonmember) and 10 (full member); for example, 7 may be used to specify how certainly an object can be said to belong to a certain class. If there are several attributes within one area, a corresponding number of additional values can be used. In an analysis context, such as overlay, fuzzy values are handled equally with other attributes that are linked to the actual areas.

15.4   GIS Analytic Models

Several theoretical models have been evolved to exploit GIS analytic functions. The three prime examples are cartographic algebra, expert systems, and linear combinations. The models must be regarded as aids to expert judgment, because the drawback in all of them is that they are based on theoretical assumptions that cannot cover all practical situations.

15.4.1   Cartographic algebra

Cartographic algebra is based on the assumption that a set of simple operations can be found and joined sequentially to constitute relatively complex modeling (Figure 15.13). This process starts with an existing set of attribute tables, which are then processed in a sequence of operations, each of which produces a new column in a table. Finally, the attribute sought is compiled. It may be represented by a flowchart in which the thematic map is processed successively until the final version is produced (Strand 1991a).

Attribute
tables

Figure 15.13
Cartographic algebra. Thematic layers A, B, and C with their own attribute tables
(a) are manipulated by simple operations *(b)* to a new thematic layer and
attribute tables *(c)*, which are, again, by a simple operation, manipulated to one
result layer *(d)* and a result attribute table *(e)*.

Cartographic algebra may require the following operations:

- Reclassification
- Averaging
- Maximizing
- Subtracting
- Adding
- Scattering
- Streaming

The success of the model is contingent on the manipulated data being of a form suitable to the operations concerned. For example, attribute values arranged in classes cannot be added or subtracted and are therefore unsuitable for cartographic algebra.

15.4.2 Expert systems

GISs are increasingly implemented with ever more "intelligence." One intelligent approach is that of the *expert system*, in which analytical results are assessed automatically in terms of criteria entered as ancillary information, or attributes. On the basis of these criteria, the system assigns priorities to various combinations of attributes and then provides an output comprising recommended choices and courses of action (Figure 15.14).

Attribute	Attribute	Priority
Cultivated	Development	9
Cultivated	Recreation	6
Cultivated	Preservation	7
Grazing	Development	3
Grazing	Recreation	1
Grazing	Preservation	8
Forest	Development	4
Forest	Recreation	2

(a)

(b)

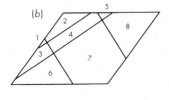

ID		Attribute		Attribute	Priority
1	1	Cultivated	1	Development	9
2	2	Grazing	1	Development	3
3	1	Cultivated	2	Recreation	6
4	2	Grazing	2	Recreation	1
5	3	Forest	2	Recreation	2
6	1	Cultivated	3	Preservation	7
7	2	Grazing	3	Preservation	8
8	3	Forest	3	Preservation	5

Figure 15.14
Expert systems. The expert's criteria in table *(a)* are combined with attribute table *(b)* for an automatic evaluation of the result of an overlay operation.

In GIS, each attribute type may comprise a number of attribute values. "Soil quality" may have three quality classes, "conservation" may have two conservation classes, and "Quaternary geology" may comprise three types of geologic deposits. The attribute values of these three attribute types may be combined to yield $3 \times 2 \times 3 = 18$ possible combinations. One or more experts may then rank the 18 combinations with regard to a particular goal, such as development. All combinations, including conservation classes, are ill suited to development; some may be unworkable for technical or other reasons and hence must be discarded.

The evaluations of the experts are then arranged in a table in which the 18 combinations are ranked qualitatively: "very suitable," "suitable," "moderately suitable," "moderately unsuitable," and "very unsuitable"—or quantitatively by priority 1–9. The table can then be entered in the GIS and will serve as a reference for every record (ITU) in the attribute table that results from GIS analyses. The expert recommendations can thus be entered as a new attribute, or as a column in the attribute table, and can be queried whenever reports are exported. It is clear that the answers provided by an expert system can be no more reliable than the professional expertise available in compiling the evaluations entered.

15.4.3 Linear combination

The linear combination model is used both in impact assessment and in suitability analysis. The starting point is a scale, such as from 0 to

5, which is used to rank features with respect to a stated application. The attribute values of all attribute types are first coded by a knowledgeable expert. Rank on the scale indicates the mutual relationships of the attribute values. In the aforementioned expert systems example, for instance, the three categories of soil quality may be assigned 1 for low quality, 3 for medium quality, and 5 for high quality. Corresponding rankings may be assigned to the conservation and Quaternary geology attributes. The result is that all attributes have rank.

In practice, the different attributes will have varying impacts on the end product of the analysis. Conservation may well weigh more heavily than geology in selecting an area for development. Consequently, the various types of attributes must be weighted to indicate their relative importance. All records (ITUs) in the attribute table are then weighted according to the attribute type involved. The result is a set of weighted features, which may then become new attribute values in the attribute table.

15.5 Practical Application of GIS Analytical Functions

15.5.1 Statement of the problem

In many towns, land use depends on natural resource data. However, urban area planning often makes poor use of such data. As a rule, extensive surveys are conducted, but further processing, including the superimposition of data and the assigning of priorities to categories of natural resources, is often neglected.

A digital map database could be established for the purpose of area planning, and natural resource data are often available on manuscript maps. The goal should be to produce maps of selected natural data and then superimpose them selectively to produce a map of priorities for wildlife, fishing, conservation, and outdoor activities. An additional goal should be to produce a composite ground resource map for use in town planning and land administration.

15.5.2 Procedure

Raw data
The raw data used to assign priorities to categories of natural resources:

- Digital basic data from 1:50,000 topographic maps: lake and watercourse contours, elevation contours, and roads
- Digital plan extract of township plan
- Digital data for precipitation area classification
- Manuscript maps for wildlife (reindeer, deer, moose, roe deer, black grouse, wood grouse, and ptarmigan), outdoor recreation, and conservation

- Extracts of computerized registers (attribute data)
- Data having IDs in common with the manuscript maps for wildlife, outdoor recreation, and conservation
- Fish registry data, including coordinates of fish monitoring stations

Relating geometry and attributes

All natural resource themes to be related to existing computerized registers were first digitized as line information and identified by codes. For example, the wildlife information on manuscript maps contained many overlapping symbols with consecutive numbers not associated with the species involved (Figure 9.3). Consequently, IDs were necessary to relate the information to attribute data and facilitate the subsequent sorting of the various species of wildlife. After digitization, the objects (points, lines, and area boundaries) were linked to the existing attribute data, which were also used to control the use of symbols in subsequent mapmaking.

Sorting themes

The attribute data for wildlife, fish, conservation, and outdoor recreation were sorted in layers so that no areas overlapped within a layer. Typical sorting criteria were species for wildlife data and conservation topic for conservation data. After sorting, the area data were edited to close all polygons. Then the attribute data originally associated with border lines were transferred to the areas themselves. This was possible because all lines contained attributes referenced to polygons on their right and left sides, and because the polygon numbers were available as numbers in the polygon topology table. The result of this neatening process formed the basis for both the thematic maps and the subsequent processing maps used to assess priorities.

Thematic maps

Data selected from the township plan were used as background data for the thematic maps. For instance, polygons representing industrial buildings or residential areas were reclassified and joined to polygons representing built-up areas. Subjects intended to overlap in final map plotting were distinguished from each other by using differing pixel values.

One goal for the fish resources map was to represent fish population distributions accurately without direct reference to the discrete stock counts made in the field. This was accomplished in the following way. First, the digital precipitation data for partial and complete watercourses permitted computation of fish population averages for lakes and watercourses, which provided status data independent of species. Thereafter, a polygon layer representing the precipitation area borders was compiled from the watercourse register. Individual fish stock

301

counts were then correlated to precipitation areas in a point-on-polygon routine. Finally, averages were computed for each area to produce the desired fish population distribution.

The first draft distributions of fish populations and those of other subjects often revealed inconsistencies or shortcomings in the original data. The professional staff in the relevant disciplines then supplemented or corrected the original data. With the new data, new thematic maps could be produced rapidly by replicating the processes used in producing the first drafts. This, in turn, required that all stages of the production process be well organized.

Use of overlays and buffer zones

Differing area categories were overlaid to identify areas with overlapping subjects, and union functions (NAND operations) were used in testing two layers at a time. Buffer zones were employed around such line information as wildlife tracks and hiking trails, to extrapolate them to areas.

Weighting

The three most prevalent wildlife species—moose, roe deer, and ptarmigan—were assigned priority weightings: 1, 2, and 3, respectively. These weightings, which were placed in a dedicated field in the attribute table, were retained through the overlaying process to produce aggregate weightings for all polygons. Wildlife tracks, represented by buffer zones of various widths, were also weighted.

Union functions were used in successive superimpositions of layers, so that the final result was a single layer containing the relevant aggregate information (Figure 15.15). Attribute fields containing the weightings for the individual thematic layers were retained in the individual polygons, so total weighting could be computed for each. High total weightings dictated that polygons should have high priority; lower total weightings dictated lower priorities.

Moose

ID	Weight
1	1
2	1
3	0
4	1
5	3

+

Deer

ID	Weight
1	2
2	1
3	1
4	0
5	1

+

Ptarmigan

ID	Weight
1	0
2	2
3	1
4	3
5	2

=

ID	Moose Weight	Deer Weight	Ptarmigan Weight	SUM
1	1	2	0	3
2	0	1	2	3
3	1	2	3	6
4	2	0	1	3
5	1	0	2	3
6	0	0	0	0
7	1	1	1	3

Figure 15.15
Use of the union function to find the combined point sum for the various units in the figure.

Unified priorities for wildlife, fish, conservation, and outdoor recreation
Weighting was also used to assign unified priorities for wildlife, fish, conservation, and outdoor recreation. Each subject was divided into two classes, assigned weightings of 1 and 2, respectively. Following the overlay process, the total weightings for the individual polygons

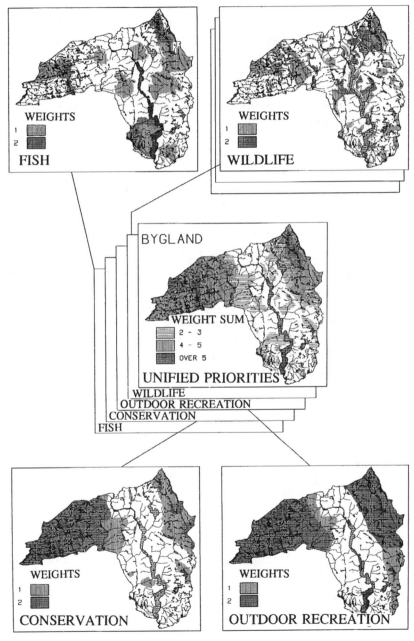

Figure 15.16
Unified priorities to natural resource data. (Courtesy of Powel GEMINI.)

provided the basis for plotting a priority map. Numerous statistical computations, including sum, average, and minimum/maximum, were performed in the attribute tables, both between various attributes for the same object and for like attributes for all objects in a thematic layer.

Adapting final results and plotting the priority map

Finally, an ordinal scale of priorities of the individual areas and polygons was compiled using the weighting sums resulting from the overlay processes. Again, the final assignments identified inconsistencies and omissions. These required revision of the foregoing weightings and computations, adjustments that were made relatively easily because the weightings of the individual subjects were retained throughout the various manipulations.

Once these adjustments had been completed, the final map was plotted (Figure 15.16). The ordinal priority ranks were used to control symbols assigned to the polygons, a project involving many themes that called for expert judgment in delineating borders. An important aspect of the entire "mechanics" of GIS processing was that human expertise was used continuously to evaluate intermediary and final results.

As is often the case in GIS applications, when expert input is used in processing, GIS obliges experts to be more concise in their specifications. A strength is that the ability of GIS to superimpose various alternatives provides rapid testing of hypotheses. GISs do not supplant professional expertise, but they do contribute to documenting it in the area planning process. The various weightings and the assignment of ordinal priorities have much in common with the processes involved in expert systems and linear combinations, as described in this chapter.

15.6 Digital Terrain Models

Digital terrain models may be used in various operations that are partly included in the operations discussed in the preceding sections.

Elevations

Using elevation data stored as a point cloud, the elevation of a random point can be computed as the weighted median of the surrounding points, with the closest points having the greatest weight. Neighboring points are searched within a specified area (Figure 15.17).

Mathematical functions more complex than the weighted median may also be used (kriging, etc.). The elevation of a new point in a grid model may be computed as the elevation of a plane passed through the closest grid points. Elevations may also be computed from the

Figure 15.17
Area search.

faces and corner elevations in a triangulated irregular network (TIN) model.

Slope and slope direction
A terrain's angle of slope and direction may be computed relative to a plane through the elevation model points. In the raster model, the angle of slope can be calculated by moving a window—for example, with a size of 3 × 3 cells—over the raster, see Figure 15.12. Based on the nine values of the cells in the window, the slope of the cell in the center can be calculated in different ways. The simplest method to calculate the angle of slope is to calculate the maximum height difference between the center cell and the neighboring cells, and divide the height difference by the distance between the centers of the two actual cells. The mean value of all the eight height slopes can also be adopted for calculating the angle of slope of the center cell. The window is moved column by column and row by row so that the slopes of all the cells in the raster are calculated, apart from the band of cells that lie on the periphery. The simple linear interpolation can be replaced by a more advanced technique that adjusts a level through the cells.

The direction of slope can also be calculated by comparing the height of the center cell with the height of the eight neighboring cells (Figure 15.18). One thus obtains eight different slope directions. The direction of the greatest height difference can be adopted as the direction of slope for the center cell.

$$tg\,\theta = \frac{b}{a}$$

Figure 15.18
The slope in a raster DTM can be calculated by simple trigonometry, based on the distance between cell centers and the corresponding height difference.

305

Figure 15.19
Slope calculation in a TIN model.

In the TIN model the angle of slope and direction of slope are constant for each triangle and are calculated from the coordinates (x, y, and z) to the corner points in the triangle (Figure 15.19).

Slope angle and direction are useful in simulations of precipitation runoff and in defining drainage areas. The triangle terrain model, which comprises sloping triangular surfaces, is ideal for simulating terrain surface conditions. Slope and slope direction are computed directly from the coordinates of the corners of a triangle. Slope is usually expressed in terms of degrees or percentage with respect to the horizontal. Slope direction is almost always expressed in degrees from north.

Contours

The contour lines of constant elevation may be interpolated directly from the digital elevation model. The best result is normally obtained when the points are divided evenly or according to the surface variations. Raster data and TIN data are therefore good starting points for interpolation. The program used identifies x and y coordinates for new points having the same interpolated elevation and then connects them to form contour lines (Figure 15.20). Many methods are used—among others, linear and spline interpolation. In the triangulated irregular network (TIN) model, triangle faces intersecting a stipulated contour elevation are identified and the points of intersection can be determined by linear interpolation. In elevation data stored in the form of

Figure 15.20
Calculation of isolines (contour lines) in a point cloud of spot heights.

point clouds, triangles are first formed, then treated similarly to those of the triangular terrain model.

In grid models, the grid points on either side of a contour line are identified. A plane through the points or linear computations can be used to locate the new point on the contour line. Height values can be linked either to the corners in each cell or to the centers. Interpolation can be earned out alternatively between the corner points or the center points and will give a somewhat different result. In all cases, the contour interval can be chosen to suit the accuracy required. The various approaches to generating contour lines can produce differing results on plotted maps.

Volumes

Accurate surveys of excavation volumes are crucial in planning roadwork, major industrial buildings, and so on. Soil borings are used to calculate cover quantities, such as down to bedrock, and replacement quantities, such as down to usable soil. From these quantities, haulage can be calculated down to the project level. Quantity surveys may be conducted in GIS by entering data in a terrain model program, which includes the extent of the excavations in elevation and ground plan, soil conditions, slopes for planned leveling, volume expansions due to blasting, and so on. The quantity survey computations in raster data are based on multiplying each cell's size by the height difference of surface in relation to which the volume is to be calculated, and then summing the volume of each cell elements.

Viewshed and shading techniques

It is often interesting to visualize the parts of the terrain that are visible from different observation points. The results of excavations may be visualized by excising a rectangular area in the terrain model and viewing it at an angle from above. Users are generally free to select the viewing point and the viewing angle that best displays the features sought. The visible areas are calculated by drawing a fan of straight lines from the point of observation. The deletion process calls for the sequential sorting of lines, starting from the observation point and working backward. The lines are tested, one by one, for visibility. By simple geometric calculations visible and invisible (vanished) parts are calculated along each line (Figure 15.21). This is a relatively time-consuming process, but is carried out most simply on raster data. Calculations are more complex using TIN data.

A terrain model may be used to automatically produce relief maps that use shading to effect the appearance of the third dimension of height. These processes demand the computation of shaded areas from an assumed solar position (Figure 15.22). The simplest approach is to color the cells of a raster model as seen and not-seen.

Figure 15.21
When calculating visible areas in digital terrain models, a number of lines have to be drawn from the observation point to test which parts of the terrain are not visible.

Visualization and draping

Profiles and grids in the area are plotted to suitable densities. Regular grids are best for the purpose, so several triangulated irregular network (TIN) models support regular grids for plotting. Drawing straight lines between successive grid points often produces a rough, serrated terrain surface. Consequently, some systems support splines, or curved-line segments, for plotting.

Perspective drawings employing profiles are usually based on isometric perspective images of terrain. That is, the units of measure are

Figure 15.22
Shadowing of a topographic map. (Courtesy of the Norwegian Mapping Authority.)

Figure 15.23
Perspective drawing. (Courtesy of Asplan Viak.)

fixed along the axes, and the foreground of the image has the same scale as the background. Central perspective produces a more realistic image of terrain but does not permit measurements, because scales vary in plots (Figure 15.23).

For the best replication of terrain, the program must delete profile lines and surfaces hidden by terrain rises as seen from a side view, as explained earlier for calculating viewshed.

Draping a digital aerial photo, satellite image, or thematic map over the surface of a terrain model produces a more realistic terrain image. However, draping two-dimensional images over three-dimensional surfaces poses problems, because a two-dimensional image does not contain enough pixels to cover the terrain surfaces. The system must then compute pixel values for surfaces not covered, which deforms the two-dimensional image.

Miscellaneous applications of terrain models

The principles of interpolation and visualization in digital terrain models may be applied to other types of three-dimensional data. For example, the third dimension in georeferenced data may be time, population data, farm productivity, geophysical measurements, or other factors. In visualizations, elevations may be replaced by any numerical values related to the observation points. Similarly, interpolation between contours may be performed or perspective drawings plotted.

15.7 Hydrologic Modeling

The use of a digital terrain model makes it possible to calculate a synthetic drainage system in the terrain, including the individual areas for

Figure 15.24
A watershed consists of watershed boundaries, subbasin, stream network, and outlets.

drainage, drainage network, and nodes through which the water runs (Figure 15.24). This kind of hydrological model can be used to simulate different drainage situations, including the establishment of parameters for calculating the dimensions of the piping network in connection with the development of a housing complex.

A grid-based terrain model represents the continuous surface of the terrain. Drainage will always be in the direction with the lowest elevation. Each cell has eight adjoining cells and eight possible drainage directions (Figure 15.25). Comparison of the basis cell's elevation value with the values of the adjoining cells thereby determines the direction. The sum of the number of cells that run into each other gives the accumulated drainage and forms the basis for calculation of the network. Cells with a high accumulated value show where the water collects and create a drainage pattern. Cells with the lowest accumulated value represent the limit for individual watersheds. The cells that represent the network (and the boundaries for the drainage areas) can be vectorized and attributes linked to the different nodes and links in the net-

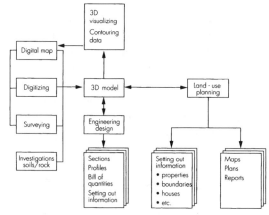

Figure 15.25
Flowchart illustrating the use of GIS in engineering.

work. The links in the network can be numbered automatically (e.g., based on a ranking system whereby the lowest number is allocated to the links with no links upstream, and the highest number to the links with the highest number of links upstream).

When calculating the dimensions of the water piping network, it is vital to know the probable water consumption, which could be represented, for example, by the number of persons who will be linked to each node. The number of planned dwellings within each section of a rainfall area may also be a suitable parameter. When dimensioning the sewage network (surface water), it is also important to know the probable volume of water that will run through each node. The drainage volume is dependent on the water flow, among other factors, which is in turn dependent on the degree of slope and type of surface. The degree of slope can be calculated from the digital terrain model. As to type of surface, bituminous roads and corresponding surfaces will result in faster drainage of cultivated land. A digital map with surface structure, with use of an overlay, can supply data on the area for sealed and permeable surfaces within each drainage field. The final calculation of piping dimensions is carried out in a special program for water and sewage networks.

15.8 Functions for Engineering GIS

Geographical data are often used in solving such engineering problems as locating basic points, surveying properties, and planning various phases of roads, water supplies, sewerage, residential areas, and so on. These tasks have previously been accomplished using stand-alone PC programs, but the trend is to integrate them into urban GISs (Figure 15.25).

Surveying and other measurements

Surveying programs usually include functions for computing coordinate geometry based on in-field measurements, with interfaces to various surveying instruments. For engineering applications, they may be extended to compute the intersections between lines, between lines and circles, and so on. They must also support computations of various geometric figures so that exact geometry may be computed.

Other operations may also be performed. The layout data of plans may be computed, and various map projections may be manipulated and transformations made between projections. Often, the systems can produce standardized deed or other maps in local or standard formats. Some systems support the method of least squares, as well as automatic quality designations in the form of error ellipses. Other systems can combine GPS vectors with conventional measurements.

Manipulating terrain

To aid engineering work, digital terrain model systems have been developed to manipulate simultaneously as many as ten separate terrain layers, such as loose fill quantity, bedrock, and various geological strata. Volumes for cut-and-fill operations are computed and documented automatically in maps, perspective drawings, profiles, and cut specifications. For example, when a new road has been aligned, it may be designed with exact curve radii, transition curves, lengths, and cross sections. Thereafter, its plans may be plotted with center lines, driving lanes, shoulders, cuts, fills, building lines, and so on. These data are then used to set out lines in terrain. The documentation concerned may be standardized, for example, to the standards of a roadwork commission.

As visualization prior to construction is desirable in many engineering applications, some GIS terrain models are integrated with three-dimensional computer-aided design (CAD) systems. Such a combination can present realistic images of the planned construction, perhaps showing a complete building, with facades, windows, and interiors surrounded by graphic symbols for trees, cars, and other extraneous objects.

CHAPTER 16

Visualization

As we have already seen, GISs are not merely automatic mapmakers but are capable of a wide range of tasks. Using GIS, the cartography of a map can be supplemented with textual information, digital images, sound, diagrams, and other miscellaneous graphical information. Today, GIS is linked with multimedia and virtual reality. Thus, the concept of visualization is extended beyond the traditional view of cartography.

Nonetheless, whether displayed on screen or printed on paper, maps are the primary GIS product for end users. However, a map will generally be only an aid in a decision-making process or in identifying a situation. The real end product may be a building permit, a new road, or a new conservation area. By using GIS, the visual product can be quite closely related to the decision-making process. As mentioned earlier, the product for the end user is a map, which has been created for a particular purpose, often with a short life span, based on data of varying origin and quality. This places new and greater demands on the map designer.

A map is a graphic rendering of information, and in GIS the cartographer's task is to transform geographical data in digital form into a visual product that is accurate and user-friendly. Regardless of the production methods used, a map is a medium for visual communication, and, like other visual communications media, its appearance affects how it is perceived and, consequently, how readily the information it contains is interpreted by the user. The transmission of information via a map is a one-way process—from the sender to the receiver, which sets particularly high demands on a map as a means of communication. Traditional cartography is a venerable craft, in which artisans train for long periods to achieve the aesthetic appeal necessary to ensure that maps communicate the necessary information to users. Even the most skilled cartographers, though, are limited by the drawing implements at their disposal. Computerized GISs are not so limited and are therefore capable of producing a far greater variety of maps than can be drawn by cartographers, even when using the best hand tools of their craft.

This capability has the enormous potential of enabling less skilled mapmakers to produce truly professional maps; it also has the drawback of permitting the production of artless maps that are at best unattractive and at worst misleading. Thus far, there are no formal rules

as to how a map should be modeled, and GIS facilities have no cartographic intelligence to guide operators in the choice of map symbols
and other graphic effects. Consequently, those who operate GIS facilities should be reasonably knowledgeable in all aspects of cartographic
communications.

The use of computers in map presentation makes it possible to divide maps into categories other than the traditional base maps and
theme maps. Today, maps are often divided into the following categories (Balstrøm et al. 1994; Brande-Lavridsen 1994):

- Hard-copy maps, drawn or printed on paper
- "Soft" maps, visible on a computer screen
- Digital map data held in an electronic database
- Real-time maps, maps that show changes temporally

16.1   Theoretical Foundation

It is prerequisite in the visualization phase that a preliminary treatment of the information has been carried out in the form of conceptual
modeling, and that the data selected are available in digital form, normally in a database. Conceptual modeling is mentioned earlier in this
book. This discussion is therefore limited to the process used to visualize the data.

16.1.1   From data to display

When visualizing geographical data, it is important to be aware that
one often operates with a series of fundamentally different types of
data, each of which must be handled in a different way. We have already mentioned that geographical data are represented in the data
model by points, lines, and areas (or raster). The same three geometric forms are essential to the presentation of geographical data in maps.
In addition, geometric volumes such as terrain surface, buildings, and
other three-dimensional objects can be represented.

Attribute data are divided into two categories: qualitative and quantitative data. *Qualitative data* describe the type of objects, such as different types of vegetation or different types of language. At the quantitative level (nominal level) the objects can only be classified into
groups. As an aid to visualization, it can be useful to subdivide qualitative data into those that are conceived of as different but associative,
and data that are conceived of as clearly different (selective) (Bertin
1983). *Quantitative data* specify number; they are, as previously mentioned, often split into three levels, according to precisely how they
can be measured: ordinal, interval, and ratio. Ordinal data are arranged
in order. Interval data can be placed in differently defined groups or
classes on a number line with an arbitrary zero point and interval. Ra-

tio data are indicated with exact numerical values measured on a scale in relation to a particular zero point and with a particular resolution. Ratio data represent the most exact or highest level of measurement (Figure 3.10). With visualization in mind, it can often be appropriate to place ordinal data in one separate group, with other quantitative data in a second, and qualitative data in a third group.

The temporal perspective is, in principle, a type of data that we wish to use to show changes over time. Generally, there is also a need to be able to show text on a map in the form of place names and attributes such as length, depth, and so on. A name often represent different types of natural phenomena such as water, rivers, mountains, centers of population, and the like, which can be indicated by use of the text on the map. Finally, we need information as to the quality of data—its accuracy, for example. As GIS is used increasingly as a decision-making tool, we now need to be able to visualize the quality of this basis for decision making.

16.1.2 Cartographic instruments

Normally, the aim of cartography is to enable the intuitive recognition and understanding of one or more messages in a map with as little uncertainty as possible. Understanding maps is a mental process that, to a great extent, is based on the human ability to recognize patterns, contexts, and differences. Communication to this end can be achieved by the use of certain cartographic instruments. Traditionally, cartographers command six graphic variables in symbolizing geographical phenomena: size, density (gray scale), color, shape, texture, and orientation (Bertin 1983) (see Figure 16.1). Of course, not every graphic symbol or effect is important in itself. A map with only one symbol, one color, or one gray tone has little value. It is the context in which the variables are presented, together with the geometric variables—points, lines, and areas—that are of importance. As far as graphic variables are concerned, it is their differences in size, color, and so on, that are associated with the differences in the measured values. The six variables differ in their application and can illustrate either the qualitative characteristics of various object types, the quantitative characteristics of various attribute values, or the ordered sequences.

Size
Variation in size is the most convenient way of illustrating variation in the quantity and, to a certain extent, order represented by symbols, whose size can then be made a direct function of the magnitude measured. However, differences in size are not always readily perceived. As with most areas of human perception, the difference in magnitude that can be perceived is related to the initial magnitude involved. Thus, the area of a circle, for example, should be increased by a factor

315

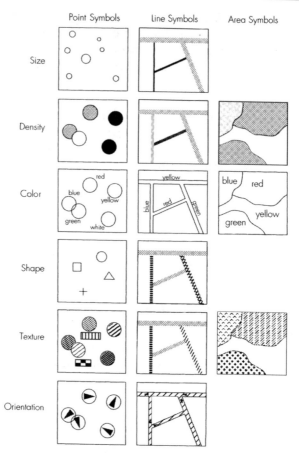

Figure 16.1
In principle, six different graphic parameters can be used to symbolize geographical phenomena: size, density, color, shape, texture, and orientation.

of 4 to give the impression of a doubling of the amount being visualized. To be able to separate area symbols with great certainty, one should not use more than five to six different symbol sizes on a map. Moreover, without a map legend or comparative symbols, it is not possible to create an impression of the absolute size relationship.

The variable instrument "size" can be used only for geographical objects such as points and lines, but not for area, because an area in itself represents a size relationship.

Lightness and gray-scale value
Variation in density (gray scale) is used mainly to illustrate ranked data; normally, darker means greater and lighter smaller. Lightness and gray values are best suited for areas, but they can also be used for point and line objects.. As for all cartographic variables, there is a lower limit to the size of the differences that can be perceived by a

user. Gray scales should be divided into no more than five to seven sections for qualitative data, lest neighboring sections be mistaken for each other. For ordered data, seven to eleven different gray values can be used (Brodersen 2000). It is also important to be aware that the human eye cannot differentiate between more than seven to eight different classes on a map without close consideration and comparison (Kraak and Ormeling 1997). In fact, the maximum number of different objects that can be observed by sight alone, without consciously connecting in the brain, is four.

The properties of density are illustrated in Figure 16.2.

The term *lightness* is closely linked to half-tone images. When each grain of pigment is two to three times less than one can see, it is known as a half-tone image. When the grains are larger, then we are talking about raster in the form of points or lines.

Maps in which symbols cover areas in a density proportional to the value of the phenomena being represented are often termed *choropleth maps*, from the Greek *khoros* (place) and *plethos* (value, number). The lightness of a single color can also be used to illustrate relative magnitude. Variations in the lightness of a single color are, in fact, often more obvious to users than differences in color hue, and display an orderly sequence quite clearly, but as we shall see in the next section, colors are relatively complicated to handle.

Color

Differences in color can be both effective and pleasing and are particularly well suited to distinguishing various qualitative (selective) phenomena from each other. In fact, color is the best cartographic variable with which to separate quantitative data, as the colors appear "attractive" and are easy to combine with other cartographic variables. Unfortunately, colors are also the most frequently misused of all cartographic variables. Although colors are often coded to scales and used to indicate quantitative properties of areas, such as population

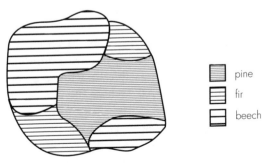

Figure 16.2
Incorrect use of density as a graphic parameter. The impression may easily be created that the map presents magnitudes, whereas in fact, the data shown are qualitative rather than quantitative. (Bjørke 1988b.)

densities, monthly temperature means, degrees of industrialization, and so on, colors themselves carry no subjective implication of rank or order. Red, blue, green, and orange might well represent the values 10, 20, 30, and 40 of some measurable quantity shown on a map, but users can only interpret the rank of a colored area correctly if the map has a color-code key that relates the colors used to the quantity involved. The use of colors also permits the exploitation of common associations and conventions. For example, blue may represent water, green may represent forest, and red, commonly associated with danger, may be used to indicate dangerous areas on a map.

Color can be created by illuminating a neutral surface (such as a computer screen) with red, green, and blue (RGB) light. This is called additive color mixing. When printing and plotting on paper, the complementary colors, cyan, magenta and yellow are used. This is known as subtractive color mixing. Black is used in printing and plotting to deepen the dark areas. From the use of these colors, we get the abbreviation CMYK.

In color theory it is usual to divide the concept of color into hue, saturation, and lightness (HLS). *Hue* is associated with the name of the color (red, green, blue, etc.) and its position in the spectrum. *Saturation* is variation in a color without the use of white or black, and *lightness* is changed by adding white or black to individual hues. One should take care to ensure that all colors used on a map have the same lightness; however, hue and saturation can vary. In general, large areas should have light colors and small areas darker colors. IHS (intensity, hue, and saturation) is a variation of HLS.

When working with digital data, changing colors with the HLS/IHS system rather than with the RGB system gives normally a better control of the design of the colors.

Shape

Various geometric shapes are best suited to indicating qualitative differences (associative data). Shape can be used only on point objects. Shapes convey no overall impression, but may be used to convey details; they are usually noticed last of all the cartographic variables. Map users are aware of the different symbols employed to make maps understandable: a cross represents a church, an airplane an airport, and so on. It is also possible to achieve a special effect by the use of more complex patterns, such as the positioning of symbols in certain groups.

Symbols with a fixed shape can be put together to form a pattern on a surface. Patterns that combine cartographic variables are frequently used to present data arranged in a particular order. In maps containing several cartographic variables, patterns differentiated by symbol size and a scale of grays are more visible than those differentiated by a single symbol shape and size.

Orientation

The differences in orientation refer to different point or line patterns and are best suited to illustrating qualitative differences. The best effect is achieved with short lines in groups placed on surfaces. In principle, oriented symbols can be used on points and lines but not on surfaces. However, different raster angles are used on surfaces, but this group of effects belongs to the texture graphic parameter.

Texture

Texture, alone or in combination, can be represented by:

- Pattern (shape)
- Raster angle (orientation)
- Grains

Pattern has generally the same qualities as shape, and raster angle the same as orientation, that means illustrating qualitative differences. Graining (half-tone density) is achieved by varying the number of points or lines per surface unit while maintaining a fixed degree white and blacking, such as 70% black and 30% white. Variable graining can give an organized sequence. Very occasionally, variations in texture are used to show qualitative differences (associative data). Graining can be used on line and surface symbols, but is little suited to point symbols.

Special observations

The three variables—size, density, and color—together cover most practical needs and are therefore also the most used variables (Figure 16.3). As mentioned earlier, shapes are usually noticed last of all the cartographic variables. In spite of this, they are used relatively often—perhaps too often. There is good reason to note how few opportunities are available to show quantitative data on a map. The only variable here is size.

Cartographic variables have varying ability both to perceive visibility for the phenomena and to distinguish between phenomena. When selecting a variable it is not enough to look at an individual variable; one must also observe how the variables work together (Figure 16.4).

Cartographic variables may be combined to display several phenomena simultaneously. For example, colors may be varied among differing size symbols to show both qualitative and quantitative data. Generally, however, care must be exercised in combining several graphical variables in one and the same symbol because map images can become complex and difficult to interpret.

In addition to the six graphic variables, several other means are also available. The geometric representations of point, line, and surface are cartographic instruments in themselves, represented on a map by dot, dash, and patch, respectively. All six graphic variables can be used in geometry.

AREA SYMBOLS

Graphic parameters	Quantitative data	Ranked data	Quantitative data
Size			
Density	Good	Good	–
Color	Good	–	–
Shape			
Orientation			
Texture			
• Orientation	Moderate	–	–
• Shape	Moderate	–	–
• Granule	Moderate	Medium	–

(a)

LINE SYMBOLS

Graphic parameters	Quantitative data	Ranked data	Quantitative data
Size	Medium	Good	Good
Density	Medium	Medium	–
Color	Medium	–	–
Shape	Moderate	–	–
Orientation	Moderate	–	–
Texture			
• Granule	Moderate	–	–

(b)

POINT SYMBOLS

Graphic parameters	Quantitative data	Ranked data	Quantitative data
Size	Good	Good	Good
Density	Good	Medium	–
Color	Medium	–	–
Shape	Moderate	–	–
Orientation	Medium	–	–
Texture	–	–	–

(c)

Figure 16.3
Relationship between visual graphic parameters and their abilities. As shown here, the three variables, size, density, and color, together cover most needs. One should note how few opportunities are available to show quantitative data on a map. The only variable available here is "size."

Maps on which several phenomena are presented often contain areas where adjacent phenomena overlap. Graphic variables must then be chosen very carefully, so that users readily comprehend the overlaps and do not interpret them as separate phenomena (Figure 16.4). In principle, the cartographic variables are suitable only for showing static data. If it is desirable to display dynamic data in addition, it will be necessary to use several maps, one for each data set. Arrow symbols can be used in this connection to indicate direction of movement.

Figure 16.4
Representation of the diversification of language areas and overlapping of language groups: *(a)* the incorrect impression that the overlap area represents a new language group; *(b)* the correct visual impression. (From Baudouin and Anker 1984.)

Visual hierarchy

A map often has to present one or more main messages on a neutral background map. This forms the basis for dividing information into a visual hierarchy. The graphic variables are used to create such a hierarchy. It is thus possible, for example, to use the fact that shape is perceived later than other graphic variables. Thick lines are perceived faster than thin lines, dark colors faster than light colors and so forth. This indicates, for example, that background data should normally have lighter colors than data representing the main message(s). As presented in the following lists, text symbols can also be used to indicate a hierarchy, with large font types representing important places and small font types representing less important places. One should remember that the list of contents should represent part of the hierarchy in the map.

Text and numbers

Text and numbers also play an important role in cartography (Kraak and Ormeling 1996). Both text and numerals should be able to:

- Display nominal differences (e.g., differentiate between different types of objects and classes)
- Display hierarchy (e.g., differentiate between more important and less important types of objects)
- Be associated with points, lines, and areas on the map (positioning on the map)

A hierarchy can be achieved by using variation in:

- Boldness
- Size
- Spacing between letters and words
- Color
- Style between upper- and lowercase letters

Nominal differences can be achieved by using variations in:

- Color
- Type

16.2 Graphic Generalization

We have previously discussed the need for generalization when the real world is transformed to GIS. In this connection we talked about generalization regarding content, or so-called conceptual generalization. In map presentation we use the term *graphic generalization* (Figure 16.5). The main function of graphic generalization is to ensure that important information on a map is registered more quickly than less important information. This is achieved by graphic effects such as simplification, adjustment, and smoothing. In concrete terms, this includes simplification of geometry, use of symbols, dividing data into groups (classification), and extension of the validity area of the data (extrapolation). The last two points, classification and extrapolation, are mainly related to statistical data. Moreover, the need for graphic generalization is closely linked to map scale, because small map scales require a greater degree of simplification, adjustment, and smoothing than larger map scales.

Typical examples of graphic generalization are the use of thick lines for important roads and thin lines for less important roads, and showing a settlement as a single area instead of using many small house

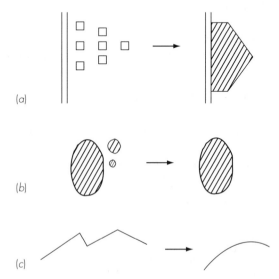

Figure 16.5
Examples of cartographic generalization: *(a)* aggregation, *(b)* amalgamation, *(c)* smoothing.

symbols. Alternatively, areas that are located close to each other and represent the same type of object are joined together in one common surface. When thick lines are used for roads, the consequence can be that other objects (buildings, railroads, etc.) may have to be moved on the map so that symbols do not conflict with each other. Graphic exaggeration may also result in other and less important objects being completely removed from the map—there may physically be no room in the map image. Correspondingly, coastlines can be simplified by removing smaller islands.

Coastlines, roads, elevation contours and similar objects are also simplified through smoothing. Different automatic smoothing routines have been developed, in which the aim is usually to preserve the main characteristics in the geometry with the help of as few points as possible. In general, an operator-controlled manual adjustment is often carried out to obtain a good map image.

The following guidelines can be applied in graphic generalization (Brodersen 1999):

1. Select and enhance principal and characteristic information.
2. Exclude and withhold less important information.
3. Enlarge and simplify important shapes.
4. Simplify small and rough shapes.
5. Combine similar object types into one group.
6. Use symbols when objects cannot be shown with a correct position.
7. Join scattered objects of the same type to one common symbol.
8. Keep the shapes that are not affected by generalization.

16.3   Selecting Map Symbols

The graphic display of information can be successful only if the variables displayed have the same mutual order, similarity, and magnitudes as the variables illustrated (Baudouin and Anker 1984). There are many choices in mapmaking, but they are seldom straightforward. The following guidelines (Palm 1988) are useful:

1. Consider:
   a. The final application of the map (tourism, documentation, planning, etc.)
   b. The probable capabilities of its users
   c. The questions to which the users want answers
   d. Select information which will give the users the answers to their questions.
   e. The grouping of information according to importance
2. Select point, line, or area symbols for the phenomena.
3. Consider the characteristics of the cartographic variables.

4. Use symbols of equal impact to represent equally important variables.
5. Use related symbols for related phenomena.
6. Consider visual phenomena and their characteristics. For example:
   a. Symmetrical symbols (roads, railways, etc.) are often visually dominant.
   b. Symbols appear smaller when they are surrounded by larger symbols, and vice versa.
   c. Whenever two areas adjoin without a border line, a light area appears lighter and a dark area appears darker.
7. Heed tradition (red means danger; the sea is blue).
8. Use existing standards.

As far as standards are concerned, one should remember that a map should always contain a title, scale, grid network, description of projection and coordinate system, as well as producer, year of publication, and source reference.

The most common errors in cartographic presentations include:

- Incorrect use of cartographic variables
- Background color too strong in relation to the most important information presented
- Excessive number of dissimilar themes presented
- Inadequate legends or instructive material

16.4 Limitations and Potentials of GIS in Cartographic Communications

Practical solutions

The ability of a GIS facility to employ cartographic variables is determined primarily by the constraints imposed by the software systems on its screens and plotters. These programs usually contain predefined symbol libraries (Figure 16.6). Users may often define their own symbols, either by digitizing symbols drawn or printed on paper or by using software tools to construct their own symbols (Figure 16.7). As shown in Figure 16.6, a large selection of different styles of lines, symbols, and text lettering is now available.

Normally, GIS will support printouts of defined attribute values, such as measured values and names of measurement stations, or permit attribute values to direct the plotting of selected symbols. The size and color of symbols may be related to the attribute value by checking with a reference (look-up) table (Figure 16.8). Density of symbols or solid colors may be used to indicate phenomena covering areas on a map. For example, colors may be assigned numerical codes, which are entered into the plotting codes for the map polygons, which in turn

Figure 16.6
Typical symbol library. (© ESRI.)

are activated when the map is plotted. Most GIS facilities support a range of letter fonts, either in basic system software or in plotter programs.

Raster displays support color assignments to each individual dot, so raster plots can include solid colors. However, raster resolution may limit the amount of detail in symbols. Similarly, variations in texture may be difficult to achieve. In general, screen resolution sets the upper limit of on-screen enlargement of a map: The greater the resolution, the more a displayed map can be enlarged.

Plotters, whether color raster plotters or film printers, seldom replicate on-screen colors exactly. This is because computer monitors have a greater spectral purity than can be achieved by plotting. This limitation may be vexing, particularly when color nuances are vital, as in the processing of remote sensing images. In such cases, extensive calibration may be required. Raster plotters are best suited for producing graphic effects, because their characteristics are similar to those of raster screens. The quality of the printed product depends on plotter resolution.

Figure 16.7
Symbol design. (© ESRI.)

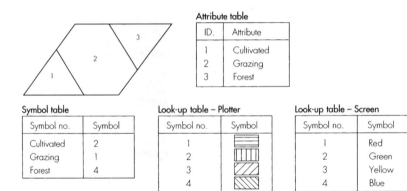

Attribute table

ID.	Attribute
1	Cultivated
2	Grazing
3	Forest

Symbol table

Symbol no.	Symbol
Cultivated	2
Grazing	1
Forest	4

Look-up table – Plotter

Symbol no.	Symbol
1	
2	
3	
4	

Look-up table – Screen

Symbol no.	Symbol
1	Red
2	Green
3	Yellow
4	Blue

Figure 16.8
Look-up table.

The process of transferring digital data to the printer is known as "RIPing," derived from the phrase "raster image processor" (RIP).

As mentioned, geometric generalization is an important tool in map design. Automatic functions have been developed that carry out certain limited parts of the generalization. Some systems support overall changes—for instance, by altering the size of symbols and the width of lines whenever the scale of a map is changed. There is no GIS available today that can carry out the entire generalization process automatically. To a great extent, therefore, the operator will still need to make decisions to the best of his or her ability.

In addition to geometric generalization, conceptual generalization is also used in map design. Conceptual generalization assumes that the operator chooses to merge or remove some object types. In this case, no machine treatment can be performed before the operator has made a choice.

As we know, reality is three-dimensional whereas maps are two-dimensional. Normally, plotting of three-dimensional objects is carried out in GIS by using perspective drawing. Some systems can display several windows of the same segment, viewed from different angles, simultaneously on the screen. There are also examples of stereoscopic presentations on screen and virtual reality, whereby the viewer can freely wander through the terrain. Virtual reality requires unusually powerful computers and is still not available in normal commercial GIS form.

Multimedia

Multimedia and virtual reality are based on the simultaneous use of sound, text, animation, and pictures/video. For example, in tourist information systems and atlases, by pointing at special mapped objects on the screen, the viewer can release information in the form of speech or the playing of music (e.g., a national anthem). Through linking, electronic atlases and other specially adapted applications are able to supply supplementary text information by selecting map objects (e.g., property title owners, historical facts on a country, and the like). By linking pictures and video to objects on a map, it is possible to increase the information value. In its simplest version, animation is based on the presentation of fixed images on the screen in an edited sequence. To obtain a live effect, 25 to 30 pictures per second must be shown. Software is also available that can create new frames by interpolating automatically changing images between two key or main frames. The most advanced method for creating animation is based on algorithms that create new frames automatically, based on the prespecification of objects, changes, and timing for changes (Kraak and Ormeling 1996). The basis for map-related multimedia is that the user can, to a greater or lesser extent, control the effects through predefined links to the map. The term *hyper maps* can be applied here because of the linking opportunities that have been developed.

16.5   Final Comments

Beautiful final products may, of course, hide poor-quality data. Data uncertainty is not easily shown in graphic form. Most experience of showing data uncertainty relates to geometric accuracy, less to showing attribute accuracy, maintenance status, completeness, and logical consistency. Uncertain boundaries between polygons can be shown by dashed lines, double lines, a line darker in its middle and lighter toward its edges, or as a zone of a particular color or of no color at all. Uncertain points can be indicated by circles of differing size and color or as vectors. The network, deformed in proportion to the uncertainty in the point basis, is also one way of showing data quality. A digital terrain model (DTM) can be used with very good effect by placing the size of the uncertainty on the z axis. It is also possible to decide to use a more powerful generalization for inaccurate data than for exact data. Attempts have been made to use blinking lights or sound signals to indicate uncertainty. Today there is no available GIS system that can automatically handle visualization of the quality of data in a satisfactory manner.

Maps are static presentations that are not particularly suited to showing temporal changes. But animation techniques can bring onscreen maps to life. For example, maps may be displayed in rapid sequence to illustrate a particular trend, such as the day-by-day spread of radioactive fallout from Chernobyl during the first two weeks following the disaster. The use of virtual reality will undoubtedly be able to open new opportunities for showing the time dimension.

In short, a GIS exploits its analytical prowess to generate a wide variety of maps rapidly from the same data. However, GIS maps do have cartographic limitations, because few aesthetic capabilities are supported in GIS. The GIS user is unable to manipulate the overall aesthetic appeal of a map as can the conventional cartographer.

One reason for this disparity is that the art of mapmaking (and map printing) has developed over the centuries; the art, like the artisan, is mature. GIS is still young; in consequence, its cartographic aesthetics are relatively crude and the disciplines involved are less mature than those of conventional cartography. The skills of a GIS user are related to the analytical *manipulation* of spatial relationships, whereas those of a conventional cartographer are related to the *representation* of spatial relationships. The GIS user relies on the technological end products of the sciences: computational methods, mathematical operations, and the like. The conventional cartographer relies more on the arts, which involve perception and are therefore integral in both aesthetics and communications.

CHAPTER 17

Choosing a GIS:
Organizational Issues

17.1 Technology and Organization

17.1.1 Historical background

In many ways, the evolution of the applications of geographic information mirrors the course of history. From the earliest seagoing charts to the most recent star atlases, maps have reflected not only a contemporary knowledge of the world but also the contemporary technologies employed to gather it (Figure 17.1). From the earliest Roman census compiled for taxation purposes, to the Domesday Book, to modern demographic studies, overviews of populations and their distribution have recorded stages in the development of commerce, trade, and government. Conversely, many of the quests for improved georeferenced information have spawned changes that have affected history. The search for a reliable method of measuring longitude during the sixteenth, seventeenth, and early eighteenth centuries triggered the first government sponsorship of scientific research projects. Similarly, research into nuclear weaponry during the 1930s and 1940s led to development of the digital computer, one of the earlier examples of civilian benefits resulting from military R&D spending.

Initially, individual mapmakers expended considerable time for a limited gain in knowledge of their immediate areas, even though maps were established as useful sources of geographic information. The desire for innovation spurred progress: the need for accurate longitudinal measurement, the development of the theodolite, machine printing, photogrammetry, aerial photography, satellite remote sensing, and a host of lesser technological breakthroughs all prompted progress. As a result, the curve of qualitative cartographic improvement versus the effort expended to gain it soared—although the curve subsequently leveled off as the effort given to development produced fewer further gains.

Map sheets produced by traditional manual means in the late twentieth century may be both more attractive and more accurate than those of the late nineteenth century, but updating them still requires a considerable investment in worker-hours. The details of a public utility network of the 2000s will probably be more involved and far more

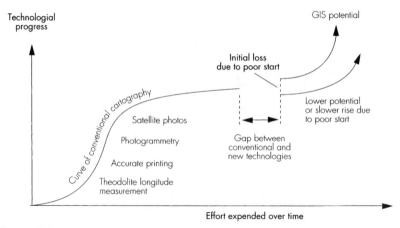

Figure 17.1
Qualitative development in cartographic disciplines.

extensive than those of the same network of the 1890s, but the need for filing of information, on paper in folders in file drawers and in computers has remained unchanged. So the human element—the speed and accuracy with which a cartographer can draw, the time required for a clerk to find and fetch a file—is clearly the ultimate constraint. Computerized GIS can change this overall picture quite dramatically because it dispenses with much of the human intervention required for the conventional acquisition, manipulation, and storage of geographic information.

An organization dealing with georeferenced information may initiate GIS to realize a gain, but, as implied in Figure 17.1, effort (in time and money) is required to switch from conventional means to GIS. Bridging the effort gap successfully is therefore the first step toward realizing the benefits afforded by GIS.

17.1.2 Bridging the gap

Bridging the gap invariably entails switching to a new technology and changing procedures (Figure 17.2). In modern business language, the first steps to initiating GIS in an organization call for a number of strategic choices. For many high-tech approaches, the benefits accrued are related to these initial choices. A GIS facility cannot be bought off the shelf; rather, it is an assemblage of items of equipment that can become a useful tool only when properly placed in an organization and supported by expertise, structured data, and organizational routines (Figure 17.3).

The principal choices concern four aspects of GIS:

1. Data
2. Hardware and software

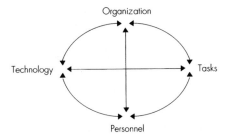

Figure 17.2
Successful initiation of GIS depends on a prudent balance of technology, organization, personnel, and tasks.

3. Expertise
4. Structuring (organization)

Definitions of these four aspects are usually interdependent and may change over time in ways that are difficult to predict. So a systematic approach is essential.

The tasks and issues involved in initiating GIS may be classified either as organizational or technological. This chapter concentrates on the organizational questions. Technical questions are dealt with in Chapter 18.

In practice, GIS projects tend to overfocus on technology and underestimate the organizational tasks. This has also been the case in other areas of information technology (IT). Studies in the United Kingdom show that 30% of large IT projects have budget overruns, owing primarily to organizational problems. In the United States corresponding studies show that 10% of the problems in IT projects are technology related and 90% are related to organizational questions. Nordic studies (Tveitdal 1987; Kylen and Hekland 1990) show that many organizations with GIS experience see, in retrospect, that their GIS activities could have been started more profitably in other ways.

Even though introduction of a GIS is very similar to the introduction of other IT, it appears that GIS is particularly difficult to handle from an organizational viewpoint. There are several reasons for this. With the introduction of GIS, many different users will normally be supplied with data from a variety of sources. This makes it necessary to develop new and complex organizational relations to obtain a good

Figure 17.3
Interlinked chain of a GIS facility.

flow of data. In addition, it can be problematic for an organization to handle the long data establishment period; this often leads to frustration because of the long period between the time of investment and the realization of the utilitarian value. In IT projects it is usual to allow a payback period of one to three years, whereas with larger GIS projects one should normally allow for six to eight years. In that sense, there is a large difference between CAD (computer-aided design) and GIS. It is possible to start using CAD as soon as the software has been installed (and training given). It is not possible to obtain any benefit from GIS until at least some of the database has been established.

17.1.3 Impact of GIS

The organizations that make choices in regard to GIS may start from widely varying points. For example, many may already have access to a computer system that includes a network, as well as to other systems that need to communicate with GIS. Some may have digital georeferenced data available and possess some expertise in the use of software, hardware, and GIS in general. Others may have a form of GIS already in operation. In certain countries, organizations tend to choose total solutions in preference to step-by-step solutions. In such cases system integrators have become an important element in composing major systems using software, hardware, and data from different sources. The extreme solution is to outsource the entire operation for a longer or shorter period.

The scope of a GIS implementation project is related to the final impact of GIS on an organization. Two categories can be established according to impact:

1. *Minimum impact.* Within the organization, GIS is seen as a useful new tool to be acquired, much as a new computer or telephone exchange. Typically, the GIS users may be within a division of a larger company or a bureau of a governmental agency.
2. *Major impact.* Initiating GIS changes the way in which the organization operates. Consequently, the problems addressed affect virtually all phases of the organizational hierarchy and operational paradigms. Typical organizations include municipal authorities, public utilities, and cartographic agencies.

The demarcation between these two categories is a blurred zone rather than a sharp line. GIS implemented in one small bureau of a large governmental agency may radically change that bureau. For the bureau, the impact is major; for the parent agency, GIS is merely a tool that enables one of its many bureaus to accomplish its task. Conversely, a large agency in a small country may need to be reorganized completely in its conversion to GIS. In terms of equivalent efforts in larger countries, however, its efforts may be small. Hence the qualifier "from the users' viewpoint" should be applied to all assessments of the impact of implementing GIS. Most of the following discussion is devoted to the implementation of GIS in organizations for which it effects a major impact. A third category of organization spawned by GIS, such as a national geographic database, is also described later in this chapter.

17.2 Phases in Organizational Issues

The introduction of GIS into an organization can be based on three different philosophies:

1. Technology-driven introduction—the best/newest technology always gives more benefit than the next best.
2. Plan-driven introduction—a good plan and its accomplishment always give the best result.
3. Process-driven introduction—there are so many unknown factors that the organization has to learn as the introduction is underway, and this common learning process always gives the best result.

However, experience has shown that the best technology does not always give the best result. Experience also shows that two approximately similar plans for introduction can produce extremely different results in different organizations. Likewise, a process-driven introduction does not always produce the learning effect and the result that were anticipated. The conclusion, therefore, is that none of these methods is always the best, and a GIS introduction should therefore include a balanced part of each method. Nevertheless, it is a fact that most organizations choose a poor plan-driven introduction.

As we have seen, GIS differs considerably from its conventional predecessors. Consequently, the effect of the introduction of GIS on an organization may result in the need for restructuring. The prudent, realistic approach usually requires a balance between remodeling the organization to accommodate GIS and shaping GIS to suit the needs of the organization.

Evolution of the organizational changes normally includes:

- Development of a business concept and identification of goals
- Appraisal of current situation
- Review of the experience of others with GIS
- Identification of user requirements
- Assessments, including cost-benefit analyses and delineation of:
  - Choice of strategies
  - Development of a plan for future conduct
  - Data structuring

The current trend is to regard information as a widely available and independent resource available in numerous ways through the integration of data processing schemes. Consequently, planning should not be limited to the organization itself, but should include external users, data suppliers, and other cooperating organizations. Hence, several organizations may participate in a single GIS project. Preparatory work should also aim to instill motivation, build competence, and promote maturity within the organization. Indeed, the organizational evolution itself should draw on the expertise of professionals who know the tasks, executives who know the organization, staff and system specialists and/or external consultants.

Evaluations of the various aspects of data processing and computer hardware and software should be aligned with activities relating to the evolution of organizational changes.

The budget for the planning phases should, of course, reflect the cost and complexity of the system envisioned. An investment in the preparatory phase of 5 to 10% of an overall project budget is usually adequate to minimize project risks.

17.3 Development of a Business Concept and the Identification of Goals

A thoroughly prepared business concept should form the basis for initiating GIS in any organization. Many different motives may lie behind a decision to introduce GIS. Management may wish to increase internal productivity or external competitive ability. There may be a professional interest in new technologies, or the organization may be obliged to respond to internal or external pressures to approach tasks in a new or more vigorous way.

The motives for choosing GIS must be thoroughly understood, for only then can goals be identified with sufficient clarity as to become attainable. This obvious requirement often means limiting the overall scope of the GIS project. The business concept forms the basis for defining tasks to be implemented by the organization, and the user requirements are a consequence of these tasks.

It is usual to define two types of goals: effect goals and result goals. An effect goal is linked to what one wishes to achieve with the project, which can be specified as greater efficiency, better service, or another outcome. A result goal is concrete and can involve the development of a certain database, automation of particular work process, or other such achievement. The aims should be formulated in such a way as to give as precise a description as possible of what results one hopes to attain through the project, and they must be realistic for the business. It is often natural to specify one short-term and one long-term aim for the introduction of GIS. In principle, this means setting objectives, then reviewing the results as compared with those objectives. For this reason, the aims should be concrete and, as far as possible, testable. For example, they can be expressed in percentage form as an improvement in efficiency, an increase in income, a reduction in costs, the number of new services and products, or increased service.

Initially, GISs must often be "sold" within an organization, and GIS salespeople—enthusiasts capable of promoting GISs—are sometimes needed to trigger the processes leading to the implementation of a GIS facility.

17.4 Appraisal of Current Tasks, Users, Data, and Data Flow

One of the fundamental starting points for appraising a new technology is a clear identification and description of the current situation. The appraisal should, of course, include *all* users, regardless of whether the system to be implemented is limited to a single bureau or agency or is to be a larger, multiuser system.

The first step in an appraisal, then, is to compile an overview of actual and potential users; the next is to elicit the following relevant information from these users:

- Tasks and products
- All production processes
- All data sources
- All files and filing systems
- Descriptions of all data (including content, accuracy, completeness, and timeliness)
- All data users
- Data flow, documents, and communications between processes, staff, and internal and external files
- How data are managed and maintained
- Descriptions of current software, computer systems, and network
- Human factors, including the working environment and individual expertise

Assessments of current tasks must include evaluation of factors having financial impact, along with problems related to the conduct of current tasks, such as proliferating delays and a decline in quality. An assessment must be made of the ideal situation sought by individual users of geographic information—which can be resolved by use of new technology—such as reflected by a demand for new products and services, ways of changing current products and services, and necessary organizational changes. The assessment of current tasks should also cover projected time or cost savings resulting from realization of the ideal situation, side effects of the new technology, and number of present and potential future users. This information can form the basis for later cost-benefit analyses.

The depth of detail should, of course, suit the goals of the assessment. For example, a detailed overview of a production process is essential only if one of the purposes of the transfer to GIS is to rationalize production. Many organizations are already computerized and thus have data processing infrastructures in place. As these infrastructures may affect the introduction of GIS facilities, they should therefore be evaluated with respect to system approach and software systems, user interfaces, databases, operating systems, programming languages, hardware, and communications facilities.

Product information is vital. In some cases, production processes may depend on the information and organizational structures required to produce specific products. In other cases, product improvement or process optimization may be important factors. A clear diagram of information flow provides an overview of the various processes and their interrelationships. However, because data flow may be complex and therefore difficult to describe, special techniques, such as structured systems analysis, may be necessary.

17.5 Review of Others' Experience with GIS

All organizations and activities have their own particular requirements. Nonetheless, a review of the experience of others is useful whenever new technology is implemented. As is the case with all technologies, operating GIS facilities falls into three broad classes:

1. *State of the art* or *leading edge*. Usually only a few organizations are at the forefront of the field, because keeping abreast of the latest developments is expensive and requires expertise to deal with unproved approaches.
2. *Proven*. In the burgeoning field of GIS, most facilities now in operation comprise off-the-shelf hardware and software adapted to specific tasks and proven in operation. Directly or indirectly, the cost-benefit ratio of such a facility is usually near optimum.
3. *Outmoded*. Aging facilities in the proven category are often reliable producers. However, as constituent parts or whole facilities become outmoded by newer designs, cost-benefit ratios are less favorable because greater benefits may be realized through upgrading.

All organizations have their own requirements, but in most cases the experience of organizations with proven facilities are the most valuable in defining user needs. The majority of facilities now in operation comprise commercial off-the-shelf (COTS) hardware and software, but realization of user needs should be balanced against the availability of

new technology. Usually, only a handful of organizations are able to stay at the forefront of the field because keeping abreast of the latest developments means that cost-prohibitive and special expertise is needed to deal with an unproven approach. It is not surprising, therefore, that no one was ever fired for buying from IBM or Microsoft. It is important, however, to balance this conservative tendency. Whenever GIS is to be implemented a review should be carried out based on a variety of sources, including user experience of state-of-the-art facilities and information from system suppliers and others familiar with the market and technological development.

In most cases, the experiences of organizations with proven facilities are the most valuable in designing a new GIS facility. State-of-the-art facilities are usually either too expensive or too exotic for organizations not involved directly in GIS R&D. Outmoded facilities are seldom of interest to an organization contemplating a new technology.

When GIS is to be employed in a totally new sector, the process described in Section 17.10 should be applied.

17.6 Identification of User Requirements

The business concept forms the basis for defining tasks to be implemented by the organization, and the user requirements are a consequence of these tasks. It is possible to identify the requirements only when user tasks have been identified and are clearly understood. Definition of user requirements is closely related to the development of the business concept and may be an integral part of building the concept.

As discussed earlier, defining user requirements is a process in which a great deal of work may take place during and even after system procurement. Traditionally, corporate GIS acquisitions for large organizations with operational responsibilities involved detailed specification of the user needs up front, often with the aid of a feasibility project. Increasingly, however, the greater part of the work to define user needs takes place during and even after system procurement has been specified, as the first stage in system integration. This exposition is based primarily on an increasing partnership between users and vendors and the view that GIS introduction is a process whereby learning is more important than detailed planning. But whether identification of user needs is done pre- or postprocurement, it is a basic activity that should be treated in a structured way. Each organization has to find the best balance between pre- and postprocurement in its definition of user requirements.

Preprocurement definition of user requirements
User requirements for most implementations of GIS in both government and utility organizations are normally identified through user

surveys, often as part of a feasibility study to establish the overall potential for GIS in the organization. A standard questionnaire should be employed in the user surveys to compile homogeneous information from varying sources, and care should be taken in selecting representative respondents if the information gleaned is to be meaningful. Because questionnaires are always subject to misinterpretation and users are not always aware of their specific needs, user studies are best conducted through personal interviews. Unfortunately, these are both time-consuming and costly. In most cases a survey should be limited to a representative sample including, of course, the major users who will benefit most from the new GIS technology. The sample should also include a broad spectrum of other users to give the best overall picture of what is required.

User surveys can provide an overview that is valid only at the time they are conducted, and results should be assessed and analyzed by GIS experts, who can more readily see the potential and the limitations.

Because GIS awareness is a process in the organization, workshops can serve as a forum for speeding up this process. Both workshops and user surveys may be carried out as integrated activities for identification of user requirements. Workshops can also be used as an approach to building a total business model for GIS, but business modeling for the sole purpose of GIS implementation is now relatively rare. Many organizations have already built such models as part of a wider program, and generic models are now also available for local governments and utilities.

Information compiled from user surveys, workshops, and reviews of documented experience will normally produce a long list of different user needs, which will have to be given a structured and neutral evaluation. The business model should form the basis for all evaluations—only requirements that are within the business model should be evaluated. (This is an important reason for business modeling.) Ideally, the ranking of needs should be based on a solid cost-benefit analysis, or at least on an evaluation in which both costs and benefits have been taken into consideration.

Postprocurement definition of user requirements

The basic assumption for the postprocurement activities of GIS is that the organization has selected a supplier of GIS software to develop applications that meet the business concept of the organization. This selection is made on the basis of vendors' responses to some relatively simple previous research. The requirements common to all such cases, irrespective of size or sector of the organization, will constitute some form of design specification. Although the terminology may vary, such design specifications normally comprise two parts: a specification of applications and a functional specification. The specification sets out

the organization or user needs for GIS in terms of the applications for which the system is required, a logical data model, and the required service levels, such as user numbers, user locations, and requirements for uptime and other performance levels. As in the preprocurement activity, user requirements may be identified through user surveys, workshops, and a review of other users' experience. The functional specification defines how the chosen GIS software will be implemented to meet these needs. It is usual for suppliers to produce the functional specification, based on their own knowledge of the software. The nature of the postprocurement activity varies according to the scale of implementation and the nature of the application, but user involvement should be relatively high inasmuch as GIS technology is mainly end-user oriented.

17.7 Financial Evaluations—Cost-Benefit Analyses

A prerequisite for obtaining funding for GIS should be that the proposals for introduction be based on solid plans and financial evaluations. It is important to know what can be achieved through GIS so that the right decisions can be reached as to what functions should be performed and how profit can be achieved. Cost-benefit analyses ensure that projects can be selected that can show profitability. User surveys and (possibly) pilot projects provide the basis for such analyses.

The purpose of financial evaluations is primarily to:

- Improve the basis for decision
- Increase profit
- Increase economy consciousness
- Improve planning

The following factors, which are important in making decisions, are affected by cost-benefit analyses:

- Ranking of prospective projects (GIS versus other systems)
- Revealing the need for further consideration/planning
- Considering the adaptation of the GIS project to increase profitability
- Ranking of solutions internally with the project
- Deciding whether projects be carried out or not

17.7.1 Theoretical basis

In business, costing tends to focus on the improved internal efficiency and increased productivity resulting from a new technology. This is usually expressed in terms of profitability. Government agencies evaluating the change to new technologies often take socioeconomic factors into consideration. This mandates a broad view that includes both internal efficiency and external effectiveness. Socioeconomic costing, therefore, is based on the inclusion of all parties likely to be affected by the change under consideration.

In a company or a municipal department, inner efficiency (benefit) can be measured either in terms of increased productivity or as savings in time and money. The external effectiveness of an organization is determined by how well it serves its users, both public and private. For users, increased effectiveness is manifested in lower costs, higher productivity, or both. However, time savings are relevant only if they can be used to equal advantage elsewhere.

Cost-benefit analyses customarily weigh various options against each other with respect to their applicability over a prescribed time period. The zero option, which retains the status quo with no change of technology, is almost always included. This requires a prediction of future events without the new technology. Other options may incorporate varying degrees of technological change, but they must be pure; that is, each individual cost-benefit assessment must be independent of the costs and benefits that would accrue without technological change.

In principle, socioeconomic evaluations and business economic evaluations comprise the same phases:

1. Define different projects/alternative actions.
2. Establish assumptions for calculation: time perspective, rates of interest, etc.
3. Chart the positive and negative effects of the various project alternatives.
4. Quantify the effects (i.e., cost and benefit).
5. Calculate the cost-benefit ratio.
6. Consider the nonquantifiable effects.
7. Carry out sensitivity analyses, if necessary.
8. Recommend or choose solutions.
9. Establish prerequisites (preconditions) for the realization of gains.
10. Work out strategies or plans for realization of gains (including budget proposals).

17.7.2 Negative effects: costs

Commercial cost accounting is usually straightforward. Yet when socioeconomic factors are involved, cost accounting becomes tenuous, because the complete effects of investments elude concise definition. For example, a national map service may accurately estimate the initial cost of a new map series, as the costs of geodetic data, mapmaking, field surveys, reproduction, and the like are easily ascertained. However, the ultimate distribution of costs among the various end products, as ordinary map sheets, digital map data, and so on, is seldom known as accurately.

Cost accounts usually include the following three main categories:

1. Planning costs
   - Internal personnel costs
   - External consultants
2. Establishment costs
   - Data conversion
   - Applications development
   - Equipment and software
   - External consultants
3. Operations and maintenance costs
   - Internal personnel costs (training, operation, user support, and administration)
   - External consultants
   - Equipment maintenance
   - Software maintenance
   - Cost of capital
   - Insurance

In addition, there may be intangible liabilities, including:

- Increased vulnerability due to reliance on computers, which may fail because of hardware or software malfunctions or failures
- Poorer working environment, due to equipment-generated noise, tedious digitalization tasks, etc.
- Increased entry-level competence, which can bar some staff members

17.7.3 Positive effects: benefits

The tangible benefits that may accrue can be divided into three main categories:

1. Benefits in resources
   - Reduction in staff hours
   - Reduced costs
2. Benefits in products and services
   - Faster processing
   - Increased income, sale of new products
3. Benefits in effect
   - Improved internal and external communication
   - Lower costs and less work for others

The intangible benefits may include:

- Higher-quality goods and services
- More conclusive decisions
- More rapid decisions

- Superior plans; strategic positioning and management
- Greater understanding of problems
- Enhanced expertise and more challenging work
- Stronger competitive ability

All vital intangible effects, positive and negative, should be described and incorporated in the decision-making process.

17.7.4 Quantification and analyses

Cost-benefit analyses appraise costs and benefits in measurable terms, according to the prime rule of the relationship

$$\text{benefits (or costs)} = \text{quantity} \times \text{unit price} \qquad (17.1)$$

A typical computation would be of labor savings expressed in worker-years, based on a productivity increase of, for example, 10% per staff member. The unit price would then be the average staff labor cost per year. It is generally recommended that relatively conservative estimates form the basis of all quantification, to avoid presenting over optimistic figures, or, alternatively, that sensitivity analyses are used (as discussed in Section 17.7.6).

Costs and benefits often occur at different times, with costs usually being incurred before benefits accrue (Figure 17.4). Both costs and benefits may be divided into nonrecurring and annual values. These differences must be resolved so that the various values may be compared directly. For example, an income of 1000 monetary units received five years hence will, because of loss of interest, be worth less than a present income of 1000 of the same monetary units. Consequently, future values must be discounted to permit assessment in terms of present values.

The usual procedure involves computing the present value (PV) of a potential investment or project. The PV is the aggregate discounted

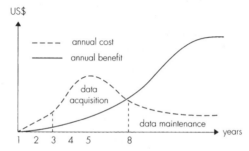

Figure 17.4
The benefit will come after a period of investment. Cost and benefit curves indicate that data acquisition and maintenance normally account for 60 to 80% of aggregate costs.

value of the expected benefits, and the aggregate discounted value of the expected costs. Net present value (NPV) is computed by the relationship

$$\text{NPV} = \sum_{t=0}^{T} \frac{B_t - C_t}{(1 + r)^t} \qquad (17.2)$$

where B is the benefit, or return, and C the cost at time t in years, r is the discount rate, and T is the time horizon (the point in time defined as the end of the economic life of the project or the time at which the assessment no longer evaluates costs and benefits). A typical computation is shown in Figure 17.5.

The final computational result depends strongly on the interest rate chosen. In general, the interest rate should be high for high-risk projects, short-term projects, or combinations thereof. Only effects that occur within the project(s) life span will be dealt with here. This means that the lifetime has to be long enough to include the main effects of the project(s). One should bear in mind that effects that occur late in the life span have relatively little influence on the current value estimates. The time horizon of most computerization projects is 5 to 8 years. However, the time horizon of GIS projects is longer, usually about 10 to 20 years. This is because the data employed in GIS are valid over longer periods of time.

Alternatives are often ranked in terms of their respective benefit-cost ratios, which are the ratios of the present values of their aggregate benefits to aggregate costs:

$$\text{benefit-cost ratio} = \frac{\text{present value of benefits}}{\text{present value of costs}} \qquad (17.3)$$

The benefit-cost ratio usually provides the best index for ranking options. However, in some cases, where options involving considerably

Year	Annual costs 1.000 US$		Annual benefit 1.000 US$	
	1991 US$	Present value (1991)	1991 US$	Present value (1991)
1991	200	200	50	50
1992	300	280	100	93
1993	300	262	200	175
1994	200	163	400	327
1995	200	153	700	534
1996	200	143	800	571
1997	200	134	800	535
SUM:	1.600	1.335	3.050	2.285

Cost	Benefit	Benefit/cost	Benefit ÷ cost
1.335	2.285	1.71 : 1	US$ 950

Figure 17.5
Typical computation of NPV, based on a discount rate of 7% and a time horizon of five years.

different levels of costs are involved, the aggregate net benefit is also useful. This represents the difference between benefits and costs:

$$\text{net benefit} = \text{present value of benefits} - \text{present value of costs}$$
$$(17.4)$$

Methods that do not include considerations of time and interest rates cannot be used as profitability criteria.

17.7.5 Prerequisites for harvesting benefits

The cost-benefit analyses involved in decision problems are invariably based on assumptions. The prerequisites that form the basis of the project will have great significance for the recommendations resulting from the cost-benefit analysis.

The main prerequisites will often be:

- Which decisions have to be made
- Which measures have to be initiated

The following aspects can form the basis for clarification of the prerequisites:

- When can the gains be taken out?
- Is there any need for new developments?
- Do we need new competence?
- Is reorganization necessary?
- Will this mean adaptation for the employees?
- What is the relationship to other projects?
- What is the relationship to other internal and external activities?
- What is the timetable for completion of the project?
- In how many budget periods will there be need for investments?
- Are there any particular cost and resource restraints?
- Who is to approve the startup of the project?

17.7.6 Sensitivity analysis

The role of the assumptions, however, is not always clear at the outset; therefore an analysis may be performed to identify the assumptions to which the outcome of the cost-benefit analysis is sensitive. This involves changing the assumptions (parameters) of the decision problem to reveal how the changes affect the outcome. The analysis involves individual computations, which may be based on:

- Pessimistic or optimistic assumptions for costs and benefits
- Postponement of various activities or investments
- Change in the order of activities
- Changes in the rate of investment

17.7.7 Review of the cost-benefit relationship

There are no exact measures for assessing cost-benefit alternatives. However, rule-of-thumb experience indicates the following:

1. Measures that result in a cost-benefit ratio of 1:2 or more should be enacted, when assessed both in commercial terms and in socioeconomic terms.
2. Measures that result in a cost-benefit ratio between 1:0.8 and 1:2 should be analyzed further with an eye to cutting costs or increasing benefits.
3. Measures that result in a cost-benefit ratio of less than 1:0.8 should be abandoned.

The cost-benefit relationship is an insufficient basis for a decision as to which projects should be carried out. We should also take into account the nonquantifiable effects.

17.7.8 Evaluation of the nonquantifiable effects

The nonquantifiable effects should be handled equally for all the projects to be ranked. To ensure that this is done, a precise description must be given. An evaluation should also be made as to whether each single effect has a great or small influence on the project and what the chances are that this effect will occur. Thus, there can be four categories, which will be weighted differently in evaluation of the projects:

- Great effect and good chances of realization
- Great effect but slight chance of realization
- Little effect and slight chance of realization
- Little effect and good chance of realization

17.8 Developing a Strategic Plan

The term *strategy* originates from the Greek *strategos* (*stratos*, army + *agein*, to lead). Today, *strategy* is normally used to describe a plan as to how different situations should be handled to reach goals that have been defined. In the planning process of a strategy, the prerequisites should be clarified for the choices to be made. Ideally, the process should result in clear directions as to what decisions and steps should be taken, how they should be carried out, and which resources should be used and by whom. As stated earlier, the cost-benefit analysis will form a solid basis for this type of evaluation.

A plan-driven introduction is based on a recipe (the plan) with a collection of ingredients. Planning normally consists of the following points:

- Establishing goals
- Identifying status

- Consideration of others' experience
- Identification of user needs
- Ranking of actions (based on cost-benefit) in respect to data, equipment, etc.
- Development of logical data model
- Production of the plan

The ingredients or variables in the plan will be competence, data, and organization, as well as equipment and software.

The technologies involved in GIS are developing rapidly and continuously, so GIS users may improvise new applications at any time. Consequently, the strategic plan should be sufficiently flexible to allow its conduct to be influenced by prevailing conditions. The ranking of alternatives in the plan is based on the results of the cost-benefit analyses, in which investments in GIS are tested against imminent or current investments elsewhere in the organization. In addition, public agencies must consider their official duties. With a suitable strategy, the optimum initiation of GIS activities may be viewed as shown in Figure 17.6.

A strategy plan should identify and illuminate:

- Method of introduction
- What is to be automated
- Start date(s)
- Mode of introduction
- Level of investment
- Geographical and thematic coverage
- Future data flow
- Organizational matters
- Project organization
- Budget and financing

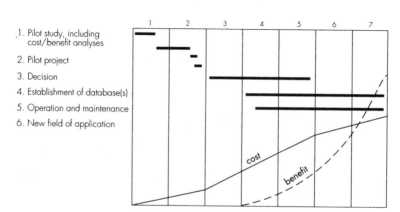

Figure 17.6
Idealized initiation of GIS, where costs and benefits accrue during a project.

Experience shows that it is important that planning be based on a realistic level of ambition. The higher the level of ambition, the greater the risk that some aims in a number of areas will not be reached. Frustration can easily occur within the organization, and the project may be viewed as unsuccessful—even though it may have had many positive effects.

17.8.1 Method of introduction

The realization of GIS projects can be based on different models. These can be divided roughly into four categories.

1. Pilot project followed by main project
2. Massive introduction in one main project
3. Gradual introduction in defined phases (part projects)
4. Simplified strategy: small, sure, seen, success (the four S's)

Many larger organizations choose to carry out a pilot project to assess organizational problems and the staffing requirements for and time consumed in transferring data from conventional to digital form. A pilot project should also help the organization to develop increased appreciation of the GIS concept, mediate organizational problems, and contribute to verification of cost-benefit analyses.

The pilot project is followed by a main project, which is planned on the basis of experience gained from the pilot project. Normally, this will result in an efficient main project, but the total introduction period is normally extended by the pilot project.

Other organizations choose a massive introduction in order to enter into the operative phase as quickly as possible. However, an organization can find this heavy to digest. The benefit of time saving can easily be consumed by negative factors—for example frustration among employees and reduced production. Some organizations therefore choose a gradual approach whereby different parts of the system are introduced in a particular order so that employees are never confronted with major changes. However, this can lead to the maintenance of parallel routines, which can also be very frustrating for employees. Today many organizations have already taken the technology into use gradually. Thus, the introduction of GIS is more an upgrading than establishment of new technology.

A simplified strategy can be described by an elementary "four S" guideline: "small, sure, seen, and success". The initial system should be *small* to minimize financial risk and to speed familiarization with the new technology. The organization should be *sure* that the tasks identified are clearly defined, necessary, and sufficiently limited that they may easily be realized. The results should be *seen* soon, both to justify investment and to encourage the staff involved. When these three requirements have been met, a measure of *success* will be at hand, relatively quickly and for a minimum investment.

347

Many ancillary GIS facilities are project oriented. Typical projects include digital terrain models used in planning residential developments, automation of township property registers, and the like. In most cases, the applications are clearly defined and the design challenge reduced to finding the most suitable system for the tasks involved. In practice, the system choices are often between the various PC systems available.

The conclusion is that there is no particular method of introduction that will always be best. What is certain is that the method of introduction should be *planned and accepted by the employees.*

17.8.2 What is to be automated?

Current data flow, production lines, products, and cost-benefit analyses form the basis for the macro view of what is to be automated. The macro view may include:

- Which work processes are to be replaced by GIS
- Which files are to be replaced by GIS databases
- Which lines of communication are to be replaced by electronic communications via data networks

17.8.3 Start date

In general, the later the startup, the longer the time before benefits are realized. As illustrated in Figure 17.7, a late start may cost no more than an early one, but may result in fewer aggregate benefits. The benefit-cost ratio is then less.

One of the more common justifications for postponing an investment in new technology is that newer and better technology will soon be available. The experience of GIS users thus far indicates that such postponements seldom prove profitable if the system is implemented properly. The continuous evolution of all the technologies involved in

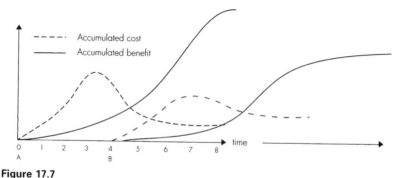

Figure 17.7

Early versus late startup of a GIS facility. An early start date (point A) results in a higher benefit-cost ratio than a later start date (point B).

GIS implies that the basic decision of whether to invest now or later remains, no matter at what time it is made.

To summarize, the factors that should influence the decision of exactly when to start include:

- The relevant technologies are available, so they no longer constitute a bottleneck.
- Postponing a start postpones benefits and thereby lessens project profitability.
- All technologies now evolve continuously, so there is no single best time to start.

17.8.4 Level of investment

The crucial choice to be made in investment is usually between constraining investments in new technologies within conventional budgets and considering more extensive investments. As is the case for almost all changes from traditional to electronic technologies, the initial investments are relatively large, yet short term. This trend should be exploited, as in establishing GIS databases quickly to realize the benefits they generate. Prolonging the establishment of a database by stretching it over phases involving both conventional analog and computer digital data only escalates costs without correspondingly increasing benefits. As shown in Figure 17.8, cautious investment often yields less than aggressive investment.

In conclusion, the guidelines for levels of investment may be summarized as follows:

- Think big but start prudently.
- Make major investments after a pilot project phase.
- In switching from conventional means to GIS, major investment should be committed in the initial implementation phase in order that the project be profitable.
- Data should be converted rapidly to digital form.

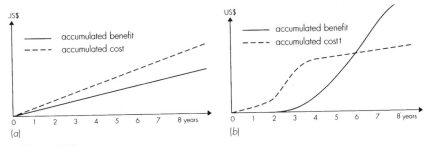

Figure 17.8
Aggressive investment often yields more than cautious investment. *(a)* Linear investment within the constraints of a conventional budget produces linear benefits, which may prove unprofitable. *(b)* A major early investment, such as in the conversion of an entire database, can yield benefits that rapidly exceed costs, for greater profitability.

17.8.5 Geographic and thematic coverage

As a rule, the basic decision to be made concerning geographical coverage within the scope of a GIS is whether to provide full digital coverage of all regions and areas involved or, rather, to digitize information fully only for areas of greater activity and more rapid change. Again, the economics of the relevant benefit-cost ratios are the best criteria.

Different user organizations have differing needs for the geographical coverage of digital information. The distinctions involved often are dictated by functional or administrative divisions (townships, regions, counties, or entire countries). In general, a need for digital geographic coverage must be fulfilled completely before the user can realize the benefits. This is because staged shifts invariably include periods when both conventional and digital data are in use, which usually increases the work involved in a given final product.

As a rule, therefore, the most cost-effective approach is for an organization to start with full geographic coverage, even though such coverage requires a greater initial investment in data acquisition, operation, and maintenance. The distinction between the coverage needs of various organizations is then not one of coverage itself, but rather of the regions or areas of responsibility involved. A managerial organization such as a major governmental agency must usually maintain geographic coverage of its entire area of responsibility. An organization dealing with projects such as roadwork need cover only those areas involved in its projects.

Once a GIS facility is planned, the question arises as to which part should be installed first. From an investment viewpoint, the part and the themes that yield the greatest proven benefit should be first. For most users, however, a part comprising several georeferenced themes concerning a single area or region provides a more comprehensive initial benefit. This is because the various themes are often synergistic. For example, a digitized property register enables GIS to provide information on properties, and a digitized road data register enables it to provide information concerning roads. However, if both digitized registers are available, the GIS can provide additional information. It can provide answers to such queries as: "Which properties front onto which roads?" and "Which properties are affected by the proposed rerouting of this road?"

The guidelines for geographic coverage may be summarized as follows:

- At least one theme covering an entire region or area of responsibility must be converted before benefits may be realized.
- Staged conversion to GIS, in which conventional and digital data may be used simultaneously, should be avoided.
- Managerial organizations have the greatest need for coverage of

entire regions or areas; organizations operating on a project basis have less need.

- Several themes pertaining to a region or area can synergistically increase benefits to users.

The potential benefit accrued (usefulness) depends to a great degree on geographic coverage, thematic coverage, and data quality. These three parameters may be regarded as a *usefulness cube,* as shown in Figure 17.9.

17.8.6 Future data flow

The description of future data flow must relate to the functions to be performed. Comparing future to current data flow helps to identify the work functions that will become superfluous, those that must be added, and those that should be modified. In a public agency, for example, the change of overall data flow caused by a change from conventional means to GIS may be regarded as being a change from series to parallel processing of work, as illustrated in Figure 17.10.

17.8.7 Organizational matters

The efficient exploitation of a new technology in an organization often mandates alterations in work routines and chains of command which, in turn, affect the overall organization (Figure 17.11). In practice, altering an organization may prove difficult because of the formal and informal positions that always exist in chains of command. Changing the organization changes relationships between personnel. Staff changes always bring in human factors that are difficult to predict or control.

Consequently, organizational matters are vital in all initial implementations of GIS facilities; the problems are often more complex and

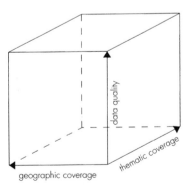

Figure 17.9
Usefulness depends on geographic coverage, thematic coverage, and data quality. These three parameters may be viewed as defining a *usefulness cube.* (Courtesy of Ken Jones.)

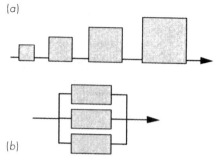

Figure 17.10
In a public agency, implementing GIS may change work processing from series *(a)* to parallel *(b)* form.

more crucial to success than are the technical problems involved. As a rule, the latter can be solved in a straightforward manner, by purchasing and installing new equipment, new software modules, and so on. Purchasing incurs costs, of course, but in a well-planned project these are anticipated. Changing and replacing staff members are less straightforward and may trigger unanticipated problems. Hence, organizational matters usually require more continuous management attention than do technical problems.

Thus, one goal of any inception study should be to delineate alternative future organization models and to recommend ways of testing their validity. New staff positions must be described in detail. In principle, new tasks and new data flows may be described independently of the persons, groups, or departments ultimately responsible. So new staff structures must be defined and management modified accordingly.

Availability of information is vital to a healthy GIS facility (Figure 17.12). Whenever information availability is restricted, GIS utility suffers. So the common tendency of an agency or department to monop-

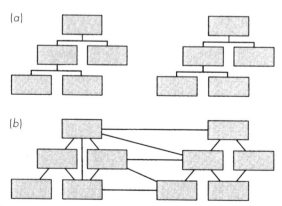

Figure 17.11
Transition from a hierarchical organization *(a)* to a network organization *(b)*.

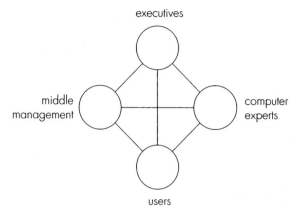

Figure 17.12
To be successful, a new technology must be accepted at all levels in an organization.

olize its own information is one of the major foes of successful GIS. Consequently, one of the problems encountered in implementing a GIS facility may involve combating the bureaucracy that blocks information flow.

Other subjective factors can further complicate the initiation of a GIS facility. Human nature apparently dictates that about a quarter of the personnel in any organization always prefer the status quo and will oppose any change whatsoever. Except in high-tech firms, executives are often indifferent to newer information technologies, partly out of ignorance and partly by being overconcerned with cutting costs rather than increasing benefits. However, as executive support is often crucial to the success of a GIS project, initial presentations may be received more readily if they contain clear executive summaries and concise examples of the utility of GIS in sectors known to be of interest to the executives concerned. Successful projects are often conducted enthusiastically by middle management, who often seek the benefits of new technologies while resisting extensive organizational changes.

Specialists are essential in the execution of a GIS project but are seldom capable of addressing the myriad details involved. For example, computer experts are usually so engrossed in their disciplines that they lack the broader view needed for dealing with organizational problems. Operators and other direct users are similarly constrained within the framework of their jobs. As shown in Figure 17.13, the introduction of a new technology in an organization often changes its power pyramid.

Project organization
Strong, independent project organization is essential to the success of any major GIS implementation (Figure 17.14). The project need not be

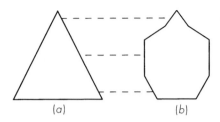

Figure 17.13
The introduction of a new technology in an organization often changes its power
pyramid from the traditional triangle *(a)* to a barrel profile *(b)*, in which fewer
employees are involved in routine tasks and middle managers assume more
comprehensive roles, both upward and downward from their location in the
traditional pyramid.

permanent, but may be disbanded when the GIS facility becomes operational. GIS, computer, and other experts are usually more readily
hired for a project than trained for the tasks involved. All projects need
committed staff and problem solvers, who need not be specialists, but
who can advantageously be recruited from among the users of the information products involved.

Organization and staffing of the project will depend on its complexity and size. In larger projects, the project manager should answer
to both a control group and a technical working group. The control
group should lay down political guidelines and provide other management support (e.g., by solving disagreements that cannot be solved
in other parts of the project organization). It should be staffed by managers from the various departments that are involved in the GIS. The
technical working group should make the important technical decisions and be the driving force behind design and development. This
group can be staffed by middle managers, who may be the same people who are leaders for the various groups carrying out the practical
work. The project manager should ensure that clear aims for the work

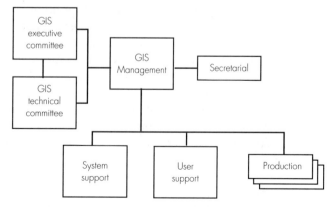

Figure 17.14
Organization of a GIS project.

are established and kept in mind at all times. He or she should also be responsible for the planning (technical and financial), performance, and follow-up of these aims. The practical performance, such as data structure and data conversion, can be organized in individual departments. Once the project has reached the operational phase, a separate group may be established for end users, to exchange experiences and define new needs.

Positioning the project in the organizational structure

There are often several alternatives for the formal positioning of a GIS project in an existing organizational structure. It can be placed directly in the line organization (e.g., in the relevant department, such as planning or water and sewage). A choice for a municipal project is to locate it either in the city manager's office or in a support department, such as IT. Many multiuser GIS projects in the United States have, in practice, been located within a department in the line organization (Somers 1994b). Even though this can create problems because of a perceived lack of neutrality, it can be counterbalanced by the advantages of clear GIS needs among the users and good accessibility to technical support and budget, thereby allowing speedy establishment within the main organization.

A more radical solution can be to establish an independent consortium for operating a joint GIS. The consortium can include participants from both the public and private sectors. Such a solution will often have a high-risk political and financial profile, but it has been carried out with success in such places as San Diego, California (Dangermond 1997).

17.8.8 Personnel and training

There will be a need for a series of specialist functions in a larger GIS project. These can include project management, system support, database and network support, user support, and production. In practice, GIS implementation projects are often understaffed. Therefore, staff requirements should be carefully assessed at the outset and the staff identified should be made available to the project. This usually requires either that staff members be relieved of their customary tasks and be assigned to the project, or that personnel be hired specifically for it.

A personnel management program for transition to a new technology should include identification of factors affecting personnel. It should include a plan for training, retraining, relocation and job rotation, and concise job descriptions, including responsibilities, employee involvement in planning, and clear definitions of positions and salaries. It is recommended that the simplest possible organizational structure be chosen, including management authority, delegation of responsibility, and division of labor. There should be clear communication to

355

and from executives between and within individual organizations. The personnel management program should also include retention of the organization's advantages and strengthening of teamwork, goal orientation, adaptability, and identification with the project. Staff proficiencies and size should be adapted to the tasks involved, and an environment of challenging work and career advancement should be created.

Many GIS projects, particularly in the public sector, suffer from a lack of training for both operators and users. Personnel management for the transition to a new technology should thus be given top priority. What type and level of expertise are to be attained? This depends on the nature of the project at the outset and on how it will be staffed, as well as on whether external services are to be involved. Ideally, all staff members who will be in contact with GIS and/or its products, either directly or indirectly, need suitable training. This also applies to executives, who should have a general understanding of both the potential and the limitations of GIS as a tool in decision making. Middle managers should have an adequate technical background to enable them to coordinate the implementation of GIS. Field staff and others who acquire data should be trained in its conversion and updating.

Most organizations elect to train their own personnel in technology conversion instead of hiring outside expertise. But who trains the trainer? The benefits of a system can be fully realized only if all functionaries, planners, salespeople, and other personnel are adequately prepared. Training should comprise on-the-job experience as well as formal education. External consultants are often most beneficial in the strategic sectors of a project, as they are not involved in the organization and can frequently resolve difficulties and disputes more easily than members of the organization's staff. Without the requisite skills, progress can be frustratingly slow; consequently, maintaining and building staff expertise should be a top-priority task.

17.8.9 Budgeting and financing

Any GIS implementation project should have a budget that allows for distribution of activities over the duration of the project and permits monitoring as well. Projects based on meticulous cost-benefit analyses are usually the healthiest. In many cases, the effort is not a single project, as several organizations may be involved in establishing a joint GIS facility. In such cases, costs are usually apportioned among the organizations according to the benefits accrued. However, the costs of facilities used by many organizations may be met in other ways, such as charging annual fees, assessing an initial outlay plus an annual fee, paying according to ability, and so on.

Government agencies must often decide whether to finance a new facility through their normal fiscal allocations or by charging user fees at market rates. As discussed later in the chapter, this question is central to the organization of a national database.

The most important organizational aspects of implementing GIS may be summarized as follows:

- Executive involvement is crucial to success.
- Organizational problems are usually greater than technical problems.
- Organizing or reorganizing should hinder monopolizing of information.
- The introduction of GIS effects changes in existing routines for information interchange between and within individual units of an organization.
- Altered work routines mandate organizational changes.
- At least one-fourth of the personnel of an organization can be expected to oppose change.
- There should be both a technical troubleshooter and an enthusiast who can market the project.
- Operator cooperation must be enlisted.
- The initial stages of implementing a GIS facility may be project oriented; organizational alterations should be tested before being finally enacted.
- Long-term organizational changes may be necessary after the initial operational phase of a new GIS facility.
- Successful GIS projects have a plan.

17.9 Developing a Logical Data Model

In the computer sciences, data modeling is considered part of the development of a data processing system. This view is common but not always correct. The development of a logical data model should not be linked to any particular technology, but should be devised by professionals who are expert in the field of application involved. It is therefore most natural to place this activity within organizational development.

Data models are discussed in detail in Chapters 3, 4, and 5. To summarize: Data models comprise objects, specified by five parameters: object type, geometry, attributes, relations, and quality. These parameters should be described in a data dictionary. The procedure for data structuring includes several topics discussed in Chapter 12:

1. Defining applications
2. Stipulating constraints, tasks, and requirements
3. Ranking according to benefit and cost
4. Descriptions of attributes
5. Coordination of conditions and definitions
6. Choice of geometric representation: raster or vector, points, lines, and surfaces
7. Descriptions of relations between objects

8. Quality requirements based on cost-benefit analysis: geometric accuracy, updating frequency, completeness and resolution, geographic coverage, logical consistency
9. Coding of data with respect to identity, object types, and attribute values

A good data model is part of the specification of a data processing system, not the result of it.

17.10 Creating National Geographic Databases and Developing New Business Sectors

Creating a national database raises numerous strategic questions on such subjects as public versus private financing, product organization, data content and quality, and so on. Many countries have addressed these questions and have created national databases of a variety of data, including property data, road data, environmental data, elevation data, statistical data, addresses, building data, and so forth.

The creation of a national database may be compared to the development of a new market—the guidelines are equally valid in the public and private sectors.

Principles of evolving strategies

All strategies start with the initial decision of whether to maintain the status quo (the *zero alternative*) or to select among alternatives to create a national database. Of course, the status quo decision requires *no* strategy. Strategies afford the means of selecting among alternative courses of action.

Strategies are best based on cost-benefit analyses, which rely on both costs and benefits being expressed in monetary units. Quantifying the costs and benefits of various courses of action is perhaps the greatest challenge of the entire strategic decision process. Strategies must also include consideration of organizational factors, such as production, storage, distribution, and updating.

All planning presupposes some experience with the subject. However, as a national geographic database is often the first of its kind in a country, the relevant experience is usually sparse. Consequently, the evolution of a strategy usually includes activities dedicated to defining the issues at hand. In general, a strategic plan may evolve in five phases:

1. Identification of current status, or zero alternative: the first user study:
   - Among users
   - Among providers (including their own organization)
2. Elaboration of a product concept defining:
   - Offerings that data users consider vital

- Characteristics of data offered: accuracy, timeliness, etc.
- Initial matrix of products offered, as in three categories of increasingly finer service
3. Formation of a cost model:
   - That simulates costs of the various product alternatives
   - That ascertains product cost chains
4. Delineation and conduct of market analyses: the second user study, which aims to identify:
   - User ranking of products and inclinations to pay for products in the various application groups
   - Size and potential of the sundry market segments, including factors that may detract from market development
5. Devising a model, based on the cost and market analyses, that promotes identification of an optimum product concept

The first step in the market analysis is to detail how users currently accomplish tasks without the benefit of the proposed product. Ideally, the users surveyed should have some experience in using their own digital data at the national level. Business criteria should be used to analyze the material compiled in the first user study. That is, cost-benefit analyses are conducted using net present values of incomes and costs. Knowledge of the behavior of current markets permits an estimate of future activities. In other words, products may be specified in terms of data content, accuracy, completeness, timeliness, distribution, costs, operation, financing, pricing, and so on.

	LOW SERVICE	MEDIUM SERVICE	HIGH SERVICE
A. GEOMETRY	Interregional All public roads No private roads	Intraregional All public roads All forest roads	All-to-all All public roads All private road
B. ATTRIBUTES	Road identity Road class Road number Public ferries	Cruising advice Speed limits Critical slopes Weight/height, etc.	Advanced advice Addresses Traffic lights Zip code etc.
C. GEOMETRIC ACCURACY	Low precision +/- 50 m x,y	Medium precision +/- 10 m x,y	High precision +/- 2 m x,y,z
D. CURRENTNESS	Revised every 5 years	Yearly Revised annually	Continuous 1 - 3 months
E. CORRECTNESS * OF ATTRIBUTES	>95% of all data	>99% within 1 yr	>99% right away
F. USER ACCESS	Magnetic medium national format	On-line access without GIS-supp.	Full GIS support on-line access
G. DELIVERY TIME TABLE	Normal cycle within 8 years	Speeded within 3 years	Crash program within 1 year
H. PRICE	Low	Medium	High

Figure 17.15
Typical product matrix for a national road database.

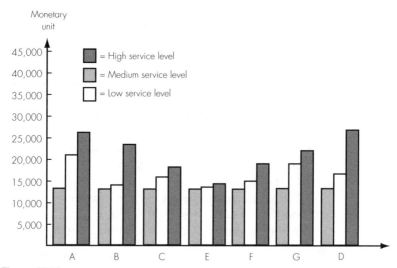

Figure 17.16
Effects on cost of a partial increase in attribute support level (see Figure 17.15).

A structured approach is used to specify products suited to identified market needs. The aspects of various products may be conveniently ordered in a matrix (Figure 17.15). A cost-benefit analysis presupposes that a cost is associated with each item in a product matrix. A realistic cost model should relate to a basic product (Figure 17.16). The model may then be used to compute the cost of the basic product and to compute the marginal costs of the alternatives.

The goal of the second user study is to measure the reactions of potential users to the various products proposed. The users involved in the first user study should be interviewed again. The user study and the supplementary treatment of the material should be carried out in a structured way. For example, a potential user may range the different alternatives according to a specific points table. This awarding of points can also be used in grouping the material.

Delineating strategies
Depending on production organization and the market, public agencies, which often operate national databases, may choose among various business approaches, such as:

1. Supplement basic data with other standard data and distribute the combination as a standard package to all users (extended map production).
2. Supplement basic data with other standard data to compile tailor-made user products.
3. Supplement basic data with user-specified data to compile tailor-made user products.
4. Coordinate data interchange without being involved in producing but with responsibility for standardization.

CHAPTER 18

Choosing a GIS: Technical Issues

We have already seen in Chapter 17 that the issues involved in initiating GIS may be classified as either organizational or technological. This chapter deals with the technical issues.

The development of a GIS pertains to all aspects of the computerization and data processing involved, and may include:

- Pilot project
- Design requirements
- Choice of hardware and software
- System implementation
- Technical database design
- Creating databases
- Operation and maintenance
- Safekeeping routines
- Evolving new applications

A GIS is usually developed partly in parallel with the relevant organizational activities. For example, design requirements may be compiled when the organizational strategy and future data flow have been defined. As stated in Chapter 17 the procurement of system can even be carried out before the user requirements are defined. The choice of a system or supplier may be made after the benefit-cost ratios have been finalized. Work routines may be delineated as the system nears completion.

18.1   Pilot Project

A pilot project entails an experimental small-scale implementation in which techniques, equipment, and subsystems for use in a full-scale GIS are tested in advance. In respect to system development, pilot projects should:

- Provide a better basis for choice of system(s)
- Test various production methods
- Identify system faults

The scope of a pilot project depends on the complexity of the organization(s) implementing the GIS facility. From a practical viewpoint, a pilot project should be budgeted in both time and money. The equipment and subsystems used should be as simple as possible, both to minimize loss should the chosen approach prove unsuitable and to ease the requirements for staff training and maintenance. As a rule, pilot projects should not take longer than one to two years. Staff involvement in a pilot project should be restricted to those persons most concerned with the GIS facility. Although widespread familiarity with GIS technologies and with the pilot project is advantageous, project progress may be complicated if an excessive number of conflicting interests are involved.

Normally, one or more limited geographic test areas should be chosen. Ideally, the test areas should be dissimilar—for example, one of high complexity and one of relatively simple content. The basic goals of a pilot project—to test approaches and prove viability—dictate maximum use of the experimental system(s) involved. Staff should be made available accordingly, lest the pilot project be delayed by lack of personnel.

Pilot project outcome and the decision to change technologies

If the outcome of a pilot project countermands GIS implementation, then the pilot phase may be extended, different approaches may be taken, or the GIS project may be abandoned completely. Before any of these steps are taken, the reasons underlying the negative outcome should be examined. As is often the case in the assessment of new technologies, a GIS pilot project can flounder if initial expectations are unduly high. Whenever suppositions of technical, financial, or organizational benefit outstrip realistic goals, disappointment invariably results. These effects can scuttle even the best of projects. Hence, workable goals at the outset of a pilot project are recommended.

If the outcome of a pilot project supports GIS implementation, work should start as soon as possible, particularly in choosing a system and starting data conversion activities. With a rapid decision and no dead time, the wave of enthusiasm generated by the pilot project can carry over into the implementation phase.

18.2   Choosing Hardware and Software for GIS

The right choice of computer hardware and software is essential. In practice, this may prove both easy and difficult. The choice is easy because it deals with comprehensible details, such as technical characteristics and prices that can be compared. Yet it is also difficult, because future applications are unknown and computer technologies change continuously. The impact of implementing GIS in an organization is uncertain, and pressure from aggressively competing system vendors also complicates choice.

Part of GIS facility	Average useful lifetime (years)	Relative cost (%)
Computer equipment	2 to 5	10
Software programs	3 to 6	20
Data	15 to 50 or more	70

Figure 18.1
Longevity of a typical GIS facility.

As in other disciplines, computerization efforts in the geographical disciplines often overemphasize equipment and software. Regrettably, data, the greatest resource and value of a GIS facility, often seem to be of lesser interest. The fallacy of focusing concern on the hardware and software of a system to the exclusion of data is illustrated clearly by the longevity of a typical GIS facility, as shown in Figure 18.1. The shorter expected useful lifetime of computer hardware and software, combined with the lesser investment involved—from 10 to 30% of the aggregate cost of a GIS implementation project—indicate that the major concern is best directed elsewhere, to the data, which are the backbone of an operating GIS.

18.2.1 Purchasing rules

Major purchases of hardware and software may be subject to company rules or guidelines, as in larger corporations or in the public sector. Today, international economic agreements* also govern public-sector purchasing. Rules are often linked to stipulated cost levels, so that a rule comes into effect when total cost exceeds a set amount.

Specific cost limits apply for different product sectors in the European Economic Area (EEA), such as goods and services, building and labor supplies and communication, and so on. Procurement notices must be published throughout the EEA for purchases in excess of these limits. It is worth noting that these limits apply to the total contract value and, to a certain extent, to total purchases within one product sector. Regulations are quite strict in regard to splitting the procurement into several contracts. In the case of a contract that extends over a number of years and involves several subcontracts, individual procurement notices may be required for each supply. Familiarity with these rules ensures that purchasing proceeds without undue delay.

*Examples include the World Trade Organization's (WTO, formerly GATT) Public Government Procurement Agreement of April 15, 1994, and the regulations agreed upon within the European Economic Area (EEA) by Directives of Public Procurement, nos. 93–96 of June 14, 1993.

18.2.2 Organization of the selection and implementation process

At this stage, responsibility for the implementation phase of the project may be defined. An organization should clearly define its own role before requests for proposals (RFPs) are sent to vendors. The assignment of responsibility for the selection and implementation process normally varies considerably, depending on the competence of the organization involved and the nature of the market supplying the goods and services. At this stage, an organization elects one of the following strategies:

1. *Retains system responsibility.* The organization, or its outside consultant, designs the system in detail and issues specifications for equipment, subsystems, and the like. The organization itself is then ultimately responsible for the system(s) being able to fulfill the design requirements.
2. *Assigns system responsibility.* The organization issues its design requirements and invites suppliers to respond with bids that include the detailed specifications of how the requirements are to be met. The supplier(s) selected carry the ultimate responsibility for the system(s) being able to fulfill the design requirements.

There are many variations of these two main approaches. An overall system may be split into subsystems. Some subsuppliers can be responsible for their sub-system functioning, and some not. The basic decision as to who performs detail design and compiles specifications and who assumes system or subsystem responsibility invariably involves trade-offs between in-house and external activities.

18.2.3 Design requirements and system specifications

The preprocurement requirements (see Section 17.6) pertaining to the various tasks, routines, sources, models, components, and subsystems constituting a GIS facility are usually incorporated in the specifications when invitations to bid are issued to selected suppliers.

In most cases, the design requirements are based on an appraisal of the current situation, the results of user surveys, cost-benefit analyses, strategic planning, and a pilot project and involve compromises in respect to what is currently available on the market. Design requirements may be compiled for:

- Organization(s) evolved for the new facility
- Tasks automated, including possible adjustment to existing systems
- Operations conducted
- Products produced
- Data employed
  - Type

- Amount
- Quality
- Data sources
- Data model(s)
- Time horizon
- Training
- Standards and their validity
- Documentation
- Data processing infrastructures already in place

Hence, the design requirements of a system stipulate how it is to function, *not* how it is to be implemented. In turn, specifications are compiled for the equipment, subsystems, routines, and other specifics of implementation that provide the various functions. Today, most organizations have software, equipment, and digital data that have to be adapted to GIS. This can form the framework for the type of technological solutions to be selected and should be clearly outlined in the specification.

The final requirement specifications are usually included in a request for proposal (RFP), which aims to provide potential vendors with adequate information for bidding. A comprehensive RFP is the basis for realistic and reliable bids. Requirement specifications fulfill several functions. Although a requirement specification tells bidders what kinds of applications a customer wants, it also acts to enhance in-house understanding of the system to be purchased. It serves as the gauge against which compliance to contract is measured. Both the reasons and goals for implementing GIS should be stated. The tasks to be addressed in the organization should be identified.

Normally, requirements vary in their importance, so they can be grouped according to:

- Requirements that must be fulfilled
- Requirements that may be fulfilled
- Future requirements that must or may be fulfilled when the technology becomes available
- Requirements that are secondary or optional, on which the vendor may bid

Cost-benefit analyses can be used to rank the importance of requirements (see Section 17.7). Design requirements can conveniently be divided into main requirements and special requirements. The functional requirements vary from organization to organization. Typical requirements include:

- Descriptions of different applications
- Data conversion that depends on the types of data processed, the technologies employed in data conversion (scanner, digitizer), and the volume of data to be converted

- Functions for importing and exporting data, including formats
- Database designs and storage functions, which depend on factors such as the data models to be preserved and the data volumes to be stored
- Printout and presentation of data, such as to meet specified standards

Requirements may also include search functions, their search criteria, and search speeds. Analytical functions are often needed for overlays and in network analyses. Many organizations already have databases (client lists, property registers, pipe network descriptions, operation and maintenance systems, etc.) that should be integrated with GIS. In such cases, the degree of integration must be described. However, in practice, experience indicates that overambitious integration plans may well degrade a GIS project that is otherwise feasible. The functions of standard operations, such as customer counter services and printouts, may also be described. The data used by all subsystems will have to be updated, so updating functions must be described. A facility may entail special functions, such as for geometrical design and for the computations of lengths, areas, volumes, and other parameters. In all cases, there should be a clear statement as to how vendors should respond to requirements that their systems do not meet but which they wish to develop.

Of course, most organizations considering implementing or updating GIS already have more or less advanced computer system(s) and networks. Therefore, the existing computer systems must be described to ensure compatibility of the new equipment and software. The descriptive parameters should include type and capacity of equipment (storage, processing speed, number of workstations, etc.), type of operating system, and other relevant software.

Many organizations have a general information technology strategy that will govern a new GIS installation. The factors involved include operating system, hardware platform, network, and so on. In addition to the general information technology strategy, special requirements may be stated for a GIS facility, such as for user interfaces and response times. The number of workstations and numbers and types of peripherals incorporated in the GIS facility must be stated. The chief parameters of the system configuration and attendant network design must be listed. Normally, the details of a new system are not described in great detail in the RFP, as technology changes so rapidly. Vendors should be encouraged to offer new technologies and to describe them in terms of their benefits for the applications involved.

The system specifications, which are based on the design requirements, include stipulations of:

- System solution
  - Computer (capacity and speed)

- Peripheral devices (number and types)
- Communication
- Software
  - Functionality
  - User-friendliness
  - Accessory modules
  - Customizing
- Maintenance
- Training
- Costs

There is no single ideal set of specifications. Organizations combine specifications in various ways to suit their individual needs. For example, organizations with ongoing products and administrative tasks, such as utilities, national map services, and municipal authorities, usually emphasize stability and service and choose systems accordingly. Agencies responsible for planning and resource management may seek greater flexibility and diversified applications. Schools and universities usually favor inexpensive elementary systems. Data security may also be a factor to be included.

18.2.4 Request for proposal

A well-prepared RFP, based on design requirements, is a prerequisite for receiving meaningful proposals. Evaluation of responses can be simplified by compiling part of the invitation to bid as a list of questions to which bidders respond. This is a form of compliance list, in which the response to a question is "comply" or "not comply."

An RFP, or an invitation to bid (ITB), should include a description of how a bid should be structured and which criteria will be used in assessing the bid. For example, the importance of fulfilling the functional and technical requirements in relation to price, the size and economic stability of the bidder, the support and training, and the documentation offered, may be described.

The manner in which queries from vendors and other relevant information are handled during bidding should be described. For example, responses may be written to all queries, with copies to all bidders. An RFP should be sent to no more than three or four selected vendors, who are expected to be able to meet the requirements of the system as well as the need for future support.

18.2.5 System evaluation

A system proposal should be evaluated in terms of how closely the bid meets the requirements stated in the RFP and the quality of the vendor's bidding. Often, requirements may be met not by a single system but by a combination of several systems. Choosing a new system that will provide integration of the new system(s) into an existing system can complicate the situation. Criteria for determining the final choice of vendors usually includes service capability, market position, company stability, price policy, references, and so on. At the end of the day, it is both the system and its vendor that are chosen.

Several objective parameters may be drawn upon to simplify choices and aid decisions. Requirements should reflect needs, not only system parameters. Systems may be ranked numerically by assigning weighted values to all selection criteria (Figure 18.2). The weighted sums for all systems can then be ranked.

Functions not available but offered as developable must be assessed separately. The fulfillment of functional requirements may conveniently be assessed separately, before the bid prices are compared. This prevents unconscious favoring of the system bid at the lowest price. Normally, no one system bid is superior in all aspects. Consequently, (and as stated in Aronoff 1989), the final choice often involves compromises.

Benchmark testing

The performance of a system *as installed* is crucial to its payoff potential; its performance at some future date is of less immediate interest. Performance may be assessed in terms of how a system executes specific functions, how rapidly it works, how simple it is for users, and how flexible it is in current applications. Such questions are best answered by testing.

Different systems are best compared by running specific test programs using uniform test data. This benchmark approach, or functional test, is valuable in assessing various systems and approaches. Each system should be tested intensively over several days.

Selection criterion	weight	rank				weighted rank			
		a	b	c	d	a	b	c	d
System cost	5	7	7	6	9	35	35	30	45
Database design	4	6	7	8	7	24	28	32	28
Human-machine interface	5	8	6	7	6	40	30	35	30
Equipment functionality	3	7	8	7	7	21	24	21	21
Own programming	3	7	5	5	6	21	15	15	18
Follow–up	4	7	6	6	8	28	24	24	32
Data interchange	5	8	7	6	5	40	35	30	35
Expansion possibilities	4	7	8	8	8	28	32	32	32
Documentation	4	7	6	6	5	28	24	24	20
Maintenance costs	4	9	5	6	6	36	20	24	24
Drawing functions	2	7	9	9	6	14	18	18	12
Unweighted sum		80	74	74	73				
Weighted sum		315	225	255	287				
Rank		1	4	3	2				

Figure 18.2
Typical comparison of four GISs (a, b, c, d), using an initial ranking of 1 to 10 and weightings of 1 to 5.

Parameters to be considered

The parameters to be considered in selecting and installing equipment include:

1. Installation details
   - Floor and other structural loads
   - Furnishings
   - Main power supply
   - Ambient temperature and humidity
   - Noise environment
2. Computer characteristics
   - Computational capacity
   - Computational accuracy
   - Communication
     - Local area network
     - Data communications protocols
     - Speeds
   - Number of peripherals supported
     - Graphic terminals
     - Alphanumeric terminals
     - Printers
     - Plotters
     - Digitizers
   - Expansions accommodated
   - Compatibility

In practice, equipment use is often impeded initially by computing capacity and the number of input/output interfaces available. The more important characteristics of the peripheral units include:

1. Graphic displays
   - Screen size
   - Image quality
   - Colors
   - National character sets
   - Ergonomics
2. Plotters
   - Accuracy
   - Resolution (raster plotters)
   - Color quality
   - Output format
   - Plotting speed
   - Plotting medium
   - Operating cost (paper, etc.)
3. Digitizers
   - Accuracy
   - Resolution
   - Size

The characteristics of tape stations, internal storage facilities, disk drives, alphanumeric terminals, printers, local area networks supported, data communications protocols available, and so on, should be listed and evaluated in a similar manner.

The software requirements should include:

- Structure of basic system and application programs
- System openness to support user-defined application programs, communications programs, etc.
- Macroprogramming capability
- User instructions and system support documentation
- Freedom to change databases
- Facilities for restricted access
- Operational records and user accounting
- Tools for restructuring databases
- Selection of languages available for entry of error messages and commands
- Programming language
- Support of data interchange with other systems (including reading and writing of different formats)
- Human–machine interface (response time, error messages, commands)
- Support and follow-up

The preceding list may, of course, be much longer, depending on the software application involved.

In general, the criteria involved in choosing a system depend on the nature and goals of the organization involved. A typical case in GIS applications is the implementation of a joint-use facility shown to have a benefit-cost ratio of 4:1. The guidelines for its specifications are:

- Start with a general-purpose system.
- Note that general systems generally entail more initial effort.
- Modular systems are easily expanded.
- Tailor-made systems constrain further development and limit the utility of databases.
- Tailor-made systems are rapidly obsolete and often expensive to maintain.
- General-purpose systems ease information interchange and interorganizational communication, which enhances benefit-cost ratios.
- Specialized systems can limit benefit-cost ratios when information is digitized.
- Evaluation should be as structured and as objective as possible.

From the early days of computer systems up to the explosive expansion of computerization triggered by the introduction of personal computers (PCs) in the early 1980s, systems were inevitably proprietary. Hence, the choice of a system was tantamount to choosing a supplier. Vestiges of this trend remain, as many systems remain proprietary. However, internationalization and intercompany tie-ups in the computer industry, the trend toward open computer architectures, increasing interprogram compatibility, and, of course, cloning and competition have all contributed to increasing the selection of suppliers capable of delivering a suitable system or subsystem. The choice among competing suppliers usually includes considerations of:

- Capabilities and service
- Market position
- Company stability
- Price policy
- References

The Open GIS Consortium (OGC) is actively developing interoperability specifications so that GIS systems can interoperate. All major vendors are focusing on OGC specification compliance and requiring that OGC compliance be a part of all RFPs.

18.3 Contracts

A mutually acceptable, clear, and complete contract is essential and benefits both purchaser and vendor. For a complete system, two contracts should be entered into, one for purchase and one for maintenance. It can also be appropriate to make separate contracts for software and hardware. The purchase agreement should contain a short description of the equipment scope and the mode of payment. Normally, the delivery will have a guarantee period, so the goods and services supplied under guarantee must be specified. Joint activities and responsibilities may be described as needed. The contract should contain guidelines for confidentiality between the contracting parties and how ownership (hardware) and use rights (software) are to be transferred to the purchaser. The treatment of delivery delays must be specified, as well as how any breach of contract is to be handled.

A maintenance agreement should contain a description of its scope and limitations and of the services to be supplied for its duration. It should also describe the financial and payment conditions and how any breach of contract is to be handled.

The final decision

The final decision to procure a GIS should be made only after contract negotiations have been completed and all financial and legal aspects have been clarified. Careful specification, bid evaluation, and contracting procedures, as discussed earlier, ensure the best decision. But only future use can determine whether the right choices have been made.

When a system has been chosen

The vendor should fulfill the contract concerning:

- Installation/integration
- Customizing, database design, etc.
- Acceptance testing

Acceptance testing may be necessary to determine whether a system functions per specifications and in conformance with the contract.

18.4 Technical Database Design

A thorough database design is essential, both for realizing the functions required and for determining:

- Scope, so that only relevant data are entered
- Ease of access to data
- Efficiency with which functions are executed
- Facility with which data are updated
- Receptiveness to new types of data
- Ease of restructuring data in the database

As discussed in Chapter 12, a data model provides the basis for a database design. The detailed design of a database depends on the system software and hardware chosen. In joint-use facilities, databases must be further designed for individual organization data or for data common to all the organizations involved.

The best approach is for each organization to design databases that best suit its specific needs, and to incorporate direct or indirect interorganizational communication via a joint solution (Figure 18.3). The

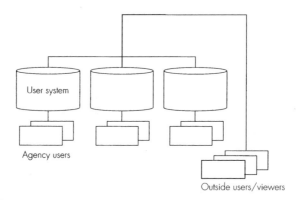

Figure 18.3

In a joint system, some information may be available to all users, while parts of the total information set are stored in user databases. The different user systems (databases) are usually separate and may be networked. Different types of user databases and storage formats may be used. Outside users may be permitted to view, retrieve and present data.

various individual databases are accessed by a data communications network. The physical separation between the databases and the communication between them are not obvious to users, who access the entire network. The distributed database approach also requires good updating routines and extensive standardization.

As the number of users having on-line access to a database or interconnected databases increases, the need to restrict access to certain parts of the data may also increase, for security, strategic, or maintenance reasons. Such considerations must also be included in database design.

Logical (software-dependent) database design

Logical database design is software dependent. Different GISs will therefore be able to offer different solutions. This is discussed in detail in Chapter 12. The principal aim is to define the most suitable storage form, such as map sheets and projects, as well as the division of types of objects into various subject layers. This implies that this design computes only geometric data. The choice of storage form and division into thematic layers determine how effectively the data can be used. Once the final choice has been made, it can be difficult to restructure the data.

Physical (hardware-dependent) database design

The physical database design is dependent on the choice of hardware. This means that it is dependent on factors such as storage medium, accessibility, and speed. This is discussed in detail in Chapter 12. In many cases, however, the constraints imposed by data storage facilities are more administrative than technical, as organizations implementing new systems frequently wish to employ an existing data processing infrastructure.

18.5 Creating a Database

Creating a database is costly, time-consuming, and a drain on facilities and staff. Therefore, the databases employed in a GIS facility should be compiled as rapidly as possible so that the data they contain can be accessible as soon as possible. Consequently, both private firms and public agencies often arrange for data conversion and entry work to be done in round-the-clock shifts. Because they know their own data and professional fields, most organizations elect to convert and enter data in-house. As the workload involved may be enormous, outside temporary staff are often hired for the routine tasks.

Identified user needs, preferably ranked according to benefit-cost ratios, should determine the data entered in a database. Clearly, data accessed by many users have the greatest utility. Therefore, database creation should start with joint data. At the bottom of the data ranking list may be information entered on the premise that it may be useful at some time. In many cases, if its use is uncertain, it need not be entered.

The practical activities required will generally include the following:

- Development of templates if needed
- Thematic map manuscript preparation
- Digitizing and editing of objects
- Attribute coding and verification

18.6   System Operation and Maintenance

Data must be maintained to retain their value. Normally, data maintenance in a major facility is considered only after the facility is fully operational, some three to seven years after the first data have been entered. However, data maintenance is vital from the very start.

Persons active in data acquisition are usually those most qualified to maintain the same data. Surveyors can both acquire and then maintain utility network data, planners can compile and then maintain zoning plans, and so on. The most direct approach is for the responsible professionals to maintain their own data by means of PCs.

Maintenance routines should include deletion of duplicate entries, which are common in conventional manual systems. This implies that the various organizations accessing common data are obliged to depend more on each other. Consequently, all organizations involved must agree upon the maintenance routines to be used.

18.7   Safekeeping and Security Routines

In many ways, computerized information is more vulnerable than conventional information. Errors in equipment, storage media, and computer programs, as well as user mistakes, can have serious consequences. Security routines are therefore necessary to guard against errors. As the collection of geographic information is very costly, damage to storage media (tapes, disks, etc.) can result in considerable financial loss. Moreover, as updating, access, and new data entry become more frequent, security must to be increased. On-line users can hardly be expected to wait several days while data are reconstructed following damage or an accident. Electrical outages or losses of data are expensive and time-consuming for both users and operating organizations. Security routines are therefore vital.

There are many causes of data loss. The storage medium may be damaged through misuse; equipment may fail; fire or other damage may occur. Data in daily use in a primary memory may be lost because of user errors, operative system errors, equipment failures, or electric power failures. Without prior warning, disk sectors or entire disks may become unreadable, or data stored on tapes may deteriorate.

Backup copies of disk files should be made periodically to minimize reconstruction work should file data be lost. A customary rule for systems in daily use is that all database changes must be automatically backed up daily from working disks onto tapes or tape cassettes. The entire database may be backed up less frequently, perhaps once a week or every other week. Some backup systems back up all transactions continuously.

Working databases independent of original databases are often used in GIS work—for example, for tasks concerning a limited geographic area. A working database should never be stored on the same disk as an original database, lest a total disk failure, *disk crash* in computer jargon, destroy both databases. Data on tapes should also be backed up frequently. Although tapes store data well, data can deteriorate on all magnetic media in time. Therefore, tapes should be recopied periodically; once a year is recommended. Backup copies should not be stored with original data or databases, lest fire destroy both, but should be stored in a separate fireproof enclosure.

Some GISs contain data that must be protected against unauthorized use, such as personal data, details of telecommunications networks, and the like. The operative system can be configured to permit only authorized access to data or parts of data stored. Complete safeguarding and security of data would be extremely expensive and time-consuming. Meticulous storage and handling routines can reduce the risk of disk damage, at a cost proportional to the measures enacted and the disk storage boxes and cabinets acquired for the purpose. Stand-by power units, uninterruptible power supplies (UPSs), and filters may be installed to guard against power outages and large fluctuations in power voltage. These are often a necessity in developing countries. However, such installations add to the cost of equipment and may be expensive, especially because each UPS unit consists of a rectifier that converts the main AC to DC for float-charging batteries, and an inverter for conversion from DC to main voltage and frequency to power the equipment.

Therefore, safekeeping and security routines must be evaluated in terms of the costs of data loss. As in other sectors and disciplines, calculated risk must be taken into account. Insurance policies may be taken out against financial loss. Normally, it is possible to insure equipment against physical damage due to fire, flood, vandalism, and theft. Insurance is also available to cover the costs of reconstructing data and the losses incurred during operational interruptions. Finally, security measures may be required to guard against the potential losses of deliberate break-in and data destruction—*data crime,* as it is now commonly called.

18.8 Evaluating New Applications

With time, most GIS users discover new needs and therefore new applications for systems and data. In many major GIS facilities that have been operational for some time, users report as many as 40 new applications implemented after the system became operational. The new applications often differ considerably from those originally anticipated for the facility and are therefore among the factors that increase the benefits GIS provides.

The procedure for assessing new applications is straightforward. After a GIS facility has been in operation for a few years, a survey of user needs may be made to chart new needs. Identified new needs may be evaluated in terms of benefit-cost ratios, just as the initial needs for which the system was originally designed should be reevaluated. Similarly, the needs for organizational alterations should be evaluated, and alterations made using the same criteria as used initially. Most new applications require programming work, and the ease with which they may be realized depends on the flexibility of the system implemented.

CHAPTER 19

Standards and Geospatial Infrastructure

Maps and other forms of geographical information constitute one of the world's oldest scientific fields and an area that has been the subject of standardization for many centuries. Long before digital methods came into use, it was generally accepted that the efficient interpretation and reading of maps was dependent on a uniform and congruous presentation. This has applied not least to sea charts, as a safe sea voyage was (and still is) based on being able to change to a new map sheet independent of regional and national borders.

The introduction of data-based methods over the past 20 to 30 years has made it even more apparent that the efficient use of geographical information is dependent on a uniform description of the data. A great many countries have therefore started work during the past 10 to 15 years on the development of national standards, especially for the exchange of digital geographical information between different computer systems. The objective is to allow the producer of data and the end user to interpret and understand the data in the same way. A standard will provide definition of data structures, data content, and rules that will:

- Increase mutual understanding of the geographic data among users
- Eliminate the technical problems of exchanging geographic data between different (geographical) information systems
- Increase integration, combination, and interoperability of geographical data and related information

Standardization can be carried out at three levels:

- *Level 1:* Generic standards (mainly IT-based; e.g., data description language, query language, encoding, transfer syntax)
- *Level 2:* GIS application independent standards (e.g., geometry, topology, quality, metadata)
- *Level 3:* GIS application specific standards (e.g., cadaster, utilities, roads, base maps, urban planning, etc.)

Levels 1 and 2 are mainly covered by international standardization and specify the framework and the general rules, whereas level 3 is a typical national task.

The uniform transfer and understanding of data is dependent on a series of factors. Transfer of spatial data involves modeling spatial data, data structures, and logical and physical file structure. To be useful, the data must also be meaningful in terms of content and quality. In addition, a common terminology has to be developed at each level. Comprehensive international work has been carried out in this field. In 1992 the decision was made to establish a committee under the European Standardization Committee—Comité Européen de Normalisation (CEN).* This work has won broad support throughout Europe and most countries participate in it. In 1994 a corresponding initiative was taken by the International Standardization Organization (ISO).† When complete, the standards will be called "ISO 19100 standards." Most of the ISO proposals for standards will probably become formal standards during 2001. Whereas CEN is most concerned with describing the more traditional aspects of geographical information such as topology, geometry, metadata, reference systems, and exchange/coding, ISO's work has, in addition, a description of the service relationship linked to geographical information (such as interface, etc.). The work of CEN and ISO is carried out independent of any application domains. The relationship between ISO and CEN is regulated by an agreement aimed at ensuring the establishment of a joint international standard.

The North Atlantic Treaty Organization (NATO), through the Digital Geographic Information Working Group (DGIWG), is carrying out international standardization work and is the first organization to develop its own standard, the Digital Geographical Information Exchange Standard (DIGEST) with an object catalog, FACC. Some years ago, a global data set was established, based on DIGEST, called the Digital Chart of the World (DCW). This data set corresponds in detail and accuracy to a scale of 1:1,000,000. The International Hydrographic Organization (IHO) has developed a standard known as S57. The file format S57 used to be called DX90. Key elements in this development are the definition of the Electronic Navigational Chart (ENC) and the Electronic Chart Display and Information System (ECDIS). Work is under way to harmonize all these standards with ISO. Both IHO and DGWG have the intention of making their standards compatible with ISO.

In recent years (1995), information technology companies, database

---

*This committee has been named CEN/TC-287—geographic information.
†This committee is known as ISO/TC-211.

system vendors, GIS vendors, and others have established the Open GIS Consortium (OGC). This organization works on the development of open solutions linked to the processing of geographical information. The purpose is to develop the necessary interface to enable total mobility between the systems, as well as the integration of data and processing into mainstream computing. The different members contribute technology that can form the basis for different specifications—abstract specifications and implementation specifications. The prominent market position of these companies gives them considerable influence on the work that is carried out within ISO. An agreement of cooperation has been entered into by OGS and ISO.

European system suppliers have developed a culture whereby they adapt to customers' and public bodies' requirements in matters concerning standardization. This differs somewhat from the situation in the United States, where powerful suppliers often control the market through de facto standards. But the Federal Geographic Data Committee (FGDC) and the U.S. Geological Survey have had the main responsibility for developing the Spatial Data Transfer Standard (SDTS). SDTS, which will serve as a national spatial data transfer mechanism for the United States, provides a solution to the problem of spatial data transfer from conceptual level to the details of physical file encoding. FGDC has also been responsible for developing a metadata standard.

In this standardization work, great emphasis is placed on the efficient transfer of GIS data from one computer platform to another, as well as on the possibility that a GIS can use data that are dispersed through a variety of systems. These standards will probably result in a much stronger integration of data and systems in the future across the bodies of different platforms, based on standards such as Open System Environment (OSE).

It is unlikely that the entire industry will move to a single standard in the near future. It is probable that there will be a few important international standards in addition to all the proprietary commercial GIS formats. In Canada, Open Geographic Data Store Interface (OGDI) has been developed to overcome these problems. OGDI is an application programming interface (API) that uses a standardized access method for raster and vector data to work in conjunction with GIS software packages and various geospatial data products, including:

- Conversion of various formats into uniform transient data structure
- Transformation of coordinate systems and projection systems
- Searching and retrieving geometric and attribute data
- Accessing different geodatabases and formats
- Using the Internet for distribution and search of geographical data

19.2 Elements for Standardization

A series of factors have to be taken into consideration when developing standards for a geographical area (town, region, country) and miscellaneous subjects.

Fundamental principles

Usually, there are a number of fundamental and general elements that are suitable for standardization. The following are some of the elements being worked on in ISO 19100:

1. Conceptual modeling and application schemas
2. Transfer format
3. Encoding
4. Spatial representation
5. Spatial referencing
6. Temporal characteristics
7. Data quality description and evaluation
8. Visualization/portraying
9. Geographic information services and interface

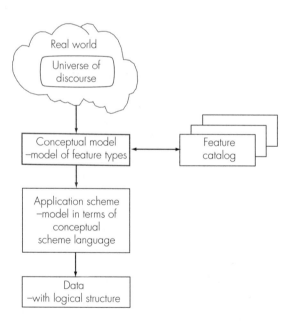

Figure 19.1
The universe of discourse is that part of the real world we want to describe—for example, features such as property boundaries, land-use areas, utilities, etc. The universe of discourse is described in a theoretic conceptual model by an object modeling technique. The model is represented in special application/conceptual schemas, which are described by a standardized conceptual schema language (adapted from ISO 19100 prototype).

10. Object catalogs
11. Metadata

As mentioned earlier, these elements can be standardized at different levels. A general description of the framework, methods, and templates is carried out on the generic level (level 1) and the GIS application independent level (level 2). This type of standardization is covered by OGC, ISO, CEN, and other organizations. Filling out the framework and templates with concrete rules and values is carried out on the GIS application level in each country or institution.

Conceptual modeling and application schemas
All standardization work should be based on a common understanding or model of how the real world should be described so that it can be represented in GIS. This type of general reference model should

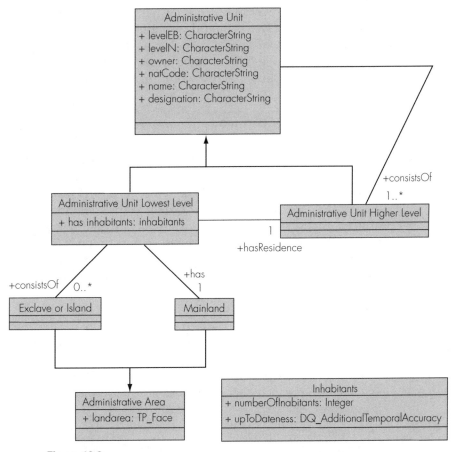

Figure 19.2
UML (Unified Modeling Language) application schema for administrative unites (ISO).

also specify all the elements that should be the object of standardization (see the elements listed earlier) and how these elements are connected. The abstract description of the real-world features is called a conceptual model. In standardization work, a formal conceptual schema is used to describe the conceptual model for the universe of discourse (Figure 19.1). Standardization work is often based on object-oriented modeling. A conceptual schema language (Unified Modeling Language—UML, etc.) is used to develop conceptual schemas or templates (Figure 19.2).

To ensure ease of data exchange between different users and different data systems, the data structures have to be fully documented in a commonly accepted way. An application schema provides the formal description of the data structures and content required by one or more application (Figure 19.7). It contains the method or the templates for a description of a data set, including both geometric data and attribute data.

The standardization of modeling work and application schemas will be very useful for everyone working with data structures, for GIS product developers, for standard developers (on application level), and for GIS users. One should note that in international standardization work, the term *feature* is normally used instead of *object type*. A feature is an abstraction of a real-world phenomenon.

19.3 Standard Transfer Formats

It is still commonplace for different GISs and DBMSs to store data in their own internal formats. The exchange of data often requires the transformation of formats and is therefore no trivial task, as a wide range of different formats are available. Were a common standard transfer format available, all system suppliers would be able to relate to this and develop regular export and import routines (Figure 19.3). Naturally, both users and producers of data have realized this, and that is why many countries have initiated the process of developing such standard formats.

The transfer format should provide transfer of geographical data sets that include metadata, spatial data, and attribute data. In general, a minimum of additional programming should be necessary for creating a standard data set (from GIS-A) or reading a standard data set (to GIS-B). Data exchange between the most common GISs and CAD systems could also be included. A standard transfer format for files should include rules for:

- Transfer of geometry
- Transfer of links between geometry and attribute data
- Attribute data transfer
- Transfer of metadata

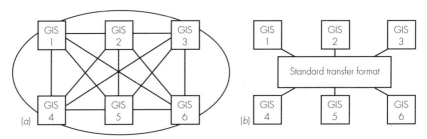

Figure 19.3
By means of standardized exchange formats, digital data can be exchanged between different GISs. A standard exchange format considerably reduces the need for translators—in this example from 15 *(a)* to 6 *(b)*.

Encoding

Encoding rules allow geographic information to be coded into a system-independent data structure suitable for transport or storage. The encoding rules specify the types of data to be coded and the syntax, structure, and coding schemes used in the resulting data structure. The resulting data structure may be stored on digital media or transferred using transfer protocols. The resulting data structure may be read and interpreted by computers, but may also be in a form that is readable by humans (ISO).

In general, a data format has the following main components:

- Head:
  - General information on format and data
  - Default setting
  - General parameters
- Index:
  - Optional information as to how one can efficiently find the data on a file
- Data dictionary:
  - Description of the structure of the data that follows (necessary in order for system to be able to interpret the data)
- Data:
  - Data elements (geometry, topology, attributes, etc.)

In a data format it is usual to split each data element into three parts: structure information, the name of the data element, and the data value itself (Figure 19.4). A special language (syntax) is often used

Figure 19.4
Each data element is often divided into structure information, the name of the element, and value.

(e.g., $<..>$ is used to enclose the names of different format elements, and $::=$ can mean "is defined as"). Special signs can be used to collect data in groups at different levels (structure). The permissible number of signs or digits has to be specified for each element. For example, one person can be described as follows, where the group level is indicated by 1 to 4 dots:

- Person
  - National identity number   H11   11—figure number
    - Name   T30   Text with 30 digits
      - Address   T30   Text with 30 digits
        - Street   T30   Text with 30 digits
        - Street number   H4   4—figure number
        - Postal code   H5   5—figure number

The description of a standard format should also include a product specification and a code list for all the elements so that the same code will always represent the same type of element or object regardless of where it is produced (see Figure 19.5).

In ASCII file format, all data are coded as ASCII signs—altogether 127 or 255 signs. ASCII covers letters in the American alphabet, numbers between 0 and 9, several special signs and different national signs. ASCII data require conversion in connection with in- and out-loading and are therefore time-consuming to read and write. ASCII format data can be moved with relative ease between different GISs, and they can also be edited in a word processing system. In the so-called binary file format, the different data are coded in different bit patterns, which can cover more than ASCII. Binary files require only partial conversion in connection with in and out loading. The supplier-specific GIS formats are often based in binary file formats.

ISO specifies the encoding rules for encoding data in Extensible Mark-up Language (XML). This standard will require the data models to be described in UML, which is a conceptual schema language. OGC has established an XML solution called Geographic Mark-up Language (GML), which works for simple features. In the United States the U.S. Geological Survey has been responsible for the development

Numerical code series	Object group		Numerical code	Object type
1 000	Survey control stations		4 001	National border
2 000	Terrain formation		4 002	County boundary
3 000	Hydrography		4 003	Township boundary
4 000	Boundaries		4 011	Property boundary
5 000	Built-up areas		4 022	National park border
6 000	Buildings and facilities			etc.
7 000	Communications			
8 000	Technical facilities			

Figure 19.5
Many countries have created national standards for coding geographical data.

of the Spatial Data Transfer Standard (SDTS), which also specifies the implementation of SDTS in terms of the ISO for a Data Descriptive File for Information Interchange. The U.S. Geological Survey has also developed a topologically structured format, Digital Line Graph (DLG). National standards are found in other countries as well, such as SAIF in Canada, NTF in Great Britain, EDIGÉO in France, ATKIS in Germany, DSFl in Denmark and SOSI in Norway.

AutoCAD's DXF/DWG format (Figure 19.6), Intergraph's SIF format, the ESRI SHAPE format, and Hewlett-Packard's HPGL format are typical industrial standards. PostScript, which is a graphic language, also has potential in the plotting, printout, and transfer of geographic data. In countries that have not established a national transfer format, industrial standards such as DXF (ASCII-based format from Auto-CAD) and pure ASCII tables are used. Vector Product Format (VPF) is a standard format, a data structure, and an organization for large geographic databases that are based on a georelational data model and are intended for direct use. VPF is defined by the National Imagery and Mapping Agency (NIMA) of the United States. The Digital Chart of the World uses VPF for distribution.

Computer Graphic Metafile (CGM) is an international standard (ISO 8632) for digital imagery. CGM files are system independent and are

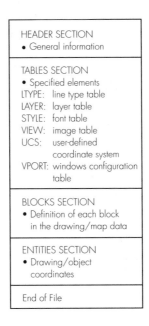

Figure 19.6
DXFs handle vector data and are in reality adapted CAD (AutoCAD format) data. The structure is based on a *header,* with general information; a *table,* which contains definitions of line table, font table, and so on; a *block,* which contains definitions for each block in the drawing; as well as *entities,* which contain the coordinates. If we only need to transfer the object's geometry, it will only be necessary to fill out the entities part.

used in GIS to unite graphics from different systems. Tagged Image File Format (TIFF; Figure 19.7) is an industrial standard developed in the United States, used primarily for transferring image data between different desktop publishing systems. TIFF allows several pixel types, such as 8-bit images and 24-bit images, and can be used together with different packing algorithms. TIFF is hardware independent but this imagery format has several limitations. It has no support for georeferencing characteristics and geodetic parameters, and user interventions are needed to a large extent to get the image in the correct position. Because of these limitations a special GIS-adjusted format has been developed, called GeoTIFF. In addition to transformation support and geodetic parameters, the image format GeoTIFF also supports geographic metadata, orientation to orthto imagery and public domain software libraries, and tools for reading and writing different formats. The development of GeoTIFF is supported by many GIS vendors. Continuous Tone Image Format (COT) is also an industrial standard developed in the United States, used in image treatment programs. These formats are also appropriate for the transfer of raster data between different GISs. Basic Image Interchange Format (BIIF) is a relatively new ISO-defined format for the exchange of raster images. Digital Terrain Elevation Data (DTED) is a standard format for elevation data, organized as regular raster grids of elevation values and divided into grids in accordance with the geographical grid network. DTED is developed by the National Imagery and Mapping Agency (NIMA) in the United States.

The standards being established by ISO and OGC will need some years to be fully implemented. Meanwhile, a pragmatic policy should

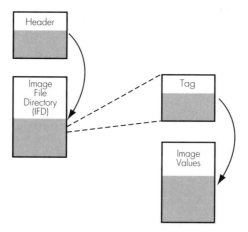

Figure 19.7
TIFF handles raster and image data. At the beginning of the file are a header and a pointer, which points at the address of the image file catalog. In the catalog are stored information on the relevant data and pointers to the data itself.

allow industrial standards to be used at national level for the transfer of data between users.

Data volume often depends on how data are formatted and designed. General formats must provide leeway for all possible situations. Consequently, in practice the data often include basic administrative data, which are not used, or space for attribute data, which is not used. A series of software modules is available for packing data for transfer.

All standardization work is time-consuming because it requires a high degree of unity and many parties are involved, all with different interests and thus using different systems. It is now possible to find solutions that can handle simultaneously many types of data warehouses with different formats.

19.4 Special Standardization Elements

Spatial representation

Geographical objects are represented spatially as either vector or raster data. Vector data consists of the geometric primitives points, lines, and areas, and the topological primitives links, nodes, and faces. Raster data is based on the division of the real world into small, uniform units (cells). At the superior level, standard schemas should be developed to describe the geometric and topologic primitives. This will increase the ability to share geographic information among applications. These schemas will be useful for geographic information system and software developers and users of geographic information in providing consistently understandable spatial data structures. Today ISO and other organizations are working on the description of conceptual schemas for the spatial characteristics.

At the application level a geometric primitive—point, line, area or raster cell—has to be specified to represent each object type. The cell size must be specified for raster data, and rules must be established for the creation of nodes and links for vector data.

Spatial referencing

Spatial referencing can be based on referencing by coordiantes or referencing by geographical identifiers. An object is not clearly georeferenced without being linked to a coordinate reference system. The most important elements in a coordinate referencing system are a datum—specified with ellipsoid, anchor point, and prime meridian—and the coordinate system itself—specified by name, the units, and the direction and sequence of the axes. There are a number of different coordinate referencing systems. The relevant coordinate referencing system should always be specified for a data set, and the description should be standardized so that misunderstandings or serious errors do not occur. At the application level, for example only the use of geographical coordinates (latitude and longitude) and UTM can be permitted. Should other systems be used, transformation parameters should be stated.

Correspondingly, indirect spatial referencing by geographic identifiers should be specified in regard to the description of reference units, such as county, city, postal code, and so on, and how the objects are linked to the units.

Temporal characteristics

ISO 8601 specifies the use of the Gregorian calendar and 24 hours local or Coordinated Universal Time (UTC) for information interchange. This should be the primary temporal reference system for use with geographic information. If other systems (Islamic calendar, Julian calendar, modern Japanese, ancient Babylon, etc.) be used, the parameters have to be stated as metadata. Specification of date and time of day can be standardized, as in the sequence year/day/hour/minute, with hours given in 24-hour format.

Data quality and quality evaluation

At the superior level, the objective of standardization of quality parameters should be to provide common principles for describing the quality of geographic data and concepts for handling quality information. Quality statements should consist of certain elements: positional accuracy, attribute accuracy, temporal accuracy, completeness, and logical consistency. On the application level, the data quality elements should be standardized with regard to specification of scope, quality measure, evaluation procedures, values, value units, and date. Data quality evaluation procedures should be standardized and describe the methodology used to apply a data quality measure. ISO and Spatial Data Transfer Standard (SDTS, U.S. Geological Survey) have specified a number of alternative methods for measuring accuracy (see Chapter 11). It is also important to note that for positional accuracy the value unit must be related to nominal ground, not to units on a map scale. Data should, in general, be given at the lowest possible unit and not aggregated. Data should also be free of bias and must be unambiguous. Quality information should be reported as metadata in a standard format.

Visualization/Portray

Different computer graphics standards (GKS, PHIGS, OpenGL, etc.) use different attributes to visualize the geometry—for instance, a line can be distinguished by thickness, color, stippling, and other features. There are also several standards for symbols. Without a visualization standard, the applications would have to set up a separate interface to each of these standards, but with a visualization standard all the supported symbol standards may be handled uniformly. This type of standardization is handled today by international standardization work (ISO, etc.). On the application and national levels, standardization can also be performed with respect to creating cartographic symbols. In standardization work the term *portrayal* is used instead of *visualization*.

Geographic information services and interfaces
Standards are needed for users to access and process geographic data from a variety of sources across a computing interface within an open information technology environment, including the Web map server interface. This means that geodata users should be able to query remote databases and control remote processing resources, and also take advantage of other distributed computing technologies, such as software delivered to the user's local environment from a remote environment for temporary use. It also means that users should have access to data acquired in a variety of ways and stored in a wide variety of relational and nonrelational databases. This type of standardization is covered by international work and carried out by ISO, OGC, and other organizations.

19.5   Standardization on the Application Level

On the application level, it is possible to standardize the different themes or applications, such as buildings, transport, utilities, land use, administration units, water, heights, control points, annotation, text, and others. At this level, concrete rules, permissible values, and so on, are specified. The work should be based on a standardized framework and templates, so that all applications are described the same way. For this purpose, one can draw on the work carried out by ISO, CEN, OGC, and similar organizations.

Application schemas
In a country or within an institution there should be common principles for defining object types and data structures. For this one can use an application schema (Figure 19.2). Rules for application schema should specify how data should be organized to reflect the particular needs of applications with similar data requirements. An application schema contains the formal rules for description of a certain data set regarding content and structure, as well as specifications of operations for manipulation and processing data.

Only the rules for organizing the data are specified in the application schema; the concrete content is described in object catalogs.

Object catalogs
An object or feature catalog should be made for each theme or application domain. An object catalog describes the data in detail and can contain:

- Objective and limitations for the standard
- Definition of terms used
- Data model, including digital representation of objects and relationship between objects
- Object attributes
- Permissible attribute values
- Spatial representation and referencing
- Theme list (codes) and, eventually, conversion tables
- Quality statements for each theme

Thus, an object catalog will contain information that is important for understanding the data. Object catalogs should have a standard organization so that they will be easy to read even though the data are produced by different institutions. Rules for visualization are usually part of a product specification, rather than part of the object catalog.

The definition of geographical data at the application level may also be done using data modeling techniques. A graphic schema [entity-relationship (ER) model] covers the definition of the main object and the relationships between object types. An example for buildings is shown in Figure 19.8.

A building is normally described by its roof outline. A small building can be represented by a center point, which will represent the spatial position of the building. Built-up areas are used for representing groups of buildings when the density at the actual scale does not make it possible to represent each building. A built-up area has its own

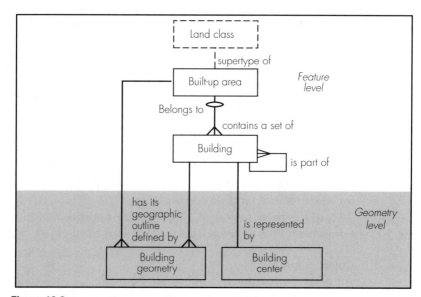

Figure 19.8
ER (entity relationship) model for built-up areas and buildings.

Object	Attribute	Explanation
Built-up area	Area_Name	Offical name of the built-up area
	Area_Type	Main classification of built-up area
Building	Building_Name	Name of the building
	Building_ID	?

Figure 19.9
Standard for object attributes.

Attribute	Legal value	Comments
Area_Type	Commercial	
	Industrial heavy	
	Industrial light	
	Residential high	
	Staff quarters	
	Unclassified	Default
Building_Type	Apartment block	
	Bus terminal	
	Car parking	
	Etc.	

Figure 19.10
Standard for attribute domain.

Geometry entity	Element	Legal values
Building center	Geometric primatives	Point
	Layer	Bouilding center
	Feature-key	Optional
Building geometry	Geometric pimative	Line, polyline
	Layer	Roof outline
		Wall outline
		Area outline
	Feature-key	Mandatory

Figure 19.11
Standard for geometry.

Map Product	Legend element	Derived from:		
		Layer	Attribute	Attribute value
1:50,000				

Figure 19.12
Standard for rules of presentation.

geometry but can also be related to a set of buildings. A building describing a part of a building complex can be related to a building "parent" describing the outline of the entire complex. This model covers the essential requirement for defining the fundamentals for georeferencing building information and for mapping buildings and built-up areas at different scales. In addition, the coordinate referencing system must be specified.

The relevant object attributes can be presented as a list, as shown for buildings in Figure 19.9. Permissible attribute values can be set up in a corresponding list (Figure 19.10). (This figure and the following examples are taken from the proposed National Standard for Zimbabwe–Digital Geographical Information.). The rules for geometry can also be set up in a list (Figure 19.11). Product-specific rules for visualization can be made as well, in tabular form, as shown in Figure 19.12. It will often be necessary to make conversion tables between a standard's code system and codes used previously.

All measurements and quantities (geometry and attribute values) can be defined as valid only by use of the International System of Units (IS units).

19.6 Metadata

Metadata are "data about data." Metadata are stored in a database and intended mainly as a tool to enable users to find out which data are available, whether they are suitable for the purpose intended, where the data are stored, whether there are any limitations linked to their access and use, and, possibly, how the data can be transferred to a suitable system (Figure 19.13). Metadatabases are intended to be an effective link between producer and user but can also be useful for internal use by producers to enable them to maintain a view of their own data over time (to avoid work duplication, among other things). Metadatabases have to be established and maintained by data producers. A decisive factor for users is that data sets should be described in a com-

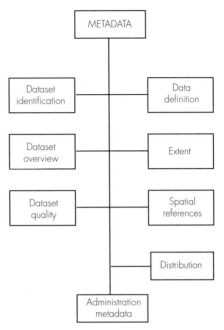

Figure 19.13
A simple data model for a metadatabase.

prehensive and standard way. The main purpose of a metadata standard is to define those data structures that should be used to describe a geographical data set. The standard for metadata does not specify how original data are organized in the computer or how data transfer will be effected in practice.

Metadata attributes
A series of elements should be included in a metadatabase for describing a data set, of which the following are the most important (ISO) (Figure 19.13):

- *Data set identification:* sufficient data proving an unambiguous identification of the data set—for example, the owner organization and the unique name of the data set within that organization, purpose, points of contact, an overall description of the dataset (e.g., a brief summary of the content, the intention of the production, language and character set used).
- *Constraint information:* information concerning the restrictions placed on the data, such as legal or security limitations.
- *Data quality:* how well the data set overall meets its nominal ground location (relative to the specification). The elements are positional accuracy, attribute accuracy, temporal accuracy, completeness, and logical consistency.
- *Maintenance information:* information about the scope and frequency of updating.

393

- *Spatial representation:* the mechanism used to represent spatial information in the data set, (e.g., direct georeferencing with the use of vector or raster, or indirectly with the use of postal codes or similar data).
- *Reference systems:* description of the spatial and temporal reference systems. For direct reference systems, the following should be specified: datum, coordinate system, map projection, height reference system, and coordinate system resolution. For indirect spatial systems, the structure has to be explained (e.g., address system, map sheet, postal code administration units). The calendar and the reference clock time must be specified for the time system.
- *Content information:* information on the content of the data set with reference to the object catalog used; object types and their attributes; description of the object type or the reference to an existing standard coding list; the domains from which attribute values may be assigned.
- *Visualization/portrayal catalog:* information identifying the visualization/portrayal catalog used by the data set.
- *Distribution:* how to acquire a data set; where and how the data set is held and how it may be transferred; identification of the distributor, copyright owner; and information about charges for the data set, including price per unit and discount. Data format and medium used for distribution (CD-ROM, Internet, etc.); information about how to access the data set eventually on-line; information about how to order the data; and future support service for users.
- *Extension information:* information about user-specified extensions.
- *Application schema information:* information about the application schema used to build the data.
- *Extent:* space covered by the dataset with respect to planes, vertical and temporal dimensions; name of the area covered, bounding area, and minimum and maximum *(x, y)* coordinates; minimum and maximum vertical coordinates; the date on which the status and description of the extent are valid.
- *Citation and responsible party information:* information for citing resources (data set, object type, publication, etc.), as well as information about the party responsible (person, organization, address, etc.).

It should also be noted that the Federal Geographic Data Committee (FGDC) has developed a more or less similar standard for metadata. According to this standard, the following seven elements should be used to describe a data set:

1. *Identification information:* basic information about the data set, including information to be used to reference the data set (citation); a characterization of the data set, including its intended use and limitations (description); time period for which the data set corresponds to the current reference (time period of content); the state of and maintenance information for the data set (status); the geographic areal domain (spatial domain); words or phrases summarizing an aspect of the data set (keyword); restrictions and legal prerequisites for accessing and using the data (access constraints and use constraints); point of contact; a graphic that provides an illustration of the data set (browse graphic); recognition of those who contributed to the data set (data set credit); security information; native data set environment information and cross-reference information.
2. *Data quality information:* a general assessment of the quality of the data set, including attribute accuracy, logical consistency report, completeness report, positional accuracy, lineage (source information, process steps, etc.), and cloud cover (for satellite data, etc.).
3. *Spatial data organization information:* the mechanism used to represent spatial information in the data set, including discrete referencing by geographical identifiers and direct referencing by use of vector or raster data models.
4. *Spatial reference information:* the description of the reference frame for, and the means to encode, coordinates in the data set, including horizontal system (resolution, geographic coordinate units, map projection, grid coordinate system, local planar system, geodetic model, etc.) and vertical coordinate system, including altitude system and depth system.
5. *Enitity and attribute information:* overviews and details about the information content of the data set, including the entity types, their attributes, and the domains from which attribute values may be assigned.
6. *Distribution information:* information about the distribution of and about obtaining the data set, including order process, the common way in which the data set may be obtained or received, file format, technical prerequistes, etc.
7. *Metadata reference information:* information on the currentness of the metadata information, and the responsible party.

Not all information is equally important for storage in a metadatabase. Priorities should be set, for example, by distinguishing information as mandatory or optional. For each metadata attribute, Figure 19.14 shows an example of data that may be used to describe geographic data sets.

Metadata attribute	Description	Con-straint	Cardin-ality	Type
Data set identification		M		
Data set title	The unique name of the data set within the owning organization(s)	M	1	string
Alternative title	Another name of the data set, which is either in the same language or another language	O	1	string
Dataset overview		M		
Abstract	A brief summary of the content of the data set	M	1	string
Purpose of the production	A summary of the intentions with which the data set was developed	M	1	string
Spatial reference system used	Data about whether geographic features in the data set are positioned in space explicitly by a coordinate system or implicitly by an indirect spatial reference system: for example, a postal address	M	N	string
Data set language	The language(s) and the character set used within the data set	M	1	string
Document reference	Reference(s) to additional documentation about the data set, including documentation language(s)	O	1	string
Data set quality		M		
Source process	A summary of the processes that the data set has undergone, including time	M	1	string
Source method	A summary of the methods that the data set has undergone	M	1	string
Source material	A summary of the original material from which the data set is produced	M	1	string

M = Mandatory
O = Option

Figure 19.14
Some standards for object attributes for metadata. Constraints: mandatory (M) or optional (O). Cardinality: one (1) or many (N). Type: string (text), numeric, date, address, point, area.

19.7 Infrastructure for Georeferenced Data

19.7.1 The principles

The primary objective of a national geospatial infrastructure is to ensure that users will be able to acquire, at the right time, adequate, complete, and consistent data sets. The infrastructure can be defined as a data series and electronically based services that satisfy the common needs of different user groups for accurate positioning and georeferenced data. Technical developments such as the Internet and GPS have made this type of work very relevant. The definition of *national spatial infrastructure* varies from country to country, but it normally includes elements such as:

- The institutional framework, which defines the policy, financial arrangements, and human resources needed and the legislative and administrative arrangements and cooperation
- Development and maintenance of geodetic network, and accurate positioning services (distribution of GPS correction signals)
- Development of national standards for geographical data
- Collection of fundamental data sets
- Management of fundamental data sets
- Technical framework, which enables users to identify, access, and use fundamental data sets
- Customer services—data distribution and application services

These types of activities have been carried out for decades in developed countries and have required a vast amount of resources, and even now the task is not completed. This is the reality that developing countries have to face.

Institutional framework

In each country there is a need for an institutional framework that can coordinate the different activities and implement policy in the field of spatial data. Traditionally, this coordinating work has been taken care of by the national mapping authorities, but their role has been challenged for several years. Because many activities are similar from country to country, benefits will be gained by international cooperation. Experience has shown that a well-functioning infrastructure is dependent on expertise, data, organization, and hardware/software. Organization can be the most demanding requirement. A prerequisite to the realization of an infrastructure is a strengthening of the human resources involved, including data producers and users, the academic sector, and the private sector.

Geodetic network and positioning services

A common and homogenous georeferencing system is a prerequisite for superimposing different data layers. Establishment of a geodetic network has for many years been a prime task for national mapping authorities. Positioning services, such as distribution of GPS corrections, are also considered basic services that should be included in a national infrastructure. Many countries still lack satisfactory geodetic networks.

Standardization

Standardization is a key to developing national infrastructure for georeferenced data. The reason is that producers of the data and the end users should interpret and understand the data in the same way. A standard can provide definition of data structures and data content, and rules that will:

- Increase mutual understanding of geographic data among users
- Eliminate the technical problems of exchanging geographic data between different information systems
- Increase integration and combination of geographic data and related data

As described earlier in this chapter, international work has been carried out in this field for several years, which the individual countries and others can draw upon. But standardization on the application level (base maps, cadaster, roads, etc.) has yet to be developed individually in each country and by each institution.

Fundamental data

In any area of the world, a number of fundamental data sets are required to assess environmental conditions, develop strategies that can lead to long-term sustainable human development, and perform other needed functions. So far no standard has been settled. According to the "Cookbook" of the organization: Global Spatial Data Infrastructure (GSDI), there are a variable number of data layers that may be considered to be of common-use status and of national or transnational importance as "framework" data. Framework layers commonly nominated in a national context include:

- Cadastral information
- Geodetic control
- Geographic feature names
- Ortho-imagery
- Elevation
- Transportation
- Hydrography
- Governmental units

It is likely that this list will grow as custodians of data identify and promote their data as necessary to increasingly advanced applications and user environments.

FGDC has started work to develop and keep current a Web-enabled clearinghouse showing existing framework data across the United States. The federal agencies' contribution to the list of layers is identical to the list from Canada, apart from geographic feature names, which are not yet included in the U. S. list. For each layer, relevant metadata will be given.

Figure 19.15 shows a possible model for establishment of a national geospatial infrastructure, which will encourage the production and use of fundamental data, and which will also be user driven. The model is based on no cost recovery for the basic spatial infrastructure and establishment of the foundation data sets, whereas maintenance of the foundation data sets should be based on cost recovery (Bernhardsen, 2001).

Technical framework
Today the Internet is the most commonly used technology for search, selection, and distribution of data. The technical framework also includes the relevant equipment such as servers, PCs, local area network, input and output devices, and software—at both the producer end and the user end. In many developing countries the technical framework still seems to be a bottleneck, particularly at the district level.

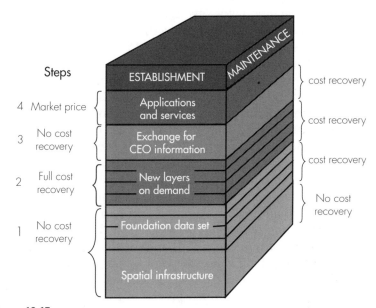

Figure 19.15
Possible model for a national geospatial infrastructure. The establishment of the basic infrastructure and the foundation data set should ideally seek no cost recovery. Production of new layers should be demand driven and based on cost recovery for both establishment and maintenance. Value-added applications should be based on market price.

Customer services

To serve the user properly, a number of national services have to be in operation. Users need to locate relevant data, and the data have to be prepared and distributed. Metadatabases are an important element in a national infrastructure. They enable the users to find out which data are available, if they are suitable for the purpose intended, where the data are stored, whether there are any limitations to their access and use, and how the data can be transferred to a suitable system. Some special services for development of customized data and products must also be in operation, for example, by value-adding the fundamental data.

19.7.2 Organizational framework

Governments throughout the world are beginning to recognize that geographic information is one of the most critical elements underpinning decision making for economic and social development, and that there is a need to allocate resources to establish an effective information infrastructure. For example, in 1994 the president of the United States issued an Executive Order that implemented a national spatial infrastructure. The European Community has developed detailed policies and strategies for a European Geographic Information Infrastructure. Similar steps are being taken in Asian countries and in Australia in cooperation with New Zealand. A number of individual countries have for several years organized/reorganized their work related to national spatial infrastructure.

In most countries, the infrastructure has traditionally been taken care of by the national mapping authorities, but today we see a need to create new alliances to promote the use, distribution and sale of geographical data. In the United States, for example, the National Spatial Data Infrastructure (NSDI) has been established as an umbrella organization to handle the generation (search, query, find, access) and use of geographic data. NSDI is composed of organizations and individuals who generate or use geographic information and manage the technologies that facilitate use and transfer of data, and the actual data. A joint organization, the European Geographic Information Infrastructure (EGII), has been established under the control of the European Space Agency (ESA) and the European Commission (EC).

In the global picture, the Global Spatial Data Infrastructure (GSDI) has taken the initiative to link national and regional spatial data infrastructures in an open process for coordinating the organization, management, and use of spatial data. The definition of GSDI encompasses the policies, organizational remits, data, technical standards, delivery mechanisms, and financial and human resources.

Because the acquisition, processing, and distribution of geodata often involve both national and local authorities as well as commercial interests, several countries have taken the initiative to establish a partnership to solve such tasks. In most developed countries, the national mapping authority is the driving force in developing a national infrastructure for georeferenced data. There is a trend for the user side to be increasingly involved and for commercial interests, such as software companies, to take an active part in building a national infrastructure. The National Performance Review (1993) in the United States has established partnership as government policy in the field. In this way, the burden on the individual institution's budget will be reduced.

Other organizations, such as brokerages and clearinghouses, have also entered the market. Data brokers ensure the efficient turnover of data being offered on the market. Clearinghouses are more involved in representing the technical solution, to allow producers to describe the data sets they have available and to allow users to find the data sets they need. Digital libraries based on the Internet and metadatabases are the most important technical factors.

19.8   Data Access and Digital Libraries

Today, the Internet is the most commonly used technology for realizing digital libraries. Several major suppliers of GIS software have developed systems that allow a certain GIS functionality on the Internet.

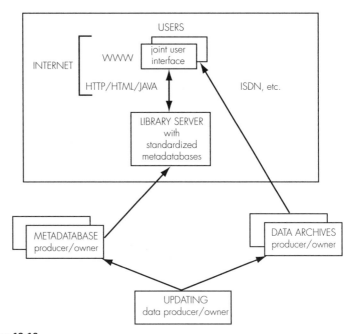

Figure 19.16
A digital library for geographical data is being offered to users on the Internet by clearinghouses.

This has made it possible to develop efficient user interfaces. A possible solution for a digital library is shown in Figure 19.16.

Normally, a digital library covers the following basic functions:

- User interface for search and selection of data
- Catalog service/metadatabase

In addition, there will often be automatic functions for ordering and invoicing.

The ultimate goal, from the user's point of view, should be to have access, through the Internet, to original databases distributed on different servers in different institutions all around the country, and ideally all around the world. By a simple selection menu, each user should be able to compose a data profile and have the data transferred automatically to his or her PC. The Open GIS Consortium (OGC) has developed a test version (Web Mapping Test Bed) in the United States which shows this capability for data search.

CHAPTER 20

Formal Problems in Establishing GIS

20.1 Ownership and Copyright

The use of geographical data often raises the question of ownership. A country's laws and statutes usually regulate claims to owning intellectual property. *Intellectual productions* is a useful term for devices whose value is not based on the ownership of a physical article. Those who can claim ownership may also regulate its use through pricing mechanisms and general licensing terms.

The legal status of geographical information is not always clear. If it fulfills the requirements of national law, it can be protected by copyright law. It can also be protected under national legislation on such issues as database protection and unfair competition. Where the data are procured in full by private firms, companies, or individuals, there would normally be no doubt that ownership belongs to those who procured the data. In most cases, however, computerized maps and registers have been partially or completely generated in the public sector. The question of ownership then depends on whether public officials regard geographical data as belonging in the public domain or as a product protected under copyright law. The ownership situation becomes even more complex in view of the fact that a number of private companies now process data from public sources, with the addition of their own data, with the object of increasing the commercial resale value.

The Bern Convention of 1951, which specified common international regulations for the protection of copyright, can be used as the basis for copyright considerations linked to mapping and geographical data for countries signatory to the convention. The new (not yet adopted) database directive from the EU gives protection to so-called "sweat of the brow" databases. This means that all data have to be registered/cataloged by one means or another in order to be protected.

Objective "facts" are not normally covered by copyright law. Certain geographical information products as well as certain elements in maps or charts, however, may not have the necessary level of importance to be copyright protected. This type of data can therefore be copied freely. For example, elevation contours are normally based on a certain amount of adaptation and are therefore protected by copyright, whereas place names alone are not protected. A database of geographical names may thus not be copyright protected, but can be protected as a catalog product (Aslesen 1996).

Basically, the same rules apply whether the case concerns data that are reproduced in digital or in analog format. Copying from one electronic storage device to another is a reproduction under copyright law. Copying a screen display is more of a problem because the display is only temporary.

Public practice in regard to claims of data ownership varies considerably from country to country, and sometimes within a country, as from state to state in the United States. In addition, the United States operates with different policies at the federal and state levels. In the United Kingdom, the public sector claims copyright. Hence, permission must be granted by the Ordnance Survey (the national map agency) for *all* digitizing of existing maps. On termination of the digitizing, a copy of the resultant digital data must be provided free of charge to the Ordnance Survey. In addition, the Ordnance Survey levies a charge proportional to the amount of digitizing work carried out. In Canada, the public sector also claims copyright over map data but not as rigorously as in the United Kingdom. Usage fees are charged whenever the value of the digitized data exceeds the costs incurred in acquisition and digitizing.

National geographic databases and GISs may be regarded as parts of future infrastructures, as with road networks and electricity supply networks. Consequently, in the United States, governmental maps and digital map data are regarded as being in the public domain, and all copying is free. That view is based on the belief that society as a whole profits from the enhanced value of products and data that initially were financed by the taxpayers.

Increasing electronic access to public geographical data through the telephone network and the Internet appears to make fixed regulations a necessity in this field. GRID—the United Nations Global Resources Information Database—offers free access to public environmental data. Many countries are now in the process of formulating this kind of regulation, but there still appears to be a certain need to obtain practical experience before approving such regulations. Because the telephone network is independent of national boundaries, an initiative has been taken at the international level to coordinate the regulations for different countries.

20.2 Cost Recovery and Pricing

Cost-recovery methods are normally used to determine commercial prices for public data. Prices are thus expected to cover such costs as data establishment, storage, and distribution. The price is usually dependent on the quality of the data with regard to detail, timeliness, completeness, topological continuity, and so on. In many countries, price is also dependent on usage (application). Thus, some national mapping authorities set lower usage fees for the public sector's use of data and higher fees for commercial uses in the private sector. Correct pricing, however, is difficult. High prices encourage users to compile their own digital map data, which then reduces the importance of the public sector. Low prices can result in data not being kept up to date because of failure to cover costs, and the value of quality data is not appreciated.

We have already seen that the United Kingdom and Canada enforce copyright regulations with corresponding fees/pricing based on the cost-recovery principle. In 1996, the United Kingdom had a total cost-recovering rate of 70 to 80% for the Ordnance Survey. A number of other European national mapping authorities (Norway, Sweden) have a cost recovery rate of 30 to 40%. In the United States, national data are available only for payment of copying and distribution costs, and cost recovery is thus quite low. The general impression today is that there is little point in a cost recovery rate of 100% for public data. There are a number of special situations in which geographical data/maps are of vital importance, such as in regard to accidents in wilderness areas. Data of this type are usually of little commercial interest, and their costs cannot therefore normally be covered. Such data would not be produced without public support. The need for public cofinancing, however, becomes most apparent when viewed from a military perspective. Historically, military needs have initiated the majority of national map series. Military requirements are often enough to justify total public financing of national data coverage at the smaller scales. However, it is probably an accepted principle in most countries that the taxpayers' burden be eased by allowing those who gain special benefit from the data to cover some of the costs. Economic theory indicates that the correct price for data produced with taxpayers' money should be based on marginal production costs, including full coverage of distribution costs and a percentage increment to cover some of the fixed costs linked to, for example, the work of gathering taxes (Groot 2001).

In addition, product price should depend on risk. If the public sector or a private company takes a high investment risk in connection with the establishment of a database, it makes economic sense to charge a higher price than if one could be sure of a certain degree of cost recovery. Joint market ventures can lower prices because the risk is spread among several partners.

The business aspects of GIS have yet to be fully developed, primarily because so far, most GIS facilities have been established to meet the needs of particular public agencies. However, commercial uses will undoubtedly become more important in the future. The development of electronic marine charts and systems for car navigation require equipment suppliers to deliver digital map data along with the equipment sold. Databases may then become commercial wares with considerable profit potential. A particularly relevant development in Europe today is that several private consortia are in the process of establishing road databases for vehicle navigation and other transport purposes. The pricing of this type of data will probably follow normal market principles, based on competition by several suppliers.

20.3 Public or Private Organization of Geodata

In the discussion regarding public or private organization of geodata, strategies must suit prevailing market conditions. But as market conditions change, prudent strategies must encompass alternative approaches with respect to organization of production and the market situation. Organization of production can be based on a decentralized, centralized, or semidecentralized approach, in which the producing organization is responsible for reasonable standardized data interchange between providers and users.

The market situation is likely to be composed of a variety of user groups with common interests only in limited (local) geographical areas and in cost sharing of centralized geodata. Market demand from a single major user would normally be inadequate to fully finance such data provision. Only a homogenization of demand would enable any uniformity of supply to be implemented.

Depending on the organization of the production side and the market, public agencies, which often operate national databases, may choose among various business approaches:

1. Supplement basic data with other standard data and distribute the combination as a standard package to all users (extended map production).
2. Supplement basic data with other standard data to compile tailor-made user products.
3. Supplement basic data with user-specified data to compile tailor-made user products.
4. Coordinate data interchange without being involved in production but with responsibility for standardization.

The question of public versus private involvement is part of the evaluation picture, as are questions of regulation and accessibility. Economies of scale and market homogeneity also influence users' choice of strategies. Financing through national budget appropriations may increase the availability of products to users, but may also hold prices artificially low so that providers cannot meet the maintenance costs. If private financing is to be involved, the market must tolerate competition and price variation, which may exclude some potential users who cannot afford the products offered.

Socioeconomic considerations dictate that public databases should be accessible to all, with few limitations. Inasmuch as the acquisition, processing, and distribution of geodata often involve both national and local authorities as well as commercial interests, several countries have taken the initiative to establish a partnership to perform such tasks. The National Performance Review (1993) has established this approach as U.S. government policy in the field. In this way, the burden on individual institutions' budgets will be reduced.

20.4 Example of Strategy for a National Map Service

Recommended strategies for national road databases in Scandinavia include the following necessary conditions (Nordic KVANTIF 1990):

1. *General strategy.* The national map service should adopt an offensive strategy to adapt to rapidly changing technologies and markets.
2. *Creating databases.* A multitheme national road database (with geometry, driving limitations, ID links to addresses, etc.) created with data from various sources should be aimed toward a small nucleus group of users willing to pay for the services provided. This will make the project feasible. This product cannot be classified as a public service, as many financially weaker groups will be excluded.
3. *Currentness.* Currentness is paramount—that is, updating every three months for important items and at least once a year for all other items.
4. *Financing.* A multitheme road database should be user financed. As the market demand is so varied, the product is not a characteristic public service. Semipublic financing may gradually increase during the updating phase as products become standardized and more users access the data. The portion of the road database dedicated to rationalization of national map production can be financed by normal public funding.
5. *Pricing.* Market pricing should involve a high entry price to cover updating costs. Otherwise, the road data will be inadequately financed, which will create problems in the updating phase and pricing problems for both the initial group and subsequent users. In the long run, the most called for data will undoubtedly become a public responsibility, a fact that may influence marginal costing in pricing.
6. *Organization.* The national map service operates both as administrator of public map policies and as producer of products for the

407

commercial market. Consequently, the map service should be divided into two units. A basic public service unit will administer the geometrical information (infrastructure). A more commercial unit will compete in the private market and should focus on value-adding activities.

7. *Access to basic data.* All private and public organizations should have equal access to basic data that are produced in connection with national map production. In this way, provider and user interests balance, and various interests may compete on an equal footing.

20.5 Data Security

Many countries have laws pertaining to the establishment and use of registers of persons. In most countries, official permission is almost always required for such registers, a restriction imposed as a means of safeguarding individual rights. In the United States, however, the passage of the Financial Services Modernization Act permits organizations to acquire and distribute substantial personal information unless an individual specifically asks to be removed from the register. Geographical data that contain information about individuals may similarly be restricted. Sensitive material such as maps of military areas or data concerning endangered species or rare plants may be considered of national interest and restricted accordingly.

In some cases, combining data from open registers may be illegal. Limited information may be openly available, but when many data sets are combined to cover larger areas or effect undesirable combinations, the result may constitute a security risk. For example, for national security reasons telecommunication administrations prohibit unauthorized persons from handling the network information contained on several map sheets. In general, the more detailed the information, the stricter the constraints on its use.

Access to computerized information may also be restricted because those who compiled it may not be confident of its quality and therefore prefer that it should not be widely used. Examples of this include data gathered ad hoc for special projects and not based on sound methods or quality controlled. As public geodatabases become increasingly accessible over the telenetwork, limitations on access will probably have to be introduced. Perhaps the level of access and the way it is granted will vary among users. For instance, certain data may be coded so that only those with access to the code will be able to interpret the data.

CHAPTER 21

A Vision for the Future

This book

In this book we have attempted to describe the most important aspects covered by the term *GIS* to give the reader a total picture of the subject. Our idea was to produce a book that would be suitable for the reader with little or no knowledge of GIS. Because this is an introductory book, many aspects have been handled relatively superficially and others are mentioned only in passing. A number of subject areas are not discussed at all. For example, there is little discussion of the different mathematical methods and formulas for handling the geometry. However, a number of textbooks are available that describe these parts of GIS theory in greater detail.

Other sources

Those who wish to learn more about the subject should refer to textbooks that give more detailed coverage of GIS. Other opportunities to study GIS are available in the form of an interactive GIS tutorial for self-study and organized courses. Courses at various academic levels are available in many countries, and these are the best means of obtaining a thorough knowledge. A number of GIS publications are available, containing both popular and academic news. The reader should also consult the References section in this book.

Status of GIS today

Just as maps have been used in many different areas of society, GIS is now gradually taking over their functions and, in addition, exploiting totally new areas. In broad terms, we can say that GIS can be used for documentation (how things used to be and how they are today) and for analyses and simulations (how things could be in the future). Although GIS has proved to be a great help in many fields, we must not ignore the fact that it is only an aid and that the results derived from GIS will have to be used in combination with good judgment and professional skill. When used correctly, however, GIS has proved to be very useful in improving many aspects of daily life both directly and indirectly, and this should be the primary reason for using this technology.

The future

The results of market research indicate that GIS will increase in volume and will be applied in new fields of activity.

Technological developments

Technological developments in the IT sector appear to continue unabated. The increase in processor speed and transfer speed in networks, expanded storage capacity, and improved databases and user interfaces are all factors that will make it easier to work with GIS in the future. Even though the GIS market in many ways can be described as technology driven (although it should, in principle, be user driven), this is a development that users welcome. We have established that today's technology is good enough to solve the main problems linked to the use of GIS, but in the future we are sure to see GIS-related solutions that are far ahead of today's boring (in many ways) monitor-oriented solutions.

We can already see the outlines of this trend in the form of multimedia (hypertext, sound, video, virtual reality) now available on the Internet. Impressive developments can be expected in this field, especially in respect to interaction between the user and the system. Although such development will benefit GIS users, it will also set new requirements for data accessibility and updating.

New applications

Geographical data and GIS are continuously finding new areas of application. To date, the GIS market has been dominated by products aimed at the professional user. New and untraditional user groups have appeared in recent years. *GIS in business,* for example, has become a common term. Looking at the number of potential users, we can see that professional users are relatively few as compared with nonprofessional consumers. As GIS and GIS-related solutions become available on the Internet, we will be better able to reach a broader group (Figure 21.1). There are users who are interested only in obtaining a quick view of the data without having to pay anything for it. The financing of geographical data on the Internet will probably be based on advertising income (e.g., from hotels and other companies willing to pay to appear on the maps).

Figure 21.1
Possible future scenario: politicians grouped around a graphic display board while discussing matters pertaining to geographical data.

As to the professional use of GIS, the greatest social significance in the future will be found in regard to environmental issues. This is a relationship that is difficult to quantify, but without systematic surveillance and documentation of the status of natural resources and analysis of future developments, large amounts will have to be invested to correct prior mistakes. The future success of GIS will also depend on how well we manage to communicate with decision makers. Once again, we have to remember that it is not the operators who are the real users, but those who make decisions on the basis of the maps and tables produced in GIS.

Standardization

International standardization (ISO and CEN) and cooperation between the system suppliers (Open GIS Consortium, etc.) will make it much easier in the future for GIS operators to bring up data from different sources. We can also expect that requirements for the standardization of data, in the form of descriptions in metadatabases, for instance, will also increase in the future. The efficient marketing of data will depend on suppliers being able to describe the data correctly and appropriately.

Data access

In the future, data will be available to a much greater extent over the Internet. In many countries today, clearinghouses and similar organizations have appeared on the scene that disseminate data via the Internet. We will probably have national and regional digital libraries for geographical data and images (aerial photos and satellite images). We will experience increased international competition in the field. For example, national mapping authorities will have to compete with both international and national companies for the supply of national data. Data on London may be bought over the Internet just as easily in Hong Kong as in Southampton.

Organization

Technological developments will always affect the way work is organized. Some may view this as a threat, others as an advantage. Thus, organization is concerned with human factors, which have remained constant for many millennia (human genes have remained more or less unchanged for at least 35,000 years). It is not likely, therefore, that it will be easier to obtain optimal organizational solutions in the future than it has been in the past.

New players in the market will change the traditional structures for the sale of data. Discussions are going on in many countries as to the extent to which the public sector, through grants to national mapping authorities, should become involved in the production and sale of data. In most countries there appears to be agreement that the public sector should be responsible for production of certain basic geographical data. This means that in the future most countries will have national mapping authorities that are responsible for carrying out national policy in this field. However, there will be wide variations in the extent to which production and any appreciation in value will be organized as a monopoly.

Research and development

Most of the research and development in the field of GIS has been linked to technological improvements. Surprisingly little attention has been given to the end user—what constitutes efficient GIS information for a decision maker, how maps should look, and so on. We hope that more research will be carried out in this aspect of the field in the future.

REFERENCES

Andersen, Ø., 1991, *Geografisk Informasjonssystem: Presisering og Problemstillinger*. Aas, Norway: Compendium.

Arnborg, S., 2000, *Databaseteknikk. I Geografisk informationsbehandling, Metoder och tillämpningar*. Eklund, L (ed.), pp 119–146. Stockholm, Sweden: Byggforskningrådet

Aronoff, S., 1989, *Geographic Information Systems: A Management Perspective*. Ottawa, Ontario, Canada: WDL.

Aslesen, L., 1996, *Paper on Copyright*. Hoenefoss, Norway: IHO.

Atenucci, J. C. (ed.), 1995, *Geographic Information Systems: A Guide to Technology*. New York: Van Nostrand Reinhold.

Balstrøm, T., Jacobi, O., Munk Sørensen, E. (eds.), 1994, *GIS i Danmark*. Copenhagen, Denmark: Teknisk Forlag, 9–21, 45–50, 61–71, 97–98.

Beck, R., 1997, Satellite imagery from earth watch and space imaging. *GIM Geomatics Info Magazine*, 10(3):60–63.

Bernhardsen, T., 1985, *Samfunnsmessig Nytteverdi av Tekniske Kart*. Arendal, Norway: Asplan Viak.

Bernhardsen, T., 1986, *A Cost/Benefit Study of GIS Methodology and Results*. Toronto, Ontario, Canada: FIG International Congress.

Berhardsen, T., 1999, *Choosing a GIS. In Geographic information systems, Volume II. Principles and Technical Issues*, eds. P. A. Longley, M. F. Goodchild, D. J. Maguire and D. W. Rhind, pp 589–600. New York, USA: John Wiley & Sons

Bernhardsen, T., and Pål Føyn Fespersen, 2001. Economics of Environmental Information Systems (EIS). NORPLAN/World Bank Report.

Berry, J., 1987, Fundamental operations in computer-assisted map analysis. *International Journal of Geographical Information Systems*, 2.

Bertin, J., 1983, *Semiology of Graphics*. Madison, Wis.: University of Wisconsin Press.

Beser, J., and Haunschild, M., 1995, Advantages of integrated GPS/GLONASS: operation GIM. *Geomatics Info Magazine*, 9(11):63–69.

Bjørke, J. T., 1988a, *Grunnleggende Datastrukturer: Eksempler på Deres Kartografiske Anvendelser*. Grimstad, Norway: paper presented at AID.

Bjørke, J. T., 1988b, *Kartinformasjon*. Grimstad, Norway: paper presented at AID.

Brande-Lavridsen, H., 1994, *Cartography*. Aalborg, Denmark: ICA paper.

Brodersen, L., 1999, *Kort om kommunikation*. Frederikshavn, Denmark: Forlaget Kortgruppen A/S

Burrough, P. A., 1986, *Principles of Geographical Information Systems for Land Resource Assessment*. Oxford: Science Publications.

Carosio, A., and Zanini, M., 1996, Landscape modelling and visualization: from 2-D surface to 3-D volumes. *GIM Geomatics Info Magazine*, 10(5):6–8.

Castle, G. H., 1993(ca), *Profiting from a geographic information system*. GIS World.

Chrisman, N., 1997, *Exploring geographic information systems*. New York, USA: John Wiley & Sons

Cullis, B. J., 1994, *A strategy for assessing organizational GIS adoption success*. GIS/LIS 1994: 208–217.

Dale, P. L., and Clarc, S. R., 1988, *Land Information Management: An Introduction with Special Reference to Cadastral Problems in Third World Countries*. Oxford: Clarendon Press.

Dangermond, J., 1997, *Address to NYC GIS Steering Committee*. Paper available on Internet, e-mail *dlashell@esri.com*.

DeMers, M., 1999, *Fundamental of geographic information systems (second edition)*, New York, USA: John Wiley & Sons

Eklundh, L., 2000, *Interpolation. In Geografisk informationsbehandling, Metoder och tillämpningar*. Eklund, L (ed.), pp 177–192. Stockholm, Sweden: Byggforskningsrådet

Eklundh, L., och Pilesjö, P. 2000, *Rumliga datastrukturer. I Geografisk informationsbehandling, Metoder och tillämpningar*. Eklund, L (ed.), pp 93–116. Stockholm, Sweden: Byggforskningsrådet

Ericsson, G. (ed.), 1987, *Community Benefit of Digital Spatial Information*. Arendal, Norway: Nordic Kvantif/Asplan Viak.

ESRI, 1995, *Understanding GIS: The ArcInfo method*. Redlands, Calif.: ESRI.

Federal Geographic Data Committee (FGDC), 1998, *Content standard for digital geospatial metadata*. Paper available on the Internett: www.fgdc.gov

Fotheringham, S. and Rogerson, P., 1994, *Spatial Analysis and GIS*. Bristol, Pa.: Taylor & Francis.

Gallardo, M., 1995, *Microsoft Windows NT*. Paper available on Internet, *http://ata.princeton.edu/falks/winnet/winnet/html*

Goodchild, M. F., and Kemp, K. K., 1990, *Introduction to GIS: Core Curriculum:* Berkeley, Calif.: National Center for Geographic Information and Analysis.

Gralla, P., 1997, *How the Internet Works—All-New Edition*. London: Prentice Hall.

Groot, R., 2001. Reform of Government and Future Performance of National Surveys. ITC, unpublished paper.

Hauska, H. och Harrie, L., 2000, *Koordinatsystem, kartprojeksjon och referansesystem. I Geografisk informationbehandling, Metoder och tillämpningar*. Eklund, L (ed.), pp 67–89. Stockholm, Sweden: Byggforskiningsrådet

Hofmann-Wellenhof, B., Lichtenegge, H., and Collins, J., 1994, *Global Positioning Systems: Theory and Practice*. New York: Springer-Verlag.

Htun, N., 1997, *The need for basic map information in support of environmental assessment and sustainable development strategies. In Framework for the World*. Rhind, D. (ed.), pp 111–119. Cambridge, UK: Geoinformation International (Pearson Professional).

Hutchinson, M. F. and Gallant, J. C., 1999. *Representation of terrain. In Geographic information systems, Volume 1. Principles and Technical Issues*, eds. P. A. Longley, M. F. Goodchild, D. J. Maguire and D. W. Rhind, pp 105–124. New York, USA: John Wiley & Sons

Huxhold, W. E., and Levinsohn, A. G., 1995, *Managing Geographic Information Systems Projects*. New York: Oxford University Press.

Iliffe, J. C., 2000, *Datums and map projections*. Scotland, UK: Whittles Publishing

Ireland, P., 1996, *Inkjet versus laser in the output game*. GIS Europe, 11(5):38.

ISO International Standard, 2001, *ISO/TC211 Geographic information*. Papers available on the Internet: www:statkart.no/isotc211

Jones, A., 1997, The new commercial satellites. *Mapping Awareness*, 11(3):14–17.

Kofler, M., and Grüber, M., 1997, Towards a 3-D GIS Database. *GIM Geomatics Info Magazine*, 11(5):55–57.

Korte, P. (ed.), 1997, *The GIS Book: The Smart Manager's Guide to Purchasing, Implementation and Running a Geographic Information System*. Entrada: OnWord Press.

Kraak, M. J., and Ormeling, F. J., 1996, *Cartography: Visualization of spatial data*, Harlow, Essex, England: Addison Wesley Longman.

Kylen, B., and Hekland, J., 1990, *Economics of Geographic Information: Organizational Impact of Technology Changes in the Road GIS Case*. Arendal, Norway: Nordic Kvantif/Asplan Viak.

Langran, G., 1992, *Time in Geographic Information Systems*. London: Taylor & Francis.

Larsen, J. M., 1991, *Brukerbehov*. Arendal, Norway: Geodatasenteret.

Laurini, R., and Thompson, D., 1992, *Fundamentals of Spatial Information Systems*. London: Academic Press: 102–106.

Lee, Y., and Plunkett, G., 1997, Desktop GIS implementation. *GIM Geomatics Info Magazine*, 11(3):33–35.

MacEachren, A. M., and Fraser, D. R., 1994, *Visualization in Modern Cartography*. Oxford: Taylor Pergamon.

Maguire, D. J., Goodchild, M. F., and Rhind, D. W., 1991, *Land applications*. In Geographical Information Systems, Vol. 1, *Principles*. Harlow, Essex, England: Longman Scientific and Technical; 251–267, 269–297, 319–335.

Mark, D. M., 1999. *Spatial representation: a cognitive view. In Geographic information systems, Volume 1. Principles and Technical Issues*, eds. P. A. Longley, M. F. Goodchild, D. J. Maguire and D. W. Rhind, pp 81–89. New York, USA: John Wiley & Sons

Martin, D. J., 1999, *Spatial representation: the social scientist's perspective. In Geographic information systems, Volume 1. Principles and Technical Issues*, eds. P. A. Longley, M. F. Goodchild, D. J. Maguire and D. W. Rhind, pp 71–79. New York, USA: John Wiley & Sons

Medycykyj-Scott, D., and Hearnshaw, H. M., (eds.), 1993, *Human Factors in Geographical Information Systems*. London: Belhaven Press.

National Committee for Digital Cartographic Data Standard (NCDCDS), 1988, Cartographic objects. Paper available on the Internet.

Newton, A. W., Zwart, P. R., and Cavill, M. E. (eds.), 1992, *Networking Spatial Information Systems*. London: Belhaven Press, 15–27.

Nielsen, F., 1994, *GIS Database Design and Organisation*. Paper available on Internet, e-mail *fofdat@inet.uni.c.dk*.

Nieuwenhuijs, S., 1995, *Oracle MultiDimension: new frontiers in spatial data management*. GIS Europe, 4(5):40–42.

Nix, M., 1997, *One-person mobile office for surveying*. GIM Geomatics Info Magazine, 11(4):32–35.

Onsrud, J. H., and Rushtom, G., 1995, *Sharing Geographic Information*. New Brunswick, N.J.: Center for Urban Policy Research.

Oosterom, P. van., 1999, *Spatial access methods. In Geographic information systems, Volume 1. Principles and Technical Issues*, eds. P. A. Longley, M. F. Goodchild, D. J. Maguire and D. W. Rhind, pp 385–400. New York, USA: John Wiley & Sons

Østensen, O., 1996, *OSO TC/211: ISO standards for GI infrastructures*. GIM Geomatics Info Magazine, 10(3):24–25.

Palm, C., 1988, *Karttecken—Kartsymbol*. Hoenefoss, Norway: Nordic Cartographic Society Summer School.

Peuquet, D. J., 1999. *Time in GIS and geographical databases. In Geographic information systems, Volume 1. Principles and Technical Issues*, eds. P. A. Longley, M. F. Goodchild, D. J. Maguire and D. W. Rhind, pp 91–103. New York, USA: John Wiley & Sons

Pilesjö, P., 2000, *Digitale höjdedata. I Geografisk informationsbehandling, Metoder och tillämpningar*. Eklund, L (ed.), pp 269–288. Stockholm, Sweden: Byggforskningsrådet

Pilesjö, P., Eklundh, L. och Olssou, L. L., 2000, *Metoder før rumlige dataanalys. I Geografisk informationsbehandling, Metoder och tillämpningar*. Eklund, L (ed.), pp 195–265. Stockholm, Sweden: Byggforskningsrådet

Sahai, R., 1996, *Inside MicroStation 95*. Entrada: OnWord Press.

Sanchez, R., 1996, *Satellite image integration with spatial data*. GIM International Journal for Geomatics, 10(10):56–59.

Schulte, V., 1996, Conversion kit. *GIS Europe*, 5(6)36–38.

Seaborn, D., 1995, *Database management in GIS: is your system a poor relation?* GIS Europe, 4(5):34–38.

Smith, N. S. and Rhind, D. W., 1999. *Characteristics and sources of framework data. In Geographic information systems, Volume II. Principles and Technical Issues*, eds. P. A. Longley, M. F. Goodchild, D. J. Maguire and D. W. Rhind, pp 655–666. New York, USA: John Wiley & Sons

Somers, R., 1994a, *Alternative GIS development strategies*. GIS/LIS, 1994:706–715.

Somers, R., 1994b, *Gis organization and staffing*. URISA, 1994:41–52.

Steinfort, A., 1996, *Is there life after GIS?* GIM Geomatics Info Magazine, 10(11):24–27.

Strand, G. H., 1991a, *Linear Combination of Models in Geographic Information Systems*. Oslo, Norway: NITO Summer School.

Strand, G. H., 1991b, *Sentrale Teknikker i Geografiske Informasjonssytemer*. Oslo, Norway: NITO Summer School.

Tomlin, C. D., 1990, *Geographic Information Systems and Cartographic Modelling*. Upper Saddle River, N.J.: Prentice Hall.

Tveitdal, S., (ed.), 1987, *Community Benefit of Digital Spatial Information: Digital Maps, Data Bases, Economics and User Experience in North America*. Norway: Arendal, Nordic Kvantif/Asplan Viak.

van der Schans, R., 1995, Differences and similarities of GIS and CAD. *GIM International Journal for Geomatics,* 9(8):43–47.

van Genderen, J. L., 1991, *Guidelines for Education and Training in Environmental Information Systems in Sub-Saharan Africa: Some Key Issues.* Delft, The Netherlands: ITC.

Veregin, H., 1995, *Developing and Testing of an Error Propagation Model for GIS Overlay Operations. International Journal of Geographical Information Systems* 92(6): 595–619

Veregin, H. V., 1999. *Data quality parameters. In Geographic information systems, Volume 1. Principles and Technical Issues,* eds. P. A. Longley, M. F. Goodchild, D. J. Maguire and D. W. Rhind, pp 177–189. New York, USA: John Wiley & Sons

von Rimscha, S., and Wolf, W., 1997, Modelling in an uncertain world. *GIS Europe,* 6(1):18–20.

Worboys, M. F., 1995, *GIS: A Computing Perspective.* London: Taylor & Francis, 61–64, 309–314.

Worboys, M. F., 1999. *Relationship databases and beyond. In Geographic information systems, Volume 1. Principles and Technical Issues,* eds. P. A. Longley, M. F. Goodchild, D. J. Maguire and D. W. Rhind, pp 373–384. New York, USA: John Wiley & Sons

Worrall, L. (ed.), 1990, *Geographic Information Systems: Developments and Applications.* London: Belhaven Press.

INDEX

Abstract data type (ADT), databases, 159
Accessibility, data quality, 232
Accuracy:
    attribute data, 225–227
    data quality, 215–234
    positional, 223–225
    surface representation, 101–102
    temporal, 227–228
Adjustments, housekeeping tools,
    261–262
Advanced data models, *see* Data model(s)
    (advanced)
Advanced spatial analysis, *see* Spatial
    analysis (advanced)
Aerial photographs:
    data collection I, 185–187
    mapping, 30–31
Affine transformation, georeferencing, 122
Ancillary features, human-machine
    interface, 151
Application program interface (API),
    databases, 159, 237
Application schemas, standards, 381–382
Area measurement, spatial analysis:
    vector, 272–273
    raster, 272–273
Areas, graphical representation, 42–43
Arithmetic operations, spatial analysis,
    267–268
ASCII, 55, 384
Asynchronous transfer mode (ATM),
    networks, 134
Attribute data:
    accuracy of, data quality, 225–227
    collection of, data collection II, 210–213
    complex operations of, spatial analysis,
        275–277
    data model(s) (basic), 87–90
    editing of, housekeeping tools, 263–264
    linkage of, map data, 91–92
    report generation from, spatial analysis,
        273–274
Automatic line tracing, quantizers, 142
Automatic Mapping/Facility Management
    (AM/FM), 11

Basic data models, *see* Data model(s) (basic)

Basic spatial analysis, *see* Spatial analysis
    (basic)
Bilinear interpolation, advanced analysis,
    296
Binary encoding, scanning and, data
    collection I, 176–182
Binary search, databases, 153
Binary system, data model, 54
Bits, data model, 54–55
Budgeting, organizational selection issues,
    356–357
Buffer zones, spatial analysis, 283
Business concept development,
    organizational selection issues,
    334–335
Byte, data model, 54

Cartographic algebra, spatial analysis
    (advanced), 297–298
Carrier-based receivers, data collection II,
    204–205
Center of gravity, data model, 47
Centroide measurement, spatial analysis,
    269
Channels, data collection I, 188
Classification and reclassification, spatial
    analysis, 277
CMYK, plotting, 143, 318
Code-based receivers, data collection II,
    203–204
Comité Européen de Noramlisation (CEN),
    378
Commands, human-machine interface,
    149
Completeness, data quality, 230–231
Compression, raster data:
    block codes, 81–82
    chain codes, 80–81
    run-length encoding, 80
Compression, vector data:
    corridor, 70
    Douglas-Peucker method, 69–70
    spline, 70
Computer:
    Geographical Information Systems (GIS)
        and, 1–4
    hardware and communications, 130–132

Computer *(Continued)*
  history of, 31–35
  recent developments, 34–35
Computer-aided design (CAD), 11–12, 164
Conceptual database design, database
    implementation, 244–248
Conceptual model, data quality, 228
Conceptual modeling, standards, 381–382
Conformal transformation, georeferencing,
    121–122
Connectivity operations, spatial analysis
    (advanced), 288–291
Consistency, data quality:
  conceptual, 228
  domain, 228–229
  format, 228
  topology, 228
Continuous-feed scanner, quantizers, 141
Continuous phenomena, data model, 48–49
Continuous systems, *see* Georeferences and
    coordinate systems
Contours, digital terrain models, 306–307
Contracts, technical selection issues,
    371–372
Conversion of height, georeferencing, 125
Coordinate reference system,
    georeferencing, 116
Coordinate systems, direct georeferencing,
    115–118
Coordinate thinning, housekeeping tools,
    262–263
Coordinate transformation, georeferencing:
  coordinate system transformation,
    122–123
  datum transformation, 121–123
  projection transformation, 122
Copyright, formal problems, 403–404
Correcting and adapting functions,
    housekeeping tools, 256–263
Correction, remote sensing, 193–195
Cost-benefit analysis:
  benefits, 341–342
  costs, 340–341
  harvesting benefits, prerequisites for,
    344
  nonquantifiable effects evaluation, 345
  quantification and analyses, 342–344
  review, 345
  sensitivity analysis, 344
  theoretical basis, 339–340
Cost recovery, formal problems, 404–406
Cubic convolution, advanced analysis, 296

Data access:
  customer services, 400
  future directions, 411
Database creation, technical selection
    issues, 373–374
Database design, technical selection issues,
    372–373
Database implementation, 235–239,
    244–250
  database design, 244–250
  database types, 235–239
  distributed databases, 239
  conceptual, 244–248
  logical, 248–249
  physical, 249–250
  map data, 240
  partitioning and indexing, 240–244
  structured query language (SQL),
    160–161
Database management systems, 152–164
  database systems, 156–164
    hierarchical database, 161
    network data structure, 161
    object-oriented database, 161–164
    relational database, 157–159
  files, 152–154
Data coding:
  raster data model, 77–78
  vector data model, 59–63
Data collection I, 170–197
  aerial photographs and interpretation,
    185–187
  digitizing maps, 171–176
  manual digitizing or scanning,
    184–185
  overview, 170
  remote sensing, 187–197
    analytic analyze methods, 195–197
    correction, 193–195
  scanning, 176–184
Data collection II, 199–214
  attribute data collection, 210–213
  photogrammetric mapping, 207–210
  satellite positioning system, 201–207
  surveying, 199–201
  text data, 214
Data compression:
  raster data model, 79–81
  vector data model, 69–70
Data-entry functions, housekeeping tools,
    253
Data model(s), 37–45. *See also* Real world
  database development

geometric representation of objects, 41–43
object attributes, 41
object relations, 43–44
objects, 41–45
quality, 44–45
shortcomings of, 46–51
conceptual generalization, 50–51
continuous phenomena, 48–49
field, 48–49
map role, 51
uncertainty, 49–50
Data model(s) (advanced), 94–111
combinations, 111
movable objects, 108
network model, 109–110
surface movement, 110–111
surface representation, 95–102. *See also*
Surface representation
three-dimensional objects, 102–105
time representation, 105–108
Data model(s) (basic), 54–93
attribute data and computer registers,
87–91
coding and entering, 88–89
storage, 90–91
map data and attribute data linkage,
91–93
overview, 54
raster data model, 73–84. *See also* Raster
data model
vector data model, 55–73. *See also* Vector
data model
vector/raster conversion, 84–86
Data processing, history of, 32–34
Data quality, 215–234, 388
accessibility, 232
accuracy, 218–221
attribute data accuracy, 225–227
completeness, 230–231
error sources, 232–234
lineage, 232
logical consistency, 228–230
overview elements, 231–232
positional accuracy, 223–225
quality control, 234
resolution, 222
rules, 234
sampling rate, 222
selection criteria, 215–218
storage precision, 223
temporal accuracy, 227–228
Data storage:
attribute data, 90–91

raster data model, 78–79
vector data model, 70–72
Data storage operations organization,
housekeeping tools, 254–256
Delaunay triangle, 98
Digital elevation model (DEM), 94
Digital Geographical Information
Exchange Standard (DIGEST), 378
Digital map, defined, 54–55
Digital representation, data model, 54–55
Digital terrain model, 304–312
contours, 306–307
elevations, 304
shading and draping, 308–309
slope and slope direction, 305–306
viewshed and shading techniques,
307–308
visualization techniques, 308–309
volumes, 307
Digitizers, quantizers, 139–142
Digitizing maps, data collection I, 171–176
Direct file, databases, 153
Direct georeferencing, continuous systems,
112–113
Discrete georeferencing systems,
described, 127–129
Discrete phenomena, data model, data
model, 48–49
Displays, hardware and communications,
137–139
Distance in raster, spatial analysis, 272
Distributed databases, database
implementation, 239
Draping, digital terrain models, 308–309
Drum scanner, quantizers, 141

Electrographic plotters, plotters and
output devices, 143–144
Elevation referencing, continuous systems,
123–125
Elevations, digital terrain models, 304–305
Ellipsoid, defined, 114
Entity-relationship (ER), database
implementation, 247–248
ETRF89, georeferencing, 115
Environmental management, Geographical
Information Systems (GIS), 15–17
Error, editing and correcting for,
housekeeping tools, 257–258
Error sources, data quality, 232–234
Euclidian distance, spatial analysis, 270
ERS-1,2, data collection I, 192

Euler number, spatial analysis, 268
Expert systems, spatial analysis (advanced), 298–299
Exploration, mapping, 27–31

Feature, data model, 40
Field, data model, data model, 48–49
Field, file, databases, 152
File, databases, 152
Files, database management systems, 152–154
Film printers, plotters and output devices, 144
Financing, organizational selection issues, 356–357
Flat-bed scanner, quantizers, 140–141
Flat-file database, 156
Flat triangles, digital terrain model, 99
Flight simulators, 35
Floating point, data model, 55
Friction values, advanced analysis, 291
Fundamental data, infrastructure, 399–400
Fuzzy analysis, spatial analysis (advanced), 49–50, 297
Fuzzy theory, data model, 49–50

General utility functions, housekeeping tools, 256
Generalization, data mode:
  cartographic, 50–51
  conceptual, 50–51
Generalization, graphic, visualization, 322–323
Geocentric coordinates, georeferencing 115, 117–118
Geographic coordinates, georeferencing, 115–117
Geographical information processing, history of, 31–32
Geographical Information Systems (GIS):
  computer and, 1–4
  computerized information benefits, 21–25
  defined, 4–10
  digital data coding for, 59–63
  diversity, 10–12
  formal problems in establishing, 403–407
    cost recovery and pricing, 404–406
    ownership and copyright, 403–404
    public/private organization, 406–407
    security, 408
  future directions, 409–412

real world and, 36–54. *See also* Data model; Real world
selection of:
  organizational issues, 329–360. *See also* Organizational selection issues
  technical issues, 361–376. *See also* Technical selection issues
socioeconomics, 12–21
  complexity, 12–13
  environmental and resource management, 15–17
  land transportation, 19–20
  military uses, 20–21
  operation and maintenance, 13–14
  planning and development, 17
  public services management, 17–18
  safety at sea, 18–19
  users of, 25–26
Geoid, defined, 124–125
Geoid-ellipsoid separation, georeferencing, 124–125
Geometric operations, spatial analysis, 268–273
Geometry, attributes and, integrated processing, spatial analysis, 278–279
Georeferences and coordinate systems, 112–129
  continuous systems, 112–127
    direct georeferencing, 112–127
    elevation referencing, 123–125
    relative georeferencing, 125–127
  discrete systems, 127–129
Global Positioning System (GPS), 20, 21, 201–207
GLONASS, data collection II, 207
Goal identification, organizational selection issues, 334–335
  effect, 334
  result, 334
Grains, visualization, 319
Graphical representation, of objects, data model, 41–43
Grid model, surface representation, 95–97

Half-tone-images, visualization, 317
Hardware and communications, 130–145
  computers, 130–132
  displays, 137–139
  networks, 133–137
  plotters and output devices, 142–144
    electrographic plotters, 143–144
    film printers, 144

ink-jet printers, 142–143
thermal plotters, 144
quantizers, 139–142
automatic line tracing, 142
digitizers, 139–140
scanners, 140–142
video, 142
technical selection issues, 361–371. *See also* Technical selection issues
Hashing, database implementation, 241–242
Hierarchical database, database management systems, 161
Housekeeping tools, 251–264
attribute data editing, 263–264
correcting and adapting functions, 256–263
adjustments, 261–262
coordinate thinning, 263
error and omission editing and correcting, 257–258
general utility, 256
line smoothing, 263
topology creation, 258–259
transformations, 259–261
data-entry functions, 253
data storage operations organization, 254–256
importing existing data, 253
overview, 251–252
Human-machine interface, software, 148–151
Hybrid system, database implementation, 235
Hydrologic modeling, spatial analysis (advanced), 309–311

Icons, human-machine interface, 150–151
Importing existing data, housekeeping tools, 253
Index file, pointer file, databases, 153
Indexing, spatial, database implementation, 240–244
Infrastructure, georeferenced data, 397–401
Ink-jet printers, plotters and output devices, 142–143
Integer, data model:
single, 54
double, 54
Introduction, organizational selection issues:
technology driven, 333–334

plan-driven, 333–334
process-driven, 333–334
method, 347–348
Institutional framework, standards, 397
Integrated data analyses procedures, spatial analysis, 285–287
Integrated processing, of geometry and attributes, spatial analysis, 278–279
Integrated system, data quality, 237
Integrated terrain units (ITU), 92–93
Intelligent Transportation System (ITS), 20
International Hydrographic Organization (IHO), 378
International Standards Organization (ISO), 378
Internet, networks, 136–137
Interpolation algorithms, digital terrain model, 98
Interpolation techniques, digital terrain model, 96
Interpolation and extrapolation, spatial analysis (advanced), 292–297
Interval data, data model, 45–46
Isoline model, surface representation, 100–101

Label point, housekeeping tools, 260
Land Information Systems (LIS), 11
Land transportation, Geographical Information Systems (GIS), 19–20
Line tracing, data collection I, 175
Lineage, data quality, 232
Linear combination, spatial analysis (advanced), 299–300
Line smoothing, housekeeping tools, 263
Lines, graphical representation, 42
Lines on polygon, spatial analysis, overlay, 282–283
Linking, data model:
by ID, 91–92
using geometry, 92–93
Local area network (LAN):
developments in, 34
hardware and communications, 133–135
Logic operations, spatial analysis, 266–267
Logical consistency, data quality, 228–230
Logical data model, development of, organizational selection issues, 357–358
Logical database design, database implementation, 248–249

Maintenance, technical selection issues, 374

Manhattan distance, spatial analysis, 270

Manual digitizing or scanning, data collection I, 184–185

Map data:
database implementation, 240
retrieval and search, spatial analysis, 274–275

Map projection coordinates, direct georeferencing, 115, 118

Map projection, direct georeferencing, 118–119

Map symbol selection, visualization, 323–324

Mapping, history of, 27–31

Menus, human-machine interface, 150

Messages, human-machine interface, 148–151

Metadata, standards, 392–397

Metropolitan area network (MAN), hardware and communications, 134

Microprocessor, history of, 32–34

Military uses:
Geographical Information Systems (GIS), 20–21
mapping, 30–31

Morton index, database implementation, 242–243

Movable objects, data model(s) (advanced), 198–111

Multimedia:
software, 164–165
visualization, 326–327

Multispectral registration, data collection I, 188

National geographic database creation, organizational selection issues, 358–360

Natural Resource Information Systems (NRIS), 11

Nearest neighbour analysis, spatial analysis, 268

Network data structure, database management systems, 161

Network model, data model(s) (advanced), 109–110

Network operations, spatial analysis (advanced), 288–290

Networks, hardware and communications, 133–137

Nominal data, data model, 45–46

Nominal data, data quality, 226

Non-qualitative data, data quality, 226

North Atlantic Treaty Organization (NATO), 378

Object catalogue, standards, 389–392

Object-oriented database, database management systems, 161–164

Objects, data model, 40–45
attributes, data model, 41
geometric representation, data model, 41–43
quality, data model, 44–45
relations, data model, 43–44
type, data model, 40

Omission, editing and correcting for, housekeeping tools, 257–258

Open System Environment (OSE), 379

Operating system, software, 148

Operation and maintenance, technical selection issues, 374

Ordinal data:
data model, 45–46
ordinal data quality, 226

Organizational framework, infrastructure, 400–401

Organizational selection issues, 329–360
appraisals, 335–336
business concept development and goal identification, 334–335
cost-benefit analysis, 339–345
benefits, 341–342
costs, 340–341
harvesting benefits, prerequisites for, 344
nonquantifiable effects evaluation, 345
quantification and analyses, 342–344
review, 345
sensitivity analysis, 344
theoretical basis, 339–340
historical perspective, 329–330
logical data model development, 357–358
national geographic database creation, 358–360
phases in, 333–334
reviews, 336–337
strategic plan development, 345–357
automation goals, 348
budgeting and financing, 356–357
geographic coverage, 350–351

introduction method, 347–348
investment level, 349
organizational matters, 351–355
personnel and training, 355–356
start date, 348–349
user requirements identification,
337–339
Overlapping polygons, data model, 46–47
Overlay:
raster data, spatial analysis, 284–285
spatial analysis, 279–283
Ownership, formal problems, 403–404

Partitioning, database implementation,
241–244
Personal computer:
hardware and communications, 130–132
history of, 32–34
Personnel, organizational selection issues,
355–356
Photogrammetric mapping, data collection
II, 207–210
Physical database design, database
implementation, 249–250
Pilot project, technical selection issues,
361–362
Pixel value transformation, data collection
I, 193
Plotters and output devices, 142–144
electrographic plotters, 143–144
film printers, 144
ink-jet printers, 142–143
thermal plotters, 144
Points:
graphical representation, 41–42
surface representation, 94–95
Points in polygon, spatial analysis,
overlay, 282
Polygon:
lines in, spatial analysis, overlay,
282–283
points in, spatial analysis, overlay, 282
Polygon overlay, spatial analysis, 278–282
Portray, standards, 388
Positional accuracy, data quality, 223–225
Precision, accuracy versus, data quality,
218–219
Precision, digital representation:
single, 55, 222
double, 55, 222
Pricing, formal problems, 404–406
Private/public organization, formal
problems, 406–407

Property rights, mapping, 30
Proximity operations:
on raster data, spatial analysis
(advanced), 296–297
spatial analysis (advanced), 292–297
Public/private organization, formal
problems, 406–407
Public services management, Geographical
Information Systems (GIS), 17–18

Quad-tree model, raster data model,
73–78
Quality data, 45–46, 222, 314
Quantitative data, 45–46, 222, 314
Quantizers, 139–142
automatic line tracing, 142
digitizers, 139–140
scanners, 140–141
screen digitizing, 142
video, 142
Queries, human-machine interface, 149
Query language, structured, database
implementation, 160–161

Radiometer, data collection I, 188
Ratio data, data model, 45–46
Raster data, proximity operations on,
spatial analysis (advanced), 296–297
Raster data connectivity, spatial analysis
(advanced), 290–291
Raster data model, 73–81. See also Data
model(s) (basic); Vector data model
creation of, 75–77
data coding, 77–78
data compression, 79–81
overview, 73–75
quad-tree model, 82–84
resolution, 74
vector data model conversion, 84–86
vector data model versus, 86–87
Raster data overlay, spatial analysis,
284–285
Real world, 36–39. See also Data model(s)
entity, 36–37
extension of concept, 52–53
Reclassification, spatial analysis, 277
Record, databases, 152
Record based structures, database
management system, 157–161
hierarchical data structures
relational database system,
network data structures, 161

Relational database, database management systems, 157–161

Relative georeferencing, continuous systems, 125–127

Remote sensing, 187–197
    analytic methods, 195–197
    correction, 193–195

Resampling, data collection I, 193–195

Research and development, future directions, 412

Resource management, Geographical Information Systems (GIS), 15–17

Sequential file, databases, 153

Safekeeping and security, technical selection issues, 374–375

Safety at sea, Geographical Information Systems (GIS), 18–19

Satellite positioning system, data collection II, 201–207

Satellite track, data collection I, 188–189

Scanners, quantizers, 140–142

Scanning:
    data collection I, 176–181

Screen digitalisation, georeferencing, 142

Security:
    formal problems, 408
    technical selection issues, 374–375

Selection criteria, data quality, 215–218

Shading, digital terrain models, 307

Sidelooking airborn radar (SAR), data collection I, 192–193

Significant figure, data model, 55

Simple list, databases, 152

Sink, digital terrain model, 96–97

Slope direction, advanced analysis, 305–306

Slope and slope direction, digital terrain models, 305–306

Software, 146–168
    computer-aided design (CAD), 164
    database management systems, 152–164
    database systems, 164–165
    files, 152–154
    human-machine interface, 148–151
    multimedia, 164–165
    operating system, 148
    technical selection issues, 362–371. See also Technical selection issues
    user requirements, 168–169
    working environment, 169
    World Wide Web, 166–168

Solar synchronous orbit, data collection I, 190

Spaghetti model:
    comments on, 73
    vector data model, 63–65

Spatial analysis (advanced), 288–312
    digital terrain models, 304–309
        contours, 306–307
        elevations, 304–305
        shading and draping, 307–308
        slope and slope direction, 305–306
        visualization techniques, 308–309
        volumes, 307
    engineering GIS, 311–312
    fuzzy analysis, 297
    GIS analytic models, 297–300
        cartographic algebra, 297–298
        expert systems, 298–299
        linear combination, 299–300
    hydrologic modeling, 309–311
    network operations, 288–290
    practical application of GIS functions, 300–304
        problem statement, 300
        procedure, 300–304
    raster data connectivity, 290–291
    spatial interpolation and proximity operations, 292–297

Spatial analysis (basic), 265–287
    arithmetic operations, 267–268
    attribute data, complex operations of, 275–277
    buffer zones, 283
    classification and reclassification, 277
    geometric operations, 268–273
        areas, vector and raster, 272–273
        centroides, 269
        distances, 270–272
        volumes, 273
    integrated data analyses procedures, 285–287
    integrated processing of geometry and attributes, 278–279
    logic operations, 266–267
    map data retrieval and search, 274–275
    overlay, 279–283
        lines in polygon, 282
        points on polygon, 282–283
        polygon, 279–282
    overview, 265–266
    raster data overlay, 284–285
    statistical operations, 268

Spatial interpolation, spatial analysis (advanced), 292–297
Spatial representation, standards, 387
Spline, advanced analysis, 294
Structured query language (SQL), databases, 160–161
SQL3, database implementation, 237
Standards, 377–396
  data access, 351–352
  elements for standardization, 380–392
    application level, 389–392
    data quality, 383–388
    encoding, 383–388
    information services, 388
    principles, 380–381
    spatial references, 387
    spatial representation, 387
    temporal characteristics, 388
    transfer formats, 382–387
  future directions, 411
  metadata, 382–396
  overview, 377–379
Standardisation, infrastructure, 398
Statistical operations, spatial analysis, 268
Structured query language, database implementation, 160–161
Subtractive color mixing, visualization, 318
Surface movement, data model(s) (advanced), 110–111
Surface representation, 94–102
  accuracy, 101–102
  generally, 94–95
  grid model, 95–97
  isoline model, 100–101
  points, 97
  practical observations, 102
  triangulated irregular network (TIN) model, 98–100
Surveying, data collection II, 199–201
Symbol selection, map, visualization, 323–324
System-ID (SID), data collection I, 174

Tagging method, discrete georeferencing systems, 127–129
Technical selection issues, 361–376
  contracts, 371–372
  database creation, 373–374
  hardware and software selection, 362–371
    design and specifications, 364–367
    new applications evaluation, 376
    pilot project, 361–362

proposal request, 367
purchasing rules, 363
safekeeping and security, 374–375
selection and implementation process, 364
system evaluation, 367–371
system operation and maintenance, 374
technical database design, 372–373
Technology, future directions, 410
Temporal accuracy, data quality, 227–228
Temporal validity, data quality, 227
Text data, data collection II, 214
Thermal plotters, plotters and output devices, 144
Three-dimensional objects, data model(s) (advanced), 102–104
Time, 105–108
  data model, 105–108
  reference system, 106
Time representation, data model(s) (advanced), 105–108
Timeliness, data quality, 227
Topology, data model, 65–69
  face, 65
  link, 65–69
  node, 65–69
Topology creation, housekeeping tools, 258–259
Training, organizational selection issues, 355–356
Transfer formats, standards, 382–387
Transformation, data collection I, 172–173
Transformations, housekeeping tools, 259–261
  coordinate system, 260
  datum, 259
  map projection, 260
Trees, database implementation, 242–243
Triangulated irregular network (TIN) model, surface representation, 98–100

Universal Transverse Mercator (UTM) Grid, coordinate systems, 119–121
Urban Information Systems (UIS), 11
Uncertainty, data model, 49–51
Undershoot, data collection I, 175
User requirements:
  coordinate system, 260
  identification of, organizational selection issues, 337–339
  software, 168–169

Vector data model, 54–73. *See also* Data
  model(s) (basic)
  area, 57–58
  comments on, 73
  data compression, 69–70
  data storage, 70–73
  digital data coding, 59–62
  geometric primitives, 64, 66
  hole, 57
  line, 57
  overview, 54–55
  point, 55–57
  ring, 57, 65
  spaghetti model, 63–65
  topology model, 65–69
Video, quantizers, 142
Viewshed, advanced analysis, 307
Virtual reality, 35
Visual hierarchy, visualization, 321
Visualization, 313–328
  cartographic communication, GIS
    potential and limitations in, 324–327
  map symbol selection, 323–324
  overview, 313–314
  theoretical foundation, 314–322
    cartographic instruments, 315–322
    data to display, 314–315
Visualization techniques, digital terrain
  models, 308–309
Volume measurement, spatial analysis,
  273
Volumes, digital terrain models, 307

Wide area network (WAN):
  developments in, 34
  hardware and communications, 134, 135
Working environment, software, 169
World Wide Web (WWW):
  developments in, 34
  software, 166–168